Abiotic Stress in Plants - Adaptations to Climate Change

Edited by Manuel Oliveira and Anabela Fernandes-Silva

Published in London, United Kingdom

Abiotic Stress in Plants - Adaptations to Climate Change
http://dx.doi.org/10.5772/intechopen.104274
Edited by Manuel Oliveira and Anabela Fernandes-Silva

Contributors
Mehdi Shahgolzari, Afagh Yavari, Elham Ghasemifar, Maria del Socorro Sánchez Correa, Adriana Montserrat Espinosa González, María el Rocío Reyero Saavedra, Edgar Antonio Estrella Parra, Erick Nolasco Ontiveros, Juan Gerardo Ortiz Montiel, Jorge Eduardo Campos Contreras, Eduardo López Urrutia, José Guillermo Avila Acevedo, José del Carmen Benítez Flores, Gladys Edith Jiménez Nopala, Victor Manuel Gonzalez-Mendoza, Jorge Rocha, Mayra de la Torre, Flávio José Rodrigues Cruz, Mohammad Javad Ahmadi-Lahijani, Saeed Moori, Ismat Nawaz, Aimen Nasir, Irum Shahzadi, Puja Agnihotri, Arup Kumar Mitra, Bushra Rashid, Saba Nawaz, Iqra Maqsood, Fatima Batool, Zainab Y. Sandhu, Sameera Hassan, Faheem Akram, Nazima Rasool, Mohammad Sayyar Khan, Irba Zulfiqar, Cláudia Pereira, José Pissarra, Susana Pereira, Miguel Sampaio, Tatiana Cardoso, João Neves, Alp Ayan, Sinan Meriç, Tamer Gümüş, Çimen Atak, Kinjal Mondal, Shani Raj, Kalpna Thakur, Anjali Verma, Mala Ram, Pooja Bishnoi, Neerja Kharwal, Animesh Chowdhury, Supratim Sadhu, Sukanya Dutta, Ayush G Jain, Saroj Choudhary

First published in London, United Kingdom, 2023 by IntechOpen
IntechOpen is the global imprint of INTECHOPEN LIMITED, registered in England and Wales, registration number: 11086078, 5 Princes Gate Court, London, SW7 2QJ, United Kingdom

British Library Cataloguing-in-Publication Data
A catalogue record for this book is available from the British Library

Additional hard and PDF copies can be obtained from orders@intechopen.com

Abiotic Stress in Plants - Adaptations to Climate Change
Edited by Manuel Oliveira and Anabela Fernandes-Silva
p. cm.
Print ISBN 978-1-83768-496-0
Online ISBN 978-1-83768-497-7
eBook (PDF) ISBN 978-1-83768-498-4

For EU product safety concerns:
IN TECH d.o.o., Prolaz Marije Krucifikse Kozulić 3, 51000 Rijeka, Croatia,
info@intechopen.com or visit our website at intechopen.com.

Meet the editors

Manuel T. Oliveira is a retired professor in the Department of Agronomy, Universidade Trás-os-Montes e Alto Douro (UTAD), Portugal, specializing in soil science and viticulture. He was previously head of the Department of Agronomy and a member of the scientific council of UTAD. He was an invited researcher at Cornell University (USA), Catholic University of Santiago (Chile), and Universidad de La Rioja (Spain). He is a member of the editorial council of *Plant and Soils* and a regular referee for *Sustainable Agriculture Research* and *Agriculture, Ecosystems & Environment*. He is a member of the Horticultural Society of Portugal, the Portuguese Society of Agricultural Science, and the Portuguese Association for National Defense.

Anabela Fernandes-Silva obtained a Ph.D. from Universidade de Trás-os-Montes e Alto Douro (UTAD), Portugal, in 2008. She is an assistant professor in the Department of Agronomy, UTAD, and a full member of the Centre for the Research and Technology of Agro-Environmental and Biological Sciences (CITAB). She specializes in agrometeorology, crops water needs, sap flow, evapotranspiration, environmental stress, and oliviculture. Her research interests include crop response to water deficit, water resources management, sustainable agriculture, crops and fruit production response to climate change, and olives and olive oil composition and quality. She is a member of the Portuguese Horticulture Society and the International Society of Horticulture Science. Dr. Fernandes-Silva leads two national research projects and is a member of the investigation team of eighteen others. She is a frequent reviewer for several journals and an editorial board member of the *Agronomy Journal*. She is also a scientific supervisor for several undergraduate and post-graduate students.

Contents

Preface **IX**

Section 1
Climatic Stressors and Their Effect on Plants **1**

Chapter 1 **3**
Understanding the Impact of Global Climate Change on Abiotic Stress in Plants and the Supportive Role of PGPR
by Puja Agnihotri and Arup Kumar Mitra

Chapter 2 **19**
Ultraviolet Radiation and Its Effects on Plants
by María del Socorro Sánchez Correa, María el Rocío Reyero Saavedra, Edgar Antonio Estrella Parra, Erick Nolasco Ontiveros, José del Carmen Benítez Flores, Juan Gerardo Ortiz Montiel, Jorge Eduardo Campos Contreras, Eduardo López Urrutia, José Guillermo Ávila Acevedo, Gladys Edith Jiménez Nopala and Adriana Montserrat Espinosa González

Section 2
Physiology: Plant Response to Stressors **41**

Chapter 3 **43**
Role of Plant Hormones in Mitigating Abiotic Stress
by Nazima Rasool

Chapter 4 **65**
Reorganization of the Endomembrane System and Protein Transport Pathways under Abiotic Stress
by Miguel Sampaio, João Neves, Tatiana Cardoso, José Pissarra, Susana Pereira and Cláudia Pereira

Chapter 5 **85**
Photosynthetic Response and Adaptation of Plants in Perspective of Global Climate Change
by Mohammad Javad Ahmadi-Lahijani and Saeed Moori

Chapter 6 105
Molecular Mechanisms and Strategies Contributing toward Abiotic Stress Tolerance in Plants
by Aimen Nasir, Irum Shahzadi and Ismat Nawaz

Chapter 7 123
Transgenic Plants in Heat Stress Adaptation: Present Achievements and Prospects
by Alp Ayan, Sinan Meriç, Tamer Gümüş and Çimen Atak

Chapter 8 155
Molecular Basis of Plant Adaptation against Aridity
by Kinjal Mondal, Shani Raj, Kalpna Thakur, Anjali Verma, Neerja Kharwal, Animesh Chowdhury, Supratim Sadhu, Mala Ram, Pooja Bishnoi, Sukanya Dutta, Ayush G Jain and Saroj Choudhary

Section 3
Stress Mitigation: Plant Management 175

Chapter 9 177
Improvement of Abiotic Stress Tolerance in Plants with the Application of Nanoparticles
by Saba Nawaz, Iqra Maqsood, Fatima Batool, Zainab Y. Sandhu, Sameera Hassan, Faheem Akram and Bushra Rashid

Chapter 10 201
Seed Nanopriming to Mitigate Abiotic Stresses in Plants
by Afagh Yavari, Elham Ghasemifar and Mehdi Shahgolzari

Chapter 11 217
Toxic Aluminum and Water Deficit Interaction in Plants: Physiological Aspects and Chemical Soil Management to Improve Root Environment in the Context of Global Climate Change
by Flávio José Rodrigues Cruz

Chapter 12 233
Microbial Mitigation of Drought Stress in Plants: Adaptations to Climate Change
by Mohammad Sayyar Khan and Irba Zulfiqar

Chapter 13 259
Plant-Growth Promoting Endophytic Bacteria and Their Role for Maize Acclimatation to Abiotic Stress
by Víctor Manuel González-Mendoza, Mayra de la Torre and Jorge Rocha

Preface

Long-term shifts in weather patterns are referred to as climate change. Natural shifts have always occurred, but during the last two centuries human activities have become a major driver of climate change.

We tend to equate climate change with rising temperature; however, its increase is only one part of the story because the planet is an interconnected system in which a change in one area affects all others. It is clear that climate change includes changes in water availability (scarcity and excess), melting ice and rising sea levels, extreme weather events, losses in biodiversity, and so on. Climate change can affect every aspect of our lives such as health, food availability, security, employment, and so on.

According to the Intergovernmental Panel on Climate Change (IPCC), the current alterations are evolving at a rate not seen in the past 10,000 years and "since systematic scientific assessments began in the 1970s, the influence of human activity on the warming of the climate system has evolved from theory to established fact."

Climate change has a perilous impact on natural ecosystems and agricultural production, threatening lives and livelihoods. Increases in global temperatures are accompanied by other abiotic stresses such as drought, flooding, and changes in soil nutrient composition that reduce crop yields. Extreme weather events are already more intense across the globe, placing additional pressure on our production systems.

Abiotic stresses tend to occur in combination rather than individually and the plant responses (physiological, metabolic, and molecular) to the combination of stresses are significantly different than those to individual stresses. Faced with the challenges of climate change, biological studies strive to decipher how plants perceive different stressors, how the signals are transduced within plants, what are the response pathways elicited by them, and how are they genetically determined.

As climate change has already introduced alterations that are irreversible, we still have to devise strategies to mitigate their impacts on our crops and food supply and protect our production systems. Thus, this book provides a comprehensive overview of stress biology to increase understanding of its factors and influences and how to mitigate it.

Manuel Oliveira and Anabela Fernandes-Silva
Department of Agronomy,
Universidade Trás-os-Montes e Alto Douro (UTAD),
Vila Real, Portugal

Section 1

Climatic Stressors and Their Effect on Plants

Chapter 1

Understanding the Impact of Global Climate Change on Abiotic Stress in Plants and the Supportive Role of PGPR

Puja Agnihotri and Arup Kumar Mitra

Abstract

Plants form the fundamental trophic level of almost all the food chains, and as such are the most significant biotic component of our ecosystems. However, there is a rising threat on the growth and well-being of these organisms due to variations in climatic conditions. Climate change conditions pose threat to plants by exposing them to various abiotic stresses, such as salinity, drought and UV-B radiation, eventually leading to oxidative stress in plant cells. Plants can put up their defence against such stressors using a number of strategies namely, adaptation, avoidance and tolerance. The action of antioxidant molecules and enzymes play a pivotal role in fighting the oxidative stress and its key player, reactive oxygen species (ROS). Plants can also develop an epigenetic memory of the stress, by modulating the expression of genes involved in stress tolerance via the epigenetic code. With the rise in environmental challenges due to climate change in recent times, it is also important to underline the helpful role played by plant growth-promoting rhizobacteria (PGPR) in building more stress-resilient plants, and the diverse array of plant genera with which these PGPR can associate.

Keywords: salinity, drought, UV-B, antioxidant, ROS, epigenetic code, epigenetic memory

1. Introduction

Presently, there is a rising concern regarding the extreme changes in climatic conditions and their related impact on the living (biotic) and non-living (abiotic) components of our ecosystems. The recent cases of flooding, drought, cyclones and hurricanes have caused an alarm and called for us to review the way in which we understand and interact with our biosphere. These situations arise not only due to anthropogenic activities but also due to natural phenomenon. A vital factor that contributes to climatic fluctuation is upsurge and accumulation of greenhouse gases which eventually cause a rise in global temperature [1–5]. There has been a significant increase in the amount of research pertaining to the mechanisms for and development of abiotic stress tolerance in plants over the past couple of decades. This is because plants are the beings most readily and deeply affected by climate change issues due to their immobile nature.

Under environmental stress conditions, arising from temperature extremes, fluctuations in rainfall and wind patterns, heat, salinity, pH variations, drought, electromagnetic radiation, etc., plants show numerous symptoms of stress-induced phytotoxicity. This ranges from modifications in metabolic and physiological activities to depletion in overall productivity. Consequently, this becomes a raging concern in case of crop plants and other plant beings that are crucial for providing ecosystem services (such as pulpwood, timber, ecotourism and natural habitat for wildlife). Due to loss in growth and productivity, there is an issue of global food shortage and also the deeply worrying aspect of oxygen depletion [6–15]. Therefore, among all the effects of the current trend of climate change on our planet, the ones on plant systems are of utmost importance.

In the backdrop of climate change, some abiotic elements of our biosphere are more susceptible to variations than the others. These include drought, salinity and UV-B. These factors influence plant life forms in solitary or in combined ways, whereby morphological, physiological and biochemical attributes of plants are effectively altered. Simultaneously, there are revelled modifications in the epigenetic codes of the plant genome, leading to what is known as chromatin-based 'epigenetic memory'. This memory may help the plant in future when it is exposed to similar kinds of stress, wherein there is elevation of plant defence activities. Plants can also respond to environmental stressors by adapting and acclimatising using various strategies. However, the climate-influenced rise in abiotic stressors continues to hamper plant growth and productivity on a larger scale and also draw in the biotic or pathogenic stress challenge into the picture.

As such, we have discussed in the present chapter some key issues related to climate change and its impact, such as the nature and origin of climate variability, its expected trend in near future, how the climate change conditions affect the sessile plant beings, the strategies adapted by plants to overcome the stress created due to the increasingly challenging environment, and the role of epigenetic mechanisms in helping the plants adapt and acclimatise better to these conditions. Moreover, the role of friendly bacteria also termed as 'plant growth-promoting rhizo-bacteria' in alleviating the negative effects of climate change on plants has also been delineated. With a clear understanding of the issue at hand, we can equip ourselves better to face the testing times that lay ahead of us.

2. Causal factors of climate change and forecast

The various physical processes in our environment at a local scale combined with responses to global climatic phenomena at a large scale form the basis of 'regional climates'. Notable examples of global-scale climatic phenomena and their respective variability in present times include Monsoon Systems, Tropical Phenomena (including the popular El Nino Southern Oscillation), Cyclones, Blocking, Atlantic Multidecadal Oscillation, Pacific South American Pattern, etc [1]. These phenomena are largely considered relevant to climate dynamics because of perceived or proven confidence that they can influence regional climate and are likely to change over time. Various physical parameters are taken into account while understanding the dynamics of regional climate, such as overall transfer of heat and moisture, and their momentum into a region [1]. According to recent updates, there continues to be a debate around the causality of natural and anthropogenic factors towards influencing climate change and also the impact of each of the causal factors. One of the latest findings has suggested that both natural and anthropogenic factors are responsible for temperature changes, contrary to what one might think, that the anthropogenic factors majorly contribute to the global rise in temperature. In addition to this, the study also indicates that there is a connection between the rise in temperature and concentration of greenhouse gases [2]. The current trend of temperature variability is arguably accredited to both natural and anthropogenic factors. For instance, sharp deviation from the present greenhouse warming trend has been noted in regions with variable volcanic activities—a period of unusually heavy activity is followed by strong cooling, while a period of low activity is accompanied with greater warming [3]. However, it is still supported by several studies that by limiting the timescale of climate analyses to more recent times, we can see that anthropogenic factors have been the major contributors to the greenhouse global warming trend. Additionally, the anthropogenic aerosols as well as greenhouse gases have influenced climate change through influencing regional temperatures and long-term changes in monsoons [3].

Despite the need to understand and precisely delineate the underlying causes of the changes in climatic conditions (such as temperature change and precipitation variability), we also need to keep an eye on the current trend of climate variability. Keeping in mind the Indian subcontinent, we have presented here the status of and predictions for climate change in South East Asia. It has been reported that S.E. Asia has observed a rise in temperature at the rate of 0.14–0.20°C per decade since the 1960s [4]. This is also accompanied by an increase in number of hotter days and warmer nights, with a simultaneous decrease in cooler weather [5, 6]. The trend for incidence of heavy and light rain episodes is positive, while that of moderate rain episodes is negative [7]. The annual rainfall on total wet day has increased at an average of 22 mm per decade, while that on extreme rain days has increased at an average of 10 mm per decade [6, 8]. It is indicated that warming is expected to persist, with extensive variation on regional basis [1]. Also, there is likely to be moderate increase in precipitation, with the exception of the part of Indonesia near to the southeast Indian Ocean. In parts of terrain, variation in precipitation is most likely to be strong [1].

3. Impact of climate change from ecological perspective

For quite a few decades now, the impact of global climate change on regional climates and the associated effects on regional as well as local ecosystems have been a matter of extensive discussion. Changes in climatic conditions in different regions across the globe have adversely affected agricultural productivity, food security, various ecosystem services, and overall composition as well as quality of flora and fauna [9–11]. Some of the major outcomes of climate changes have been temperature extremes and uncertainty or unevenness in rainfall patterns, which eventually pose a threat to agricultural crops [12–14]. Variability in temperatures and precipitation has also been found to influence cropping patterns, crop yields, and phenology, i.e. leaf development, anthesis, asynchrony between anthesis and pollinators, increased respiration, decrease in pollen germination, shorter grain filling period, and lesser biomass production [12, 15, 16].

Greenhouse gas emission and/or concentration pose a threat to the flexibility and adaptability of natural ecosystems, through influencing climate change as well as ocean acidification [17]. The recent 2018 International Panel on Climate Change (IPCC) Special Report on 1.5°C alerts that drastic climate change impacts will ensue if the planet is allowed to warm beyond 1.5°C, and such impacts include drought, flood, heat waves and sea-level rise [17, 18]. Such effects would not only harm man-kind and the lifestyle we are presently accustomed to but also the natural biodiversity in general. The previously agreed upon temperature target was 2°C; however, the half-degree variation was considered vital to avert the risk of Arctic and Coral Reef Ecosystems' degradation [17, 18]. A vital lesson to learn from this Special Report is that there is an estimated 12 years of time to reduce the net carbon emissions by half in order to avert the severe impacts mentioned earlier; however, achievement of this target would still potentially result in continued global warming as well as the associated impacts [17].

Among the most notable effects of global climate change on ecology and biosphere, salinity is number one. Several researchers have noted that one of the main reasons behind the rising levels of soil salinity across the globe is global climate change and its associated impacts, such as increasing temperature, lower precipitation, higher evapotranspiration, consequent aridisation of susceptible regions and rise in sea levels [19, 20]. Although salinity or sodicity in soil mostly originates due to natural factors such as weathering and there is some amount of it always existing in soils, the influence of climate change conditions is also substantial, whereby the amount of salinity exceeds beyond the threshold.

An important stress factor for living organisms, in particular plant life forms, is water-deficit or drought. Drought or drought-like situation arises for plants when there is inadequate water supply near the roots. This is caused by several factors such as natural climatic or geographical conditions, irregular rainfall pattern, high environmental temperature, high light intensity, spells of dry wind, water-retaining capacity of soil and water-deficit due to high transpiration rate [21, 22]. Although agricultural drought is not a big threat in itself, since it is a common natural phenomenon, and is often preceded by meteorological drought, it can still be seen as a rising abiotic stressor for plants due to the wasteful and careless anthropogenic practices. The alarming rise in greenhouse gas concentration in environment and the subsequent global warming (due to the tendency of these gases to be well mixed in atmosphere) has led to an upsurge in soil and surface water temperatures, leading to drought-like conditions. Over the past two centuries, the concentrations of carbon dioxide and methane have increased to 30% and 150%, respectively, and have thus influenced climate change through global warming and alterations in rainfall pattern [22, 23].

A direct impact of global change in climatic conditions of temperature and accumulation of greenhouse gases is thinning of the protective ozone layer of Earth's stratosphere, a phenomenon that is being studied for almost three decades now [24]. An immediate concern arising from this observation was the impact of solar UV radiation on animal as well as plant life forms. Both UV-A and UV-B are potentially harmful to biological molecules and cellular systems, and it has been noted that the interaction of UV-B with several other climate change factors (such as temperature, drought/precipitation and greenhouse gas like CO_2) can further complicate its effect, as depicted in **Figure 1** [24].

It is thus understandable that climate change is leading to enhanced abiotic stress for organisms, in particular, the sessile plant beings. The ongoing section shall throw

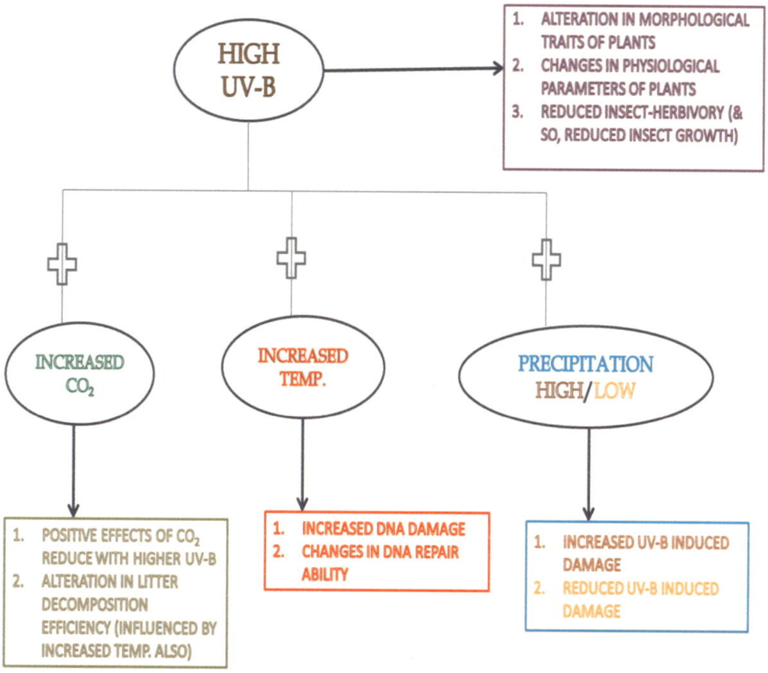

Figure 1.
Schematic overview of how UV-B affects life forms at different levels, alone as well as through its interaction with various climate change factors (derived from Caldwell et al. *[24]).*

light on how the abiotic stress factors catalyse their detrimental effects on plants and the strategies employed by plants to tackle these effects.

4. The challenge of abiotic stress for plants: Harms and defence strategies

As discussed in the preceding section, salinity is one of the significant stressors for plants and is currently being elevated by climate change conditions. It is crucial to discuss the impact of salinity stress because according some reports, and it is estimated that salt-affected land is leading to a loss of approximately 12 billion USD annually, and future predictions for agricultural production highlights the significance of working efficiently under high saline conditions [25]. Some of the principal ways by which salinity manifests its adverse effects on the physiology and biochemistry of plant systems are as follows:

i. Heightened Na^+ accumulation in plant cells causes efflux of K^+ and Ca^+ ions, which eventually leads to imbalance in cellular homeostasis, nutrient deficiency, oxidative stress, growth retardation and cell death [19]. Additionally, higher concentration of Na^+ ions in soil may lead to reduced uptake of K^+ and Ca^+ ions by the plants, further causing hindrance to proper cellular functioning and enzymatic activities [19].

ii. Impairing photosynthesis through several means such as stomatal closure, decline in primary and accessory photosynthetic pigments concentrations and damage to chloroplast ultrastructure [19, 26]. Additionally, reduction in photosynthetic pigment concentrations and photosynthetic efficacy together under increasing saline stress are attributed to loss of photosynthetic-membrane integrity, destruction of proteins and enzymes in photosynthetic pathway, dehydration of cell membrane leading to reduction in CO_2 permeability, enhanced senescence, alteration in enzymatic activities due to morphed cytosolic integrity and negative feedback by reduced sink activity [26].

Apart from salinity, abiotic stressors such as drought and UV-B radiation are also currently on the rise due to climate change conditions. Some of the principal effects of drought and excess UV-B light on plant systems are as follows:

i. Water-deficit conditions lead to loss of turgor in plant cells, reduction in plant water potential, disruption of enzymatic activities and reduced energy supply from photosynthesis. These factors eventually affect vital physiological processes such as cell division, elongation and differentiation, thereby arresting plant growth and development [27–31].

ii. Drought stress can also induce changes in morphology and anatomy of plants, i.e. reduced leaf size, lower aperture in and reduced number of stomata, thickened cell wall, cutinisation of leaf surface, enhancement in conductive system (viz. large vessels), etc. [21]. Additionally, the total biomass of plant greatly reduced under drought conditions, with an increase in root-to-shoot ratio [21].

iii. Photosynthetic efficacy is highly affected under water-deficit stress due to some notable reasons: decrease in chlorophyll pigment concentration, reduced leaf surface (due to arrested growth and development), disruption in activity and/or concentration of enzymes like RuBisCO, PEP carboxylase, fructose-1,6 bisphosphatase, sucrose phosphate synthase as a result of reduced water potential, and decline in the efficacy of both the cyclic and non-cyclic types of electron transport in the photosynthetic light reactions [32–34].

iv. Impact of drought stress is also seen on plant cell membranes, whereby the association of lipids with proteins, activity of bound enzymes and transport capacity all are hampered [21].

v. UV-B radiation influences its adverse effects on plant systems by targeting biomolecules and metabolic pathways. For instance, impairment of photosynthetic electron transport chain and increased activity of membrane localised NADPH oxidases and peroxidises both eventually lead to an overproduction of reactive oxygen species (ROS) which is the main effector molecule of oxidative stress. additionally, and UV-B stress is found to be associated with impaired pathogen resistance and alteration in the antioxidant machinery (i.e. pathways of glutathione, phenylpropanoids, cinnamates, flavonoid, respectively, and pyridoxine biosynthesis pathways) [35–37]

In addition to the points mentioned above, all the abiotic stress factors discussed herein are also potential contributors to the rise in cellular concentrations of ROS. As rightly pointed out, all kinds of stress eventually lead to a rise in ROS

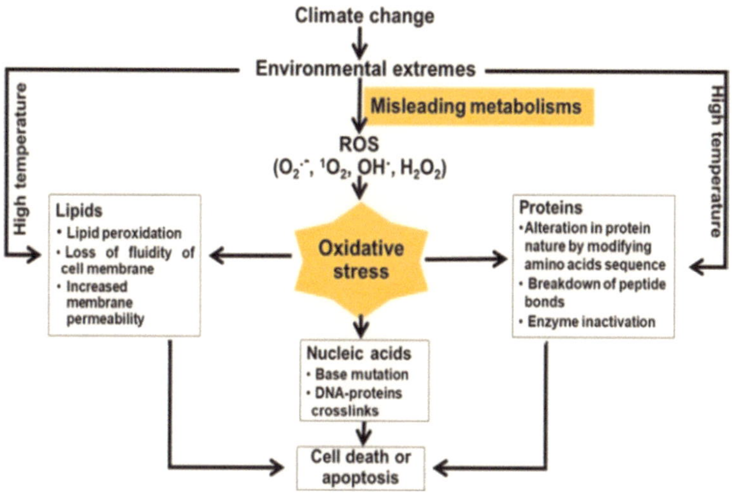

Figure 2.
Schematic overview of oxidative stress effects induced due to ROS overproduction (Taken from Dutta et al. *[39]).*

concentrations beyond their threshold value, thereby manifesting symptoms of oxidative stress [38]. The same has been depicted in **Figure 2**. Moreover, climate change conditions have also witnessed a rise in biotic stress for plants, in particular for the agronomically important plants, for instance, heightened pathogenic and pest stress, and weed stress [39].

Even though there is an upsurge of stressing conditions for plants, these immobile yet versatile organisms prove their resilience by strategically responding to the environmental stressors. Some of the major defence strategies employed by plants against the abiotic stress factors have been depicted in **Figure 3**.

5. Climate change conditions leave their imprints in plants through abiotic stress: the epigenetic effect

It is clear to us that the sessile plant beings are capable of displaying various defence strategies in response to stress factors. In nature, the plants are almost always exposed to more than one kind of stress factors, implying that response strategies are also versatile. The diverse array of signalling pathways and molecules that are involved in environmental stress defence result from reprogramming of gene expression patterns, which in turn are beautifully regulated. From research spanning the last two decades, there is growing evidence that an epigenetic regulation of gene expression also takes place under different kinds of abiotic stress [40–43]. Following the consensus, we are using the definition of 'epigenetics' as the changes in gene expression activity due to alterations that are outside of the DNA sequence of the gene, and that these changes may be meiotically or mitotically inheritable but are largely displaying the non-Mendelian feature of reversibility [40, 42, 44].

The epigenetic response of plant to different environmental stimuli is essential not only for abiotic stress tolerance but also for various other essential processes such as

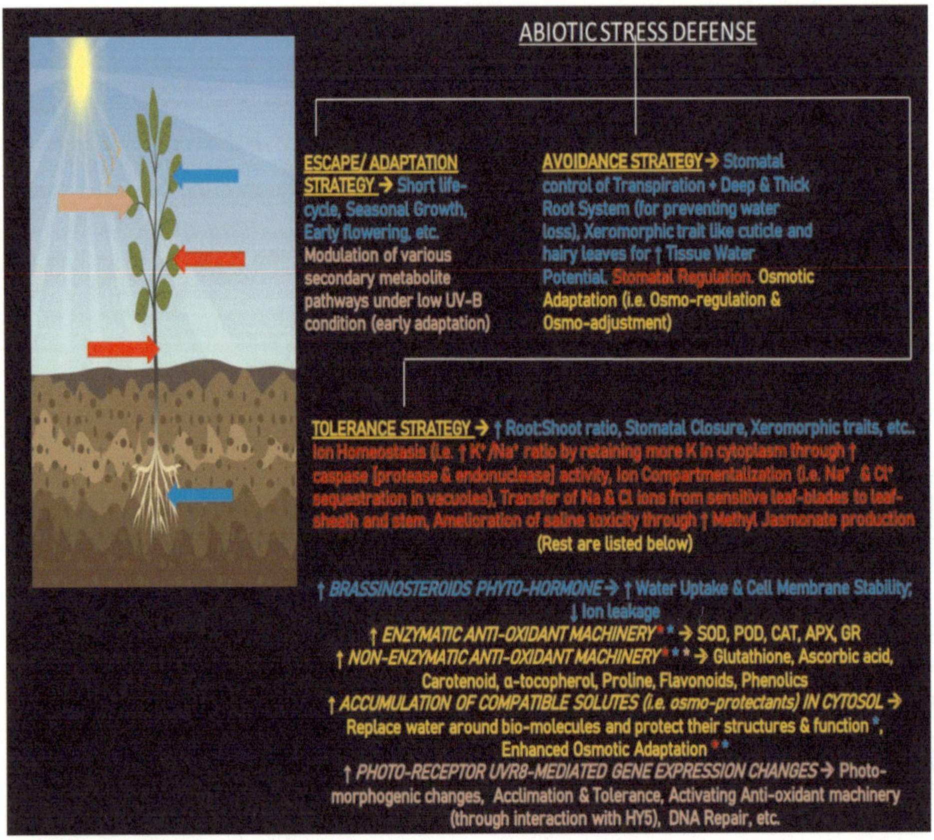

Figure 3.
Schematic representation of some major defence strategies employed by plants in response to various abiotic stresses. The strategies are categorised as 'escape/adaptation', 'avoidance' and 'tolerance', respectively. Colour codes—Blue: Drought stress; Red—Salinity stress; Pink—UV-B stress; Yellow-Common to different kinds of stress. Symbols—'↑' indicates 'increase in'; '↓' indicates 'decrease in'; Bold Arrows indicate the part of plant affected by each type of abiotic stress, i.e., leaf, stem, and root, respectively. (Extracted from Müller-Xing et al. *[40], Salehi-Lisar* et al. *[21], Kamran* et al. *[19], Podolec* et al. *[41]) (Figure is slightly modified from the literature for simplicity).*

leaf development, floral transition and bud dormancy [42]. In case of environmental stress, it has been noted that the marks of epigenetic changes display stability or transgenerational inheritance, leading to what is popularly known as 'epigenetic memory' [40, 42]. This also becomes of source of 'phenotypic plasticity' (represented as a simplified equation below).

NATURAL GENETIC VARIATION + EPIGENETIC VARIATION =
PHENOTYPIC & FUNCTIONAL DIVERSITY (origin of 'phenotypic plasticity')

However, not all epigenetic marks are stable; some of them are transient, i.e. in case of DNA repair and/or cell cycle phases. To delineate the epigenetic mechanisms of regulating gene expression, it is important to understand how the epigenetic changes work. Briefly, the genomic DNA in eukaryotes is packaged into organised structure called the chromatin. The chromatin is composed of basic units called nucleosome (as shown below):

DOI: http://dx.doi.org/10.5772/intechopen.109618

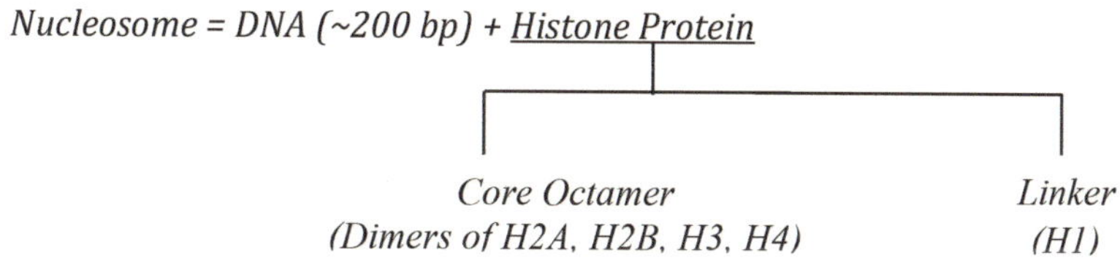

The targets for epigenetic regulation are as follows:

i. DNA methylation, wherein the 5th C of cytosine residues of the DNA backbone get attached with methyl group.

ii. Histone modification, wherein the histone proteins are subject to post-translational modifications such as acetylation, methylation, phosphorylation, sumoylation and ubiuitination, mostly at the N-terminal region of the core complex. Largely, the lysine residues are targets of histone modifications.

iii. Incorporation of histone variants.

These mechanisms, also sometimes known as the 'epigenetic code' directly regulate the activity of a gene by influencing the arrangement of nuleosome and consequently the compactness of chromatin. If the chromatin is tightly packed ('heterochromatin'), it is less accessible for expression, and if the chromatin is loosely packed ('euchromatin'), it is readily accessible by RNA Pol II for carrying out gene transcription. It has been indicated that under abiotic stress, plants can display three different kinds of epigenetic memories (**Figure 4**) [40, 45].

There is growing evidence that plants from diverse genera, including forest trees, respond to abiotic stresses of drought, salinity and UV through epigenetic modulation of gene expression [40, 46]. Additionally, these epigenetic imprints are also linked with adaptation, acclimation as well as acclimatisation under stressful conditions.

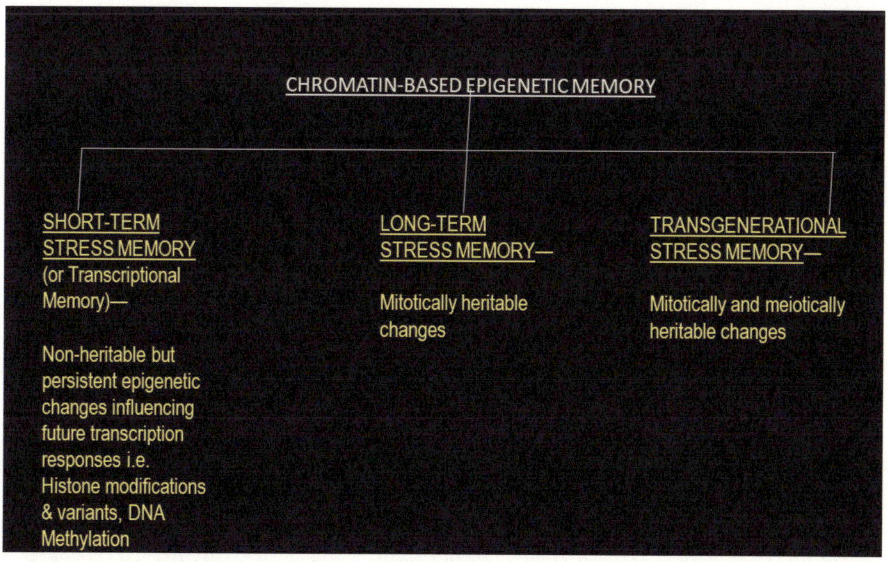

Figure 4.
Different types of epigenetic memories (Extracted from Müller-Xing et al. *[40]).*

6. Plant growth-promoting rhizobacteria as a vital tool under abiotic stress

Micro-organisms are an integral part of all ecosystems. They perform varied functions that are not only vital to their own survival but are also beneficial to different other life forms, including plants. Be it the free-living or symbiotic N_2 fixing bacteria, or the mycorrrhizal associations of fungi and plants, micro-organisms are famous for their service to the most fundamental trophic level of all food chains, i.e. plants, and consequently to the entire planet as well. For several decades now, the environmental significance of bacteria has been a matter of utmost interest, and as such scientists from all over the world have kept a close watch on the developments in this field. Bacteria are exploited to amend environmental problems arising from both natural and anthropogenic challenges, and this includes environmental pollution with toxic heavy metals or pesticides, and abiotic stress factors such as salinity or drought. Bacteria are also utilised to assist the sessile plant beings in overcoming such environmental challenges, and one of the versatile number of tools with which they help their plant hosts is collectively termed as 'plant growth promoting properties'. Some essential plant growth-promoting properties of bacteria include N_2 fixation, phosphate or potassium or other mineral solubilisation (through organic acid production), siderophore production, auxin (indole acetic acid) production and ACC deaminase activity [47–49]. If these bacteria can colonise in the rhizosphere of the plant, they form a group called 'plant growth-promoting rhizobacteria' (PGPR in short). The PGPR can also assist the plants in enhancing their antioxidant defence machinery and adapt better to the growing stress [50, 51]. In recent times, the role of PGPR in assisting plants under various abiotic stresses is being extensively explored. For instance, the efficient role of PGPR of *Bacillus* genus in salt tolerance in tall fescue, alleviating drought stress in maize and wheat through PGPR of genera *Bacillus* and *Enterobacter*, using PGPR *Bacillus subtilis* to mitigate drought stress in potatoes by suppressing oxidative stress in the plant and enhancing antioxidative enzymes, improvement in essential oil production by the medicinally important rosemary plant under salinity stress through treatment with PGPR *Pseudomonas fluorescens*, and many more [52–55]. Moreover, given the fact that bacterial species have high potential for adaptation and plasticity in metabolism, their applicability can be dynamic. The utility of PGPR thus covers a wide range of plants, from forest trees, grassland, agricultural crops to medicinally as well as aesthetically important plants, making them well applicable for cultivating more stress-resilient plants in the current backdrop of climate change.

7. Conclusion

The present trend of climate change and the resulting variation in temperature and precipitation indicate a global rise in abiotic stress factors for plants, mainly drought, salinity and UV radiation. These factors affect the plant in an integrated way, eventually leading a loss in the plant productivity. This becomes an alarming concern from agricultural as well as ecological perspective. The versatile plant beings have a number of mechanisms to articulate their defence against these stress-inducing factors, which are controlled genetically via a well coordinated cascade of signalling events. Aside from this, there is also interplay of these mechanisms with epigenetic memory, which makes the plant more resilient and better adapted to climate change conditions. However, even though plants have devised strategies of their own, it is still desirable to provide them with added assistance, given the fact that plants in the open are often exposed to multiple threats or stressors simultaneously. In this regard, the role of PGPR is imperative, because not only are they a source of beneficial activities that aid in the overall growth of plants but also a decisive tool for boosting the plants' defence mechanism against ROS and abiotic stress. Thus, the impact of climate change on plant life is manifolds, and we need to address it more resolutely with an understanding that friendly micro-organisms also play a vital role in this battle.

Acknowledgements

The authors like to extend their gratitude to Rev. Fr. Dr. Dominic Savio, S.J., principal of St. Xavier's College (Kolkata), for his continuous support. We would also like to thank the editors and the entire team of IntechOpen for this wonderful opportunity.

Conflict of interest

None declared.

Author details

Puja Agnihotri* and Arup Kumar Mitra
Department of Microbiology, St. Xavier's College, Kolkata, India

*Address all correspondence to: puja.agnihotri001@gmail.com

References

[1] Christensen JH, Kanikicharla KK, Aldrian E, An SI, Cavalcanti IF, de Castro M, et al. Climate phenomena and their relevance for future regional climate change. In: Climate Change 2013 the Physical Science Basis: Working Group I Contribution to the Fifth Assessment Report of the Intergovernmental Panel on Climate Change. Cambridge, UK and New York, USA: Cambridge University Press; 2013. pp. 1217-1308

[2] Stern DI, Kaufmann RK. Anthropogenic and natural causes of climate change. Climatic Change. 2014; **122**(1):257-269

[3] Hegerl GC, Brönnimann S, Cowan T, Friedman AR, Hawkins E, Iles C, et al. Causes of climate change over the historical record. Environmental Research Letters. 2019;**14**(12):123006

[4] Tangang FT, Juneng L, Ahmad S. Trend and interannual variability of temperature in Malaysia: 1961–2002. Theoretical and Applied Climatology. 2007;**89**(3):127-141

[5] Manton MJ, Della-Marta PM, Haylock MR, Hennessy KJ, Nicholls N, Chambers LE, et al. Trends in extreme daily rainfall and temperature in Southeast Asia and the South Pacific: 1961–1998. International Journal of Climatology. 2001;**21**(3):269-284

[6] Caesar J, Alexander LV, Trewin B, Tse-Ring K, Sorany L, Vuniyayawa V, et al. Changes in temperature and precipitation extremes over the Indo-Pacific region from 1971 to 2005. International Journal of Climatology. 2011;**31**(6):791-801

[7] Lau KM, Wu HT. Detecting trends in tropical rainfall characteristics, 1979–2003. International Journal of Climatology: A Journal of the Royal Meteorological Society. 2007;**27**(8): 979-988

[8] Alexander LV, Zhang X, Peterson TC, Caesar J, Gleason B, Klein Tank AM, et al. Global observed changes in daily climate extremes of temperature and precipitation. Journal of Geophysical Research: Atmospheres. 2006;**111**(D5): D05109, 1-22

[9] Field CB, Barros V, Stocker TF, Dahe Q. Managing the Risks of Extreme Events and Disasters to Advance Climate Change Adaptation: Special Report of the Intergovernmental Panel on Climate Change. Cambridge, UK and New York, USA: Cambridge University Press; 2012

[10] Rezaei EE, Webber H, Gaiser T, Naab J, Ewert F. Heat stress in cereals: Mechanisms and modelling. European Journal of Agronomy. 2015;**64**: 98-113

[11] Leal Filho W, Azeiteiro UM, Balogun AL, Setti AF, Mucova SA, Ayal D, et al. The influence of ecosystems services depletion to climate change adaptation efforts in Africa. Science of The Total Environment. 2021; **779**:146414

[12] Ullah A, Ahmad A, Khaliq T, Akhtar J. Recognizing production options for pearl millet in Pakistan under changing climate scenarios. Journal of Integrative Agriculture. 2017;**16**(4): 762-773

[13] Ahmed I, Ahmed S, Hussain J, Ullah A, Judge J. Assessing the impact of climate variability on maize using simulation modeling under semi-arid environment of Punjab, Pakistan.

Environmental Science and Pollution Research. 2018;**25**(28):28413-28430

[14] Lobell DB, Field CB. California perennial crops in a changing climate. Climatic Change. 2011;**109**(1):317-333

[15] Ahmad I, Wajid SA, Ahmad A, Cheema MJ, Judge J. Assessing the impact of thermo-temporal changes on the productivity of spring maize under semi-arid environment. International Journal of Agriculture and Biology. 2018; **20**(10):2203-2210

[16] Ahmed I, Ullah A, Rahman MH, Ahmad B, Wajid SA, Ahmad A, et al. Climate change impacts and adaptation strategies for agronomic crops. In: Climate Change and Agriculture. London, UK: IntechOpen; 2019. pp. 1-14

[17] Malhi Y, Franklin J, Seddon N, Solan M, Turner MG, Field CB, et al. Climate change and ecosystems: Threats, opportunities and solutions. Philosophical Transactions of the Royal Society B. 2020;**375**(1794): 20190104

[18] Hoegh-Guldberg O, Jacob D, Bindi M, Brown S, Camilloni I, Diedhiou A, et al. Impacts of 1.5 C global warming on natural and human systems. Global Warming. 2018;**2018**:175-311

[19] Kamran M, Parveen A, Ahmar S, Malik Z, Hussain S, Chattha MS, et al. An overview of hazardous impacts of soil salinity in crops, tolerance mechanisms, and amelioration through selenium supplementation. International Journal of Molecular Sciences. 2020;**21**:148. DOI: 10.3390/ijms21010148

[20] Kumar K, Kumar M, Kim SR, Ryu H, Cho YG. Insights into genomics of salt stress response in rice. Rice. 2013; **6**(1):1-5

[21] Salehi-Lisar SY, Bakhshayeshan-Agdam H. Drought stress in plants: Causes, consequences, and tolerance. In: Hossain MA et al, editors. Drought Stress Tolerance in Plants. Vol. 1. Switzerland: Springer International Publishing; 2016. DOI: 10.1007/978-3-319-28899-4_1

[22] Iqbal MS, Singh AK, Ansari MI. Effect of drought stress on crop production. In: New Frontiers in Stress Management for Durable Agriculture. Singapore: Springer; 2020. pp. 35-47

[23] Friedlingstein P, Houghton RA, Marland G, Hackler J, Boden TA, Conway TJ, et al. Update on CO2 emissions. Nature Geoscience. 2010; **3**(12):811-812

[24] Caldwell MM, Ballaré CL, Bornman JF, Flint SD, Björn LO, Teramura AH, et al. Terrestrial ecosystems, increased solar ultraviolet radiation and interactions with other climatic change factors. Photochemical & Photobiological Sciences. 2003;**2**:29-38. DOI: 10.1039/b211159b

[25] Flowers TJ, Galal HK, Bromham L. Evolution of halophytes: Multiple origins of salt tolerance in land plants. Functional Plant Biology. 2010;**37**(7): 604-612

[26] Yadav S, Atri N. Impact of salinity stress in crop plants and mitigation strategies. New Frontiers in Stress Management for Durable Agriculture. 2020;**2020**:49-63

[27] Jaleel CA, Manivannan PA, Wahid A, Farooq M, Al-Juburi HJ, Somasundaram RA, et al. Drought stress in plants: A review on morphological characteristics and pigments composition. International Journal of Agriculture and Biology. 2009;**11**(1):100-105

[28] Keyvan S. The effects of drought stress on yield, relative water content, proline, soluble carbohydrates and chlorophyll of bread wheat cultivars. Journal of Animal Plant Science. 2010; **8**(3):1051-1060

[29] Osakabe Y, Osakabe K, Shinozaki K, Tran LS. Response of plants to water stress. Frontiers in Plant Science. 2014;**5**:86

[30] Rahdari P, Hoseini SM. Drought stress: A review. International Journal of Agronomy and Plant Production. 2012; **3**(10):443-446

[31] Shao HB, Chu LY, Jaleel CA, Zhao CX. Water-deficit stress-induced anatomical changes in higher plants. Comptes Rendus Biologies. 2008;**331**(3): 215-225

[32] Chernyad'ev II. Effect of water stress on the photosynthetic apparatus of plants and the protective role of cytokinins: A review. Applied Biochemistry and Microbiology. 2005; **41**(2):115-128

[33] Farooq M, Wahid A, Kobayashi NS, Fujita DB, Basra SM. Plant drought stress: Effects, mechanisms and management. In: Sustainable Agriculture. Dordrecht: Springer; 2009. pp. 153-188

[34] Lisar SY, Motafakkerazad R, Hossain MM, Rahman IM. Causes, effects and responses. Water Stress. 2012;**25**(1):33

[35] Jenkins GI. Signal transduction in responses to UV-B radiation. Annual Review of Plant Biology. 2009;**60**: 407-431

[36] Li J, Yang L, Jin D, Nezames CD, Terzaghi W, Deng XW. UV-B-induced photomorphogenesis in Arabidopsis. Protein & Cell. 2013;**4**(7):485-492

[37] Hideg É, Jansen MA, Strid Å. UV-B exposure, ROS, and stress: Inseparable companions or loosely linked associates? Trends in Plant Science. 2013;**18**(2): 107-115

[38] Polle A, Rennenberg H. Significance of antioxidants in plant adaptation to environmental stress

[39] Dutta P, Chakraborti S, Chaudhuri KM, Mondal S. Physiological responses and resilience of plants to climate change. In: New Frontiers in Stress Management for Durable Agriculture. Singapore: Springer; 2020. pp. 3-20

[40] Müller-Xing R, Xing Q, Goodrich J. Footprints of the sun: Memory of UV and light stress in plants. Frontiers in Plant Science. 2014;**5**:574

[41] Podolec R, Lau K, Wagnon TB, Hothorn M, Ulm R. A constitutively monomeric UVR8 photoreceptor confers enhanced UV-B photomorphogenesis. PNAS. 2021;**118**(6):e2017284118. DOI: 10.1073/pnas.2017284118

[42] Chinnusamy V, Dalal M, Zhu JK. Epigenetic regulation of abiotic stress responses in plants. Plant Abiotic Stress. 2013;**2013**:203-229

[43] Chinnusamy V, Zhu JK. Epigenetic regulation of stress responses in plants. Current Opinion in Plant Biology. 2009; **12**(2):133-139

[44] Available from: https://www.cdc.gov/genomics/disease/epigenetics.htm

[45] Ding Y, Fromm M, Avramova Z. Multiple exposures to drought 'train' transcriptional responses in Arabidopsis. Nature Communications. 2012;**3**(1):1-9

[46] Carbó M, Iturra C, Correia B, Colina FJ, Meijón M, Álvarez JM, et al.

Epigenetics in forest trees: Keep calm and carry on. In: Epigenetics in Plants of Agronomic Importance: Fundamentals and Applications. Champions: Springer; 2019. pp. 381-403

[47] Agnihotri P, Maitra M, Mitra A. Isolation, Characterization and Identification of an As (V)-Resistant Plant Growth Promoting Rhizobacterium Associated with the Rhizosphere of *Azolla microphylla*. Journal of Microbiology, Biotechnology and Food Sciences. 2022;**2022**:e4728

[48] Fadiji AE, Santoyo G, Yadav AN, Babalola OO. Efforts towards overcoming drought stress in crops: Revisiting the mechanisms employed by plant growth-promoting bacteria. Frontiers in Microbiology. 2022;**13**

[49] Rodriguez H, Gonzalez T, Goire I, Bashan Y. Gluconic acid production and phosphate solubilization by the plant growth-promoting bacterium Azospirillum spp. Die Naturwissenschaften. 2004;**91**(11): 552-555

[50] Bhat MA, Kumar V, Bhat MA, Wani IA, Dar FL, Farooq I, et al. Mechanistic insights of the interaction of plant growth-promoting rhizobacteria (PGPR) with plant roots toward enhancing plant productivity by alleviating salinity stress. Frontiers in Microbiology. 2020;**11**:1952

[51] Bharti N, Barnawal D. Amelioration of salinity stress by PGPR: ACC deaminase and ROS scavenging enzymes activity. In: PGPR Amelioration in Sustainable Agriculture. India: Woodhead Publishing; 2019. pp. 85-106

[52] Li Y, You X, Tang Z, Zhu T, Liu B, Chen MX, et al. Isolation and identification of plant growth-promoting rhizobacteria from tall fescue rhizosphere and their functions under salt stress. Physiologia Plantarum. 2022; **2022**:e13817

[53] Jochum MD, McWilliams KL, Borrego EJ, Kolomiets MV, Niu G, Pierson EA, et al. Bioprospecting plant growth-promoting rhizobacteria that mitigate drought stress in grasses. Frontiers in Microbiology. 2019;**10**:2106

[54] Batool T, Ali S, Seleiman MF, Naveed NH, Ali A, Ahmed K, et al. Plant growth promoting rhizobacteria alleviates drought stress in potato in response to suppressive oxidative stress and antioxidant enzymes activities. Scientific Reports. 2020;**10**(1):1-9

[55] Dehghani Bidgoli R, Azarnezhad N, Akhbari M, Ghorbani M. Salinity stress and PGPR effects on essential oil changes in *Rosmarinus officinalis* L. Agriculture & Food Security. 2019;**8**(1):1-7

Chapter 2

Ultraviolet Radiation and Its Effects on Plants

María del Socorro Sánchez Correa,
María el Rocío Reyero Saavedra, Edgar Antonio Estrella Parra,
Erick Nolasco Ontiveros, José del Carmen Benítez Flores,
Juan Gerardo Ortiz Montiel, Jorge Eduardo Campos Contreras,
Eduardo López Urrutia, José Guillermo Ávila Acevedo,
Gladys Edith Jiménez Nopala
and Adriana Montserrat Espinosa González

Abstract

Ultraviolet radiation is a portion of the electromagnetic spectrum ranging from 10 to 400 nm, classified into three main categories: UV-A (320–400 nm), UV-B (280–320 nm), and UV-C (100–280 nm). The UV radiation from the sun that crosses the atmosphere and reaches the earth's surface is composed largely of UV-A radiation (95%) and, to a lesser extent, UV-B (5%), which is normally filtered by stratospheric ozone. With the thinning of the ozone layer, UV-B radiation penetrates deeper into the earth's surface, where it becomes dangerous due to its high energy content that acts at the molecular level, affecting the cycles of carbon, nitrogen, and other elements, thus, having a direct impact on global warming. On the other hand, UV radiation alters numerous essential organic compounds for living organisms. Since its discovery, it has been established that e UV-B causes alterations in plant development and metabolism, both primary and secondary. In this chapter, we summarize the current knowledge about the effects of UV radiation on the morphological, biochemical, and genetic processes in plants.

Keywords: UV radiation, secondary metabolites, oxidative stress, photomorphogenesis, photosynthesis, UV transcription factors

1. Introduction

As sessile organisms, plants are constantly exposed to a wide variety of stress factors, such as desiccation, environmental pollution, temperature changes, and UV radiation. Ultraviolet radiation is a part of the nonionizing radiation region of the electromagnetic spectrum and comprises about 9% of the emitted solar radiation; according to the ISO 21348 standard, it is divided into three types: UV-C (200–280 nm), UV-B (280–315 nm), and UV-A (315–400 nm) [1, 2].

The ozone layer (O^3) efficiently filters much of the shortwave UV radiation (UV-C). However, this absorption decreases rapidly for radiation with wavelengths greater than 280 nm, reaching a rate of 0% absorption for wavelengths greater than 330 nm. Factors, such as elevation above sea level, cloud cover ground reflectance, geographic latitude, and ozone gradient and can affect the amount of UV-B and UV-A radiation that reaches the Earth's surface [3].

In normal conditions, the ozone layer filters around 80% of UV-B radiation, but human activities have caused a decrease in the stratospheric ozone concentration through the emission of compounds such as chlorofluorocarbons (CFCs), carbon tetrachloride (CCl_4) and hydrochlorofluorocarbons (HCFCs). Therefore, UV-C radiation and an increased percentage of UV-B radiation can pass through [3, 4].

Although UV radiation is a minor fraction of solar energy that reaches the earth's surface, it significantly affects plants. UV-B radiation affects important biomolecules directly, including nucleic acids and proteins; these molecules absorb UV radiation easily when presenting π electrons, and this absorption can lead to metabolic, biochemical, and morphological alterations, as well as alterations in the genetic material [5, 6]. UV-A radiation produces similar effects, although they are part of the constitutive regulation of plant metabolic and morphological processes, such as photosynthesis, biomass production, and synthesis of pigments and antioxidant compounds [7].

Since the discovery of the thinning of the ozone layer, the consequent penetration of UV-B radiation into the atmosphere and its undisputable contribution to global warming of the planet, the effects of UV radiation on plants have been closely studied. Plants can use sunlight not only as a source of energy to produce carbon compounds but also as a source of environmental information; that is, they can detect it as a signal and trigger different systemic responses related to photosynthesis, phototropism, photoperiodicity, and photomorphogenesis. These same processes can be affected by the abnormal incidence of UV radiation in the atmosphere; therefore, the impact of its damage has been studied in recent decades [8]. This assessment has led to the creation of initiatives such as the Montreal Protocol, which aims at mitigating the negative effects of climate change-derived increased UV exposure through international policies [9].

In addition, the analysis of the causes of the morphological alterations shown by plants under UV light stress is difficult because they can be affected simultaneously by other environmental factors such as temperature, salinity, or drought, which together can modify development at the cellular level. The objective of this chapter is to describe the effects of UV radiation on different biochemical, morphological, and genetic processes in plants.

2. Morphological alterations

Photomorphogenesis (light-regulated plant development) in the presence of UV light has been extensively studied [10]. Plants of several species modify the development of their organs in the presence of UV light; for example, the length of the stems tends to shorten, although they form a greater number of axillary buds, while the roots tend to be longer and more abundant, akin to the development of plants that grow in conditions of low light radiation [11].

One of the stages of plant development most susceptible to the incidence of light is germination, which is also greatly affected by UV-B radiation. In *Arabidopsis thaliana* seedlings irradiated with UV light, the growth of the hypocotyl was slower [12] compared to seedlings germinating under normal conditions; even the growth of the hypocotyl is lower in etiolated plants developed in the shade but irradiated with UV light [13]. On the contrary, in this same species, it has been observed that, under these conditions, the cotyledons tend to expand, even with short periods of UV light exposure [14].

Leaves also modify their structure, tending to decrease their surface area and increase their thickness in many broadleaf plant species that have been tested for their response to UV light. Apparently, this change in morphology depends on the imbalance between cell proliferation and elongation among the different leaf tissues, which can cause a decrease in leaf area, abnormal thickening, or rolling, resulting in slow plant development [15]. While searching for modifications at the cellular level that explain the alterations in the morphology of plants under UV stress, Krasilenko et al. demonstrated in 2013 [16] that UV radiation can cause depolymerization or fragmentation of microtubules in *A. thaliana* cells, causing the reorganization of the cytoskeleton and the cell in general, so that elongation and cell division are reduced, resulting in the formation of shorter leaves, which affects the development and the complete morphology of the plant.

It is currently accepted that some plant species avoid excess light radiation by forming a waxy cuticle on the epidermis. Exposure to UV radiation-induced deposition of wax in plants of species, such as *Coffea arabica*, *Coffea canephora*, *Hordeum vulgare*, *Cucumis sativus,* and *Phaseolus vulgaris*, which results in an increase in the thickness of the cuticle. Additionally, molecules such as phenolic acids and flavonoids can accumulate in the cuticle, functioning as photoprotectors against UV light or as UV light attenuators, respectively [17].

Stomata, the structures where gas exchange occurs, are also affected by the presence of UV light. High UV irradiation causes loss of stomatal opening and closing control in response to environmental stimuli, apparently due to an altered guard cell conductance. Since the stomatic function is vital for CO_2 fixation in the light-independent reactions of photosynthesis, its deregulation can deeply affect plant development and physiology [18].

UV light plays an important role in plant development, but extreme exposure can be detrimental. Unable to relocate, plants must balance the positive and negative effects of UV radiation mostly through intracellular mechanisms, as described in the following sections.

3. Photosynthetic alterations

Photosynthesis is a light-dependent process, so it is almost inevitable that it be affected by the presence of UV radiation. There are several reports about the damage caused by UV radiation in specific sites of the photosynthetic apparatus of green plants (**Figure 1A**) [17]. Much of the damage is caused by the enhanced production of reactive oxygen species (ROS) that are involved in UV-induced responses, both as signaling agents within normal cellular processes and as damaging agents. ROS can cause damage to the proteins that make up the light-harvesting complexes of the photosystems or to those found in the protein complexes where the electron carriers of photosynthesis are concentrated, their accumulation is even known to cause the

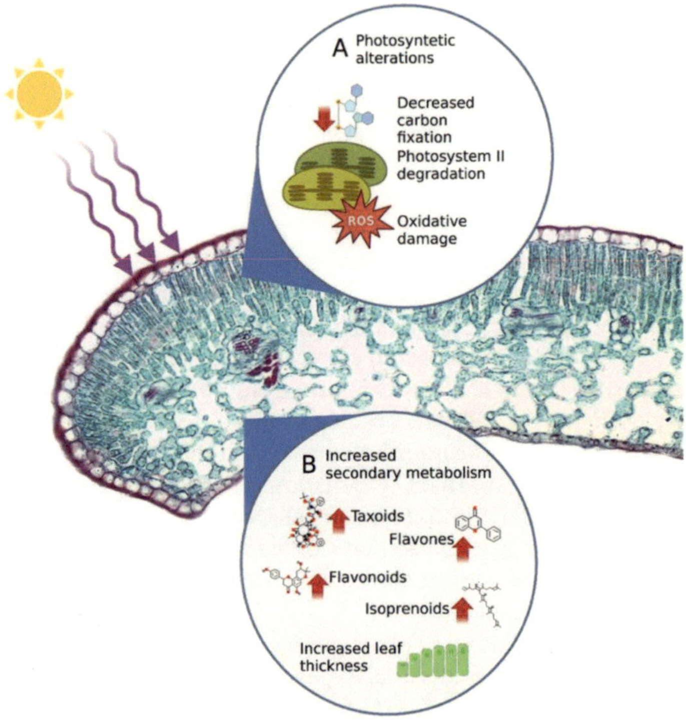

Figure 1.
Effects of UV-B light on plants and alterations caused by UV-B radiation in photosynthetic metabolism (A) and secondary metabolism (B).

destruction of ribulose bisphosphate carboxylase/oxygenase [19], and, therefore, a decrease in atmospheric carbon fixation and plant biomass occurs. Another important damage caused by ROS is the oxidation of fatty acids in the membranes, which, in combination with peroxidation and photooxidation because of UV light, breaks the essential integrity of the thylakoid membranes in the chloroplast, generating alterations in the organization of the membrane-embedded photosynthetic complexes, decreasing their photosynthetic capacity [20].

Ultraviolet light also causes damage to plant proteins; in fact, one of the effects on photosynthesis is the damage, it exerts on the enzymes that synthesize pigments such as chlorophylls [21]. In addition, pigments are also degraded by UV light, especially chlorophyll b and carotenoids, so exposure to this type of radiation can cause an imbalance in the proportion of pigments, with the consequent alteration of the photosynthetic apparatus, as has been observed recently in maize. After being exposed to UV radiation for 19 days, fluorescence and chlorophyll concentration decreased in several maize lines, although in different proportions in a line-dependent manner [22].

Several elements at Photosystem II, the site where photosynthesis begins, are sensitive to UV radiation. This complex is formed by the association of pigments and proteins, and many of these proteins are part of electron transport centers; therefore, their alteration or degradation affects the electron transport chain of photosynthesis, reducing their levels under UV light stress. In an elegant work, Ihle [23] reported that proteins D1 and D2, which are found in the reaction center of photosystem II, are especially susceptible even to low intensities of UV radiation (1 μmol m^{-2} s^{-1}) [24]. The degradation of proteins D1 and D2 adds to the alteration of the manganese (Mn) oxidizing group of water, which together cause the loss of function of the reaction center and, therefore, the inhibition of electron transport [25]. Damage at the beginning of the electron transport chain of photosynthesis makes it difficult to investigate downstream transporters; however, some reports indicate the change in the ratio of photosystems II and I due to the decrease in absorption at 700 nm—absorbed by Photosystem I—observed after prolonged exposure to UV light [26].

Plants are highly susceptible to the presence of ultraviolet light. Through research over the past four decades, it has been possible to discover the mechanisms related to damage in plant morphology, development, and metabolism. However, many questions remain to be investigated until the problem of the penetration of UV radiation into the atmosphere is resolved.

4. Oxidative stress by UV light induction

Ultraviolet radiation is an important stress in plants that elicits protective mechanisms such as the accumulation of secondary metabolites in the cell (**Figure 1B**) [27–29] and an increase in leaf thickness [30]. Interestingly, UV radiation is a hormetic stimulus, that is severe exposure is harmful, but exposure to lower sub-acute levels can stimulate protective mechanisms [31]. Consequently, plants can become resilient to UV after repeated exposure [32].

The changes in the secondary metabolism of plants from all taxa under exposure to UV radiation have been widely documented. For instance, in the moss *Pohlia nutans*, UV-B radiation enhanced flavone biosynthesis through increasing type I flavone synthase activity [33]. In *Taxus cuspidate*, UV-B radiation (3 W/m^2) provoked the accumulation of toxoids and flavonoids [34]. Also, the flavonoid contents in *Scutellaria baicalensis* reached the maximum concentration (41.86 mg/g^{-1}) after seven days under UV-A radiation [35]. In *Pisum sativum* leaves, exposure to UV-B radiation increased the nicotinamide and trigonelline content; the nicotinamide induction is an oxidative stress reaction [36]. In an analysis performed on two different ecotypes of the *Paubrasilia echinata* tree, it was shown that UV-B radiation inhibited stem growth, biomass accumulation, CO_2 assimilation, and photochemical efficiency in a shade-tolerant ecotype inhibition; in contrast, a sun-tolerant ecotype showed a positive response: UV-B increased flavonoids, lignin, and antioxidant properties, but reduced cell respiration [37]. In *Pinus radiata,* UV radiation provoked an early response reducing photosystem activity and accumulation of photoprotectors; even the primary metabolism was rearranged to minimize ROS production, also the isoprenoids compounds like carotenoids, tocopherols, phytol, and gibberellins were decreased [38]. Under exposure to UV-B radiation followed by dark treatment, the number of flavonoids and coumarins in *Clematis terniflora* increased significantly; while proteins related to photorespiration, the tricarboxylic acid cycle, and mitochondrial permeability showed differential expression profiles, indicating that UV-B radiation induces a reduction in energy consumption and maintains energy balance [39]. *Nymphoides humboldtiana* increased antioxidant activity and production of flavonoids like phloroglucinol, chlorogenic acid, epicatechin, quercetin, and ferulic acid after 13 days of exposition of UV-B radiation [40]. *Colobanthus quitensis* under UV-B radiation increased the biosynthesis of flavonoids, particularly flavone C-glycosides, metabolites located within the most metabolically active cells [41]. *Melisa officinalis* showed changes in the glycolysis and phenylpropanoid pathway under UV radiation stress with differential recovery times [28].

Studies on algae have shown similar mechanisms, as expected by their phylogenetical relation to plants. For example, UV-B radiation-induced ROS production in peroxisomes and chloroplasts in *Ulva prolifera* provokes irreversible damage under 5 W m^{-2} [42]. In *Chlamydomonas reinhardtii,* UV-C radiation stress increased ROS levels and production of antioxidant polyphenols, a phenolic including caffeic acid, cinnamic acid, coumaric acid, salicylic acid, and protocatechuic acid, among others [43].

A notable example among plant-derived compounds is the alkaloid mimosine, present in the seedlings of *Leucaena leucocephala* spp. Glabrata is particularly interesting due to its therapeutic uses as anti-cancer, antifungal, and antimicrobial, which increase its economic interest. Acute UV-C exposure of *L. leucocephala* seedlings induced a strong accumulation of mimosine, which could be implicated in general oxidative stress modulation [44].

The effect of UV-radiation stress has also been extensively studied in plants used in traditional medicine. For example, *Morus alba*, used in traditional Chinese medicine, reduces its growth and secondary metabolism after exposure to UV-B [45]. *Gingko biloba* leaves, after long-term exposure to UV-B radiation, increase flavonoids biosynthesis, and these are beneficial as therapeutic active ingredients [29]. Two different species of the Chinese herb Astragalus modified their secondary metabolite production under UV-B radiation; *A. mebranaceus* produced increased hydroxycinnamic acid derivates, while *Astragalus mongholicus* accumulated myricitrin and isoflavones, showing different tolerance to UV-B stress [30]. The flowers of *Lonicera japonica* are used as a medicinal herb in Asian countries. Under UV radiation, *L. japonica* increases the levels of oxidative pentose phosphate and secondary metabolites such as secologanic acid, secoxyloganin, and isochlorogenic acid [46]. In *Adhatoda vasica*, also used in Asiatic traditional medicine, UV-B radiation (7.2 kj m^{-2} day^{-1}) induces a reduction of superoxide radical production while increasing hydrogen peroxide production [47]. Finally, *Centella asiatica*, used in Asian and African traditional medicine, accumulated saponins and epidermal flavonols under UV-B radiation in younger leaves with high levels of saponins; in contrast, in older leaves, sapogenins were the most abundant metabolites [48].

As shown in these studies, UV light has forced algae, bryophytes, and plants to modify their metabolism—particularly the secondary metabolism—to increase their ecological success rate. However, this also has important consequences for plants of commercial interest, as seen below.

5. UV radiation as functional quality of plant foods of commercial interest

Historically, economically important plants have been exhaustively studied; recent studies have focused particularly on UV light stress, searching for alterations in organoleptic properties and secondary metabolism. In modern horticulture, plants of economic interest have been irradiated with UV light during the flowering/fruiting period, with the purpose of stimulating oxidative stress pathways as well as antioxidant production [49]. In tomato juice production, the stress caused by UV radiation in plants decreased pectolytic enzymes, improving and preserving tomato characteristics for a longer period of time [50]. Also, in a tomato cultivar, UV-A and B radiation produced higher ripening synchronization and smaller fruits.

DOI: http://dx.doi.org/10.5772/intechopen.109474

Exposure to UV-A radiation-induced accumulation of phenolics and flavonoids, making these fruits more appealing to consumers [51]. Furthermore, in tomato seedlings under UV-B radiation, carotenoid content increased as well as antioxidant enzyme activities [52].

Another plant of great economic importance is soybean (*Glycinine max*), which increases the isoflavone content of the sprouts under UV radiation [53]. In soybean seedlings, nitric oxide is induced as a protection against UV-B stress [54]. Meanwhile, on germinated soybean, UV-B radiation increased the contents of linoleic acid and erucic acid content, as well as isoflavones, phenolic acids, vitamin C, folate, and chlorophyll, improving nutritional and functional qualities [55]. Conversely, excessive UV-B exposure damaged cells and decreased the amount of isoflavones within them [56]. In cultured soybean, UV-C radiation increased the amount of genistein-O-glucoside and genistein-O-glucosyl-malonate, suggesting *in vitro* culture to obtain a high level of metabolites [57]. Moreover, in germinated soybean under UV-B radiation, total protein content and endogenous H_2O_2 were increased [58].

Cereals and ornate flowers also have responses to UV radiation. Wheat seedlings under UV stress showed an increase of phenylalanine ammonia-lyase only in the roots, indicating that UV-B radiation has a positive or negative impact, depending on the type of secondary abiotic stress factor observable in the production of phenolic compounds [59]. Also, germinated wheat under UV-B radiation increased phenols, ferulic acid, and coumaric acid. Exogenous Ca^2+ positively affected free and bound phenolic accumulations [60]. In amaranth (*Amaranthus cruentus* L.), UV-C radiation improved postharvest quality by increasing levels of quercetin, kaempferol, copene, lutein, β-carotene, and caffeic acid derivates [27]. In lily bulbs, UV-C radiation increased total phenolic content and antioxidant activity, indicating that UV-C radiation is a safe alternative for processing lily bulbs in storage [61].

Likewise, the effects of UV radiation have been studied in economically important herbs. In spinach cultivars, UV-C induced a hormetic effect that increased total phenolic compounds and reduced the presence of the parasite fungi *Alternaria* ssp. in the crops [62]. In barley seedlings, UV-B radiation up-regulated enzymatic activity, resulting in the accumulation of phenolic acids [63]. *Mentha aquatic* responded to UV-B radiation on a morphological level, increasing glandular trichomes, and on a biochemical level, increasing oxidative metabolism and overexpressing genes implicated in terpene biosynthesis, particularly volatile oils as camphene, β-pinene, and germacrene [64]. In wounded carrots under UV-A and C radiation, ROS increased, acting as a signal for ethylene synthesis, which activated the synthesis of jasmonic acid leading to the accumulation of phenolic compounds [65]. In fresh-cut carrots, UV-C doses inhibited ascorbic acid, total carotenoid, respiration, total phenols, lignin, malondialdehyde, and ethylene production; all data collected indicated extended shelf-life and overall quality maintenance [66]. In parsley, UV-C doses resulted in an increase of antioxidants such as phenylpropanoid and phenolic compounds, as well as enzymes involved in the synthesis of phenylpropanoid [49]. The effect of UV radiation induces the production of 6″-0-malonylapiin, which is a flavone glycoside, as well as the 12-oxo-phytodienoic acid [67]. UV-B radiation (1.5 kJ m^{-2}) maintained the color of broccoli florets during storage, and induced glucosinolates and hydroxylcinnamates, raising their antioxidant properties. These findings suggested that UV-B radiation is likely to induce the indole glucosinolate pathway [31], maintaining the quality of broccoli florets in low-temperature storage [68].

Fruits are also of economic interest and respond differentially to UV. Grape berries (Jumeigui variety) decreased sugar content under UV-C, promoting the accumulation of stilbenes and some flavonoids [69]. In contrast, berry clusters (red table emperor) under UV-A and B radiation decreased the amount of quercetin 3-O-glucoside and quercetin 3-O-glucuronide, suggesting that UV radiation induces postharvest changes in phenolic metabolites [70]. In fresh-cut strawberries, UV-C increased phenolic compounds, anthocyanin, cyanidin 3-glucoside, pelargonidin 3-glucoside, and cyanidin 3-glucoside-succinate, activating the phenylpropanoid pathway, thus improving antioxidant capacity without losing fruit quality [71]. In two blueberry cultivars (*Vaccinium corymbosum*), exposure to UV radiation showed that the amount of phenylpropanoid compounds was higher in the Legacy cultivar than in the Bluegold cultivar, which indicates that UV-B acclimation is different between cultivars [72]. Moreover, in highbush blueberry leaves (*V. corymbosum* L. cv. Brigitta and Bluegold), photosynthesis decreased in the Bluegold variety under UV-B radiation; in contrast, the Brigitta variety increased the photosynthesis rate as well as antioxidant activity [73]. In fragrant pear, postharvest UV-C radiation controlled blackhead disease through chitinase, β-1,3-glucanase, peroxidase, superoxide dismutase, catalase, ascorbate peroxidase, and phenylalanine ammonia-lyase [74]. In nectarine, UV-C radiation induced an increase in anthocyanin biosynthesis and promoted the antioxidant system, stimulating the phenyl propane pathway. Together, these compounds exerted antifungal action against *R. stolonifera* [75]. In young leaves of *Vitis vinifera*, low UV-B radiation increase sitosterol, stigmasterol, and lupeol, probably as an acclimation response. In contrast, diterpenes, tocopherol, phytol, E-nerolidol, monoterpenes as careen, α-pinene, and terpinolene were present in high amounts in mature leaves; these results showed that the synthesis of terpenes is an adaptive response to UV-B radiation stress [76]. In postharvest lemon fruits after UV-B radiation, phenolic compounds increased in flavedo, indicating that lemon peel modifies enzymatic activities involved in sucrose metabolism [77].

Furthermore, in *Olea europaea*, UV-B radiation increases secoiridoids and 2″-methoxyoleuropein metabolites, while decreasing oleuropein as an antioxidant defense against UV [78]. The peach (*Prunus persica*) diminishes the synthesis of anthocyanins and phenolic compounds under UV-B exposure, but after 36 h, it increases anthocyanins, cyanidin, and delphinidin compounds [79]. In Luffa seedlings, the oxylipins such as methyl jasmonate and 12-Oxo-phytodienoic acid mitigated the UV-B stress via improved photosynthetic and nitrogen metabolism, respectively [80].

Even economically important algae respond to UV radiation. In several Spirulina species, mild stress by UV-B radiation has been useful in increasing physiological and nutritional competencies in growth, rendering UV radiation useful in producing this functional food [81].

Although most of the above-mentioned economically important species appeared to benefit from UV exposure, it has been detrimental to some species. Rice (*O. sativa*) plants treated with UV-C had less palatability and were easily infested by the weevil *Sitophilus oryzae*, which provoked lower consumer acceptance and purchase intention [82]. Also, sweet cherry fruits under UV-C radiation diminished respiration, but increased rhamnose, mannose, galactarate, threonate, and aspartate contents [83].

This evidence highlights the importance of studying UV stress in plants of economic interest, as it can lead to higher yields and thus higher profits. However, care must be taken before implementing UV irradiation as a production-boosting resource because some species might be impacted negatively, as evidenced by the effects of the increased exposure to UV derived from climate change. Plant litter decomposition, especially in regions with low annual rainfall and reduction of photosynthetically active radiation (PAR), further strain crop production [84].

6. Genetic response

Several studies have focused on changes in gene expression in plants exposed to long- and short-wave UV-B radiation to identify the cellular components that regulate response to UV [85]. The results showed that UV-B radiation triggers cell growth and morphogenesis pathways [86]. UV-B response signals are also transmitted from cell to cell and are usually organ-specific [87].

Genetic approaches for phenotypic responses to UV-B are based on models of increased tolerance or aberrant responses (e.g., changes in hypocotyl growth) to UV-B irradiation [86]. Transcriptomic analysis from *Arabidopsis* seedlings exposed to different UV-B radiation intensities showed that more than 20% of the genes that modified their expression are transcription factors [85]. These approaches allowed the identification of mutants that lacked or overexpressed photoprotective compounds or inhibited hypocotyl growth in response to UV-B [14, 88].

UV-B radiation induces changes in the expression of genes that affect growth and development, as seen in *UV-B light insensitive* (*uli*) mutant plants, which present reduced hypocotyl growth relative to wild-type after UV-B exposure. Also, UV-B affects *chalcon synthase (CHS)* expression [14]; low levels of irradiation activate this gene, which is key in the biosynthesis of phenylpropanoids [89]. *CHS,* along with transcription factors, allows plants to protect themselves against UV-B.

LONG HYPOCOTYL5 (HY5) is a bZIP transcription factor that regulates morphogenesis in response to UV-B. *HY5* gene expression is a component of the UV-B-induced signaling network. Transcriptomic analysis in *Arabidopsis thaliana* showed the importance of HY5-dependent regulation in response to low-level UV-B irradiation [85]. If HY5 is lost, transcriptional induction of the UV-B response genes is impaired [86] and cells undergo programmed death [90].

HY5 is a light-induced transcription factor required for many light-responsive genes; in the dark, it is degraded by the proteasome [91]. This transcription factor is key for phytochrome and cryptochrome regulation networks [85, 92, 93]. So, it seems that HY5 does not respond to UV-B radiation exclusively, which opens the door for research on other components that specifically drive plants' response to UV-B.

Several genes are induced by UV-B independently from traditional photoreceptors, such as phytochromes and cryptochromes, through the activity of the LONG HYPOCOTYL5 (HY5) transcription factor [85]. This independence suggests that there must be a specific UV-B photoreceptor that activates HY5; however, the identity of this putative element is still unclear [94].

CONSTITUTIVELY PHOTOMORPHOGENIC1 (COP1) is an E3 ubiquitin ligase that participates in the UV-B response [95]. COP1 has three functional domains, a RING finger (ligase activity), a coiled-coil for dimerization, and a WD40 repeat domain with binding activity [93]. COP1 targets bZIP transcription factors and is required to activate *HY5* gene expression. Both proteins are localized in the nucleus and regulate photomorphogenesis under UV-B conditions in a specialized pathway [95]. COP1 was identified as a photomorphogenesis repressor in darkness and light [93, 96]. Visible light inactivated COP1 and separated it from HY5 [93], allowing HY5 stabilization and, therefore, activation of light-responsive genes [91, 97] through the interaction of phytochromes and cryptochromes [98, 99]. Phytochromes and cryptochromes interact with the SUPPRESOR OF PHYTOCHROME A (SPA) proteins, causing light-dependent COP1 inactivation. COP1 response to UV-B radiation is independent of the SPA proteins [95, 100]; rather, COP1 responds to UV-B through the interaction with the UV RESPONSE LOCUS 8 (UVR8) protein [96].

UVR8 is a seven-bladed b-propeller protein that forms a homodimer in its inactive state [101, 102] and is capable of UVR-B perception [103, 104]. In contrast to other photoreceptors (phytochrome and cryptochrome), UVR8 does not employ a bound chromophore; instead, it uses a tryptophan residue localized in the b-propeller blade [101, 102, 105]. Upon UV-B absorption, the UVR8 dimer destabilizes and the monomeric form interacts with COP1 [104]. The UVR8-COP1 heterodimer activates the transcription factor HY5, consequently activating downstream genes that are implied in metabolic and morphological alterations [104, 106]; this mechanism activates UV-B acclimation and tolerance [96]. UVR8 is usually located in the cytoplasm, while COP1 is in the nuclear bodies of hypocotyl cells. When plants are irradiated with UV-B, UVR8 translocates to the nucleus [107] and colocalizes to the COP1-rich nuclear bodies [96]. After UV-B exposition, the UVR8 dimer COP1 prevents HY5 degradation, so that HY5 can exert its transcriptional activation function [108].

In Arabidopsis plants, *uvr8* mutants do not respond when grown under UV-B radiation; they lack a photomorphogenic signal and therefore do not display the damage usually found in wild-type plants [96]. UV-B-induced gene expression is important for UV acclimation and survival. When *urv8* and *cop1* mutants are initially grown in weak UV-B exposure and later moved to high UV-B irradiance, the mutants do not show an acclimation effect. When exposed to a natural spectral balance, the *uvr8* mutant shows leaf damage. Also, *HY5* or *CHS* gene expression is undetectable

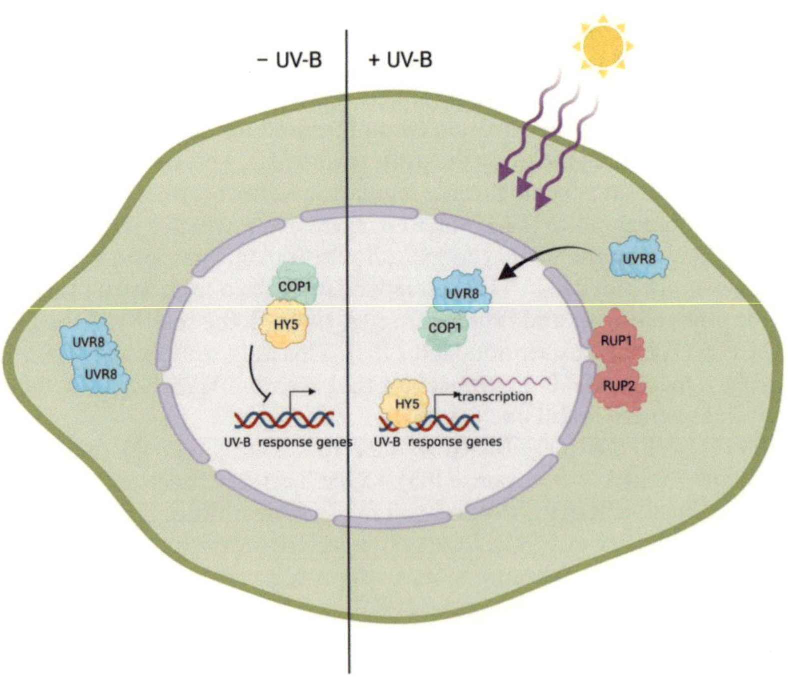

Figure 2.
ELONGATED HYPOCOTYL5 (HY5) transcription factor activation.

in both mutants. Consequently, the interaction between COP1 and UVR8 proteins is required for the regulation of UV-B response and confers UV-B protection. On the other hand, overexpression of *UVR8* leads to UV-B photomorphogenic hypersensitivity, presenting inhibition of hypocotyl growth, activation of *HY5* and *CHS* gene, and accumulation of anthocyanin [96].

COP1 is related to the repression of photomorphogenesis but it seems that UVR8 provides UV-B-specific signaling and that the interaction COP1-UVR8 occurs within minutes [96]. UVR8 is reverted to homodimer (inactive form) through the REPRESSOR OF UV-B PHOTOMORPHOGENESIS proteins (RUP1 and 2). RUP1 and RUP2 are two highly related WD40-repeat proteins that interact directly with UVR8 promoting its homodimerization, thus acting act as negative regulators [104, 108, 109]. *RUP 1* and *RUP 2* are induced by UV-B but act downstream of UVR8-COP1 forming a negative feedback loop that balances UV-B defense [109] (**Figure 2**).

Inhibition of the transcription factor HY5 by binding with COP1 in the absence of UV-B light, the expression of the response genes remains inactive (left). In the presence of UV-B light, the monomeric form of UVR8 enters the nucleus, binds to COP1, and activates HY5 (right).

7. Conclusions

The effects that UV radiation causes on plants have been extensively investigated from different perspectives. Studies on alterations in photomorphogenesis, primary metabolism, particularly photosynthesis, secondary metabolism, or gene expression have been carried out with model plants such as *Arabidopsis thaliana*; however, several other aspects still need to be addressed. Current technologies, such as omics tools, allow for the study of plants in their natural environments, considering all their complexity, and will, undoubtedly, lead to a better understanding of the impact of UV radiation in plants, an important constituent of climate change. Looking into the future, an integral view of plant responses to UV radiation has broad applications in agriculture and conservation, while providing scientific foundations for upcoming international regulations.

Acknowledgements

Consejo Nacional de Ciencia y Tecnología [CONACyT, Ciencia Básica Grant No. A1-S-14605].

Author details

María del Socorro Sánchez Correa[1], María el Rocío Reyero Saavedra[2], Edgar Antonio Estrella Parra[3], Erick Nolasco Ontiveros[3], José del Carmen Benítez Flores[4], Juan Gerardo Ortiz Montiel[5], Jorge Eduardo Campos Contreras[6], Eduardo López Urrutia[7], José Guillermo Ávila Acevedo[3], Gladys Edith Jiménez Nopala[2] and Adriana Montserrat Espinosa González[3]*

1 Faculty of Higher Studies-Iztacala (FES-I), Scientific Investigation I Laboratory, National Autonomous University of Mexico (UNAM), México City, Mexico

2 Center for Genome Sciences (CCG), Functional Genomics of Eukaryotes, National Autonomous University of Mexico (UNAM), Cuernavaca, State of Morelos, Mexico

3 Faculty of Higher Studies-Iztacala (FES-I), Laboratory of Phytochemistry, UBIPRO, National Autonomous University of Mexico (UNAM), México City, Mexico

4 Faculty of Higher Studies-Iztacala (FES-I), Laboratory of Histopathology, UMF, National Autonomous University of Mexico (UNAM), México City, Mexico

5 Faculty of Higher Studies-Iztacala (FES-I), Laboratory of Plant Tissues Culture, UMF, National Autonomous University of Mexico (UNAM), México City, Mexico

6 Faculty of Higher Studies-Iztacala (FES-I), Laboratory of Molecular Biochemistry, UBIPRO, National Autonomous University of Mexico (UNAM), México City, Mexico

7 Faculty of Higher Studies-Iztacala (FES-I), Laboratory of Functional Genomic, UBIMED, National Autonomous University of Mexico (UNAM), State of México, Mexico

*Address all correspondence to: adriana.espinosa@iztacala.unam.mx

References

[1] Hollósy F. Effects of ultraviolet radiation on plant cells. Micron. 2002;**33**(2):179-197. DOI: 10.1016/s0968-4328(01)00011-7

[2] Gill SS, Anjum NA, Gill R, Jha M, Tuteja N. DNA damage and repair in plants under ultraviolet and ionizing radiations. The Scientific World Journal. 2015;**2015**:250158. DOI: 10.1155/2015/250158

[3] Sharma S, Chatterjee S, Kataria S, Joshi J, Datta S, Vairale MG, et al. Review on responses of plants to UV-B radiation related stress. In: Singh VP, Singh S, Prasad SM, Parihar P, editors. UV-B Radiation. USA: John Wiley & Sons Inc. 2017. pp. 75-97. DOI: 10.1002/9781119143611

[4] Vanhaelewyn L, Van Der Straeten D, De Coninck B, Vandenbussche F. Ultraviolet radiation from a plant perspective: The plant-microorganism context. Frontiers of Plant Science. 2020;**15**(11):597642. DOI: 10.3389/fpls.2020.597642

[5] Carrasco-Ríos L. Efecto de la radiación ultravioleta-B en plantas. Idesia (Arica). 2009;**27**(3):59-76. DOI: 10.4067/S0718-34292009000300009

[6] Carvalho SD, Castillo JA. Influence of light on plant-phyllosphere interaction. Frontiers of Plant Science. 2018;**9**:1482. DOI: 10.3389/fpls.2018.01482

[7] Verdaguer D, Jansen MA, Llorens L, Morales LO, Neugart S. UV-A radiation effects on higher plants: Exploring the known unknown. Plant Science. 2017;**255**:72-81. DOI: 10.1016/j.plantsci.2016.11.014

[8] Robson TM, Urban KK, Jansen MAK. Plant morphological responses to UV-B. Plant, Cell & Environment. 2015;**38**:856-866. DOI: 10.1111/pce.12374

[9] Barnes PW, Williamson CE, Lucas RM, Robinson SA, Madronich S, Paul ND, et al. Ozone depletion, ultraviolet radiation, climate change and prospects for a sustainable future. Nature Sustainability. 2019;**2**(7):569-579. DOI: 10.1038/s41893-019-0314-2

[10] Podolec R, Demarsy E, Ulm R. Perception and signaling of ultraviolet-B radiation in plants. Annual Review of Plant Biology. 2021;**72**:793-822. DOI: 10.1146/annurev-arplant-050718-095946

[11] Bussell JS, Gwynn-Jones D, Griffith GW, Scullion J. Above-andbelow-ground responses of *Calamagrostis purpurea* UV-B radiation and elevated CO_2 under phosphorus limitation. Physiologia Plantarum. 2012;**145**:619-628. DOI: 10.1111/j.1399-3054.2012.01595.x

[12] Dukowic-Schulze S, Harvey A, Garcia N, Chen C, Gardner G. UV-B irradiation results in inhibition of hypocotyl elongation, cell cycle arrest, and decreased endoreduplication mediated by miR5642. *Photochemistry and Photobiology*. 2022;**98**(5):1084-1099. DOI: 10.1111/php.13574

[13] Biever JJ, Brinkman D, Gardner G. UV-B inhibition of hypocotyl growth in etiolated *Arabidopsis thaliana* seedlings is a consequence of cell cycle arrest initiated by photodimer accumulation. Journal of Experimental Botany. 2014;**65**(11):2949-2961. DOI: 10.1093/jxb/eru035

[14] Suesslin C, Frohnmeyer H. An Arabidopsis mutant defective in

UV-B light-mediated responses. The Plant Journal. 2003;**33**:591-601. DOI: 10.1046/j.1365-313X.2003.01649.x

[15] Robson TM, Aphalo PJ, Bana's AK, Barnes PW, Brelsford CC, Jenkins GI, et al. A perspective on ecologically relevant plant-UV research and its practical application. Photochemical and Photobiological Sciences. 2019;**18**:970-988. DOI: 10.1039/c8pp00526e

[16] Krasylenko YA, Yemets AI, Blume YB. Plant microtubules reorganization under the indirect UV-B exposure and during UV-B-induced programmed cell death. Plant Signaling & Behavior. 2013;**8**(5):e24031. DOI: 10.4161/psb.24031

[17] Liu L, Wang X, Chang C. Toward a smart skin: Harnessing cuticle biosynthesis for crop adaptation to drought, salinity, temperature, and ultraviolet stress. Frontiers in Plant Science. 2022;**13**:961829. DOI: 10.3389/fpls.2022.961829

[18] Jansen MAK, Van Den Noort RE. Ultraviolet-B radiation induces complex alteration in stomatal behavior. Physiologia Plantarum. 2000;**110**:189-194. DOI: 10.1034/j.1399-3054.2000.110207.x

[19] Nasibova AN. Advances in Biology and Earth Sciences. Azerbaijan: Jomard Publishing; 2022. p. 84. ISSN 2520-2847

[20] Swarna S, Lorente C, Thomas AH, Martin CB. Rate constants of quenching of the fluorescence of pterins by the iodide anion in aqueous solution. Chemical Physics Letters. 2012;**542**:62-65. DOI: 10.1016/j.cplett.2012.06.013

[21] Ranjbarfordoei A, Samson R, Damme PV. Photosynthesis performance in sweet almond [*Prunus dulcis* (mill) D. Webb] exposed to supplemental UV-Bradiation. Photosynthetica. 2011;**49**:107-111. DOI: 10.1007/s11099-011-0017-z

[22] Jovanić B, Radenković B, Despotović-Zrakić M, Bogdanović Z, Barać D. Effect of UV-B radiation on chlorophyll fluorescence, photosynthetic activity and relative chlorophyll content of five different corn hybrids. Journal of Photochemistry and Photobiology. 2022;**10**:100115. DOI: 10.1016/j.jpap.2022.100115

[23] Ihle C. Degradation and release from the thylakoid membrane of photosystem II subunits after UV-B irradiation of the liverwort *Conocephalum conicum*. Photosynthesis Research. 1997;**54**:73-78. DOI: 10.1023/A:1005871729513

[24] Booij-James IS, Dube SK, Jansen MAK, Edelman M, Mattoo AK. Ultraviolet-B radiation impacts light-mediated turnover of the photosystem II reaction center heterodimer in Arabidopsis mutants altered in phenolic metabolism. Plant Physiology. 2000;**124**(3):1275-1284. DOI: 10.1104/pp.124.3.1275

[25] Kataria S, Jajoo A, Guruprasad KN. Impact of increasing ultraviolet-B (UV-B) radiation on photosynthetic processes. Journal of Photochemistry and Photobiology B: Biology. 2014;*137*:55-66. DOI: 10.1016/j.jphotobiol.2014.02.004

[26] Renger G, Voss M, Gräber P, Schulze A. Effect of UV irradiation on different partial reactions of the primary processes of photosynthesis. In: Worrest RC, Caldwell MM, editors. Stratospheric Ozone Reduction, Solar Ultraviolet Radiation and Plant Life. NATO ASI Series. Vol. 8. Berlin, Heidelberg: Springer; 1986. pp. 110-116. DOI: 10.1007/978-3-642-70090-3_13

[27] Gogo EO, Förster N, Dannehl D, Frommherz L, Trierweiler B, Opiyo AM,

et al. Postharvest UV-C application to improve health promoting secondary plant compound pattern in vegetable amaranth. Innovative Food Science & Emerging Technologies. 2018;**45**:426-437. DOI: 10.1016/j.ifset.2018.01.002

[28] Kim S, Lee H, Kim KH. Metabolomic elucidation of recovery of *Melissa officinalis* from UV-B irradiation stress. Industrial Crops and Products. 2018;**121**:428-433. DOI: 10.1016/j.indcrop.2018.05.002

[29] Zhao B, Wang L, Pang S, Jia Z, Wang L, Li W, et al. UV-B promotes flavonoid synthesis in *Ginkgo biloba* leaves. Industrial Crops and Products. 2020;**151**:112483. DOI: 10.1016/j.indcrop.2020.112483

[30] Liu Y, Liu J, Wang HZ, Wu KX, Guo XR, Mu LQ, et al. Comparison of the global metabolic responses to UV-B radiation between two medicinal Astragalus species: An integrated metabolomics strategy. Environmental and Experimental Botany. 2020;**176**:104094. DOI: 10.1016/j.envexpbot.2020.104094

[31] Duarte-Sierra A, Munzoor Hasan SM, Angers P, Arul J. UV-B radiation hormesis in broccoli florets: Glucosinolates and hydroxy-cinnamates are enhanced by UV-B in florets during storage. Postharvest Biology and Technology. 2020;**168**:111278. DOI: 10.1016/j.postharvbio.2020.111278

[32] Klem K, Oravec M, Holub P, Šimor J, Findurová H, Surá K, et al. Interactive effects of nitrogen, UV and PAR on barley morphology and biochemistry are associated with the leaf C:N balance. Plant Physiology and Biochemistry. 2022;**172**:111-124. DOI: 10.1016/j.plaphy.2022.01.006

[33] Wang H, Liu S, Wang T, Liu H, Xu X, Chen K, et al. The moss flavone synthase I positively regulates the tolerance of plants to drought stress and UV-B radiation. Plant Science. 2020;**298**:110591. DOI: 10.1016/j.plantsci.2020.110591

[34] Jiao J, Xu XJ, Lu Y, Liu J, Fu YJ, Fu JX, et al. Identification of genes associated with biosynthesis of bioactive flavonoids and taxoids in Taxus cuspidata Sieb. Et Zucc. Plantlets exposed to UV-B radiation. Gene. 2022;**823**:146384. DOI: 10.1016/j.gene.2022.146384

[35] Miao N, Yun C, Shi Y, Gao Y, Wu S, Zhang Z, et al. Enhancement of flavonoid synthesis and antioxidant activity in *Scutellaria baicalensis* aerial parts by UV-A radiation. Industrial Crops and Products. 2022;**187**(Part B):115532. DOI: 10.1016/j.indcrop.2022.115532

[36] Berglund T, Kalbin G, Strid A, Rydström J, Ohlsson AB. UV-B- and oxidative stress-induced increase in nicotinamide and trigonelline and inhibition of defensive metabolism induction by poly(ADP-ribose) polymerase inhibitor in plant tissue. FEBS Letters. 1996;**380**(1-2):188-193. DOI: 10.1016/0014-5793(96)00027-0

[37] Faustini CGR, Novo-Gama V, Valandro-Zanetti L, Terra-Werner E, Macedo-Pezzopane JE. UV-B effects on growth, photosynthesis, total antioxidant potential and cell wall components of shade-tolerant and sun-tolerant ecotypes of Paubrasilia echinata. Flora. 2020;**271**:151679. DOI: 10.1016/j.flora.2020.151679

[38] Pascual J, Cañal MJ, Escandón M, Meijón M, Weckwerth W, Valledor L. Integrated physiological, proteomic, and metabolomic analysis of ultraviolet (UV) stress responses and adaptation mechanisms in *Pinus radiata*. Molecular and Cellular Proteomics.

2017;**16**(3):485-501. DOI: 10.1074/mcp.M116.059436

[39] Tao M, Zhu W, Han H, Liu S, Liu A, Li S, et al. Mitochondrial proteomic analysis reveals the regulation of energy metabolism and reactive oxygen species production in *Clematis terniflora* DC. Leaves under high-level UV-B radiation followed by dark treatment. Journal of Proteomics. 2022;**254**:104410. DOI: 10.1016/j.jprot.2021.104410

[40] Nocchi N, Duarte HM, Pereira RC, Konno TUP, Soares AR. Effects of UV-B radiation on secondary metabolite production, antioxidant activity, photosynthesis and herbivory interactions in *Nymphoides humboldtiana* (Menyanthaceae). Journal of Photochemistry and Photobiology B: Biology. 2020;**212**:112021. DOI: 10.1016/j.jphotobiol.2020.112021

[41] Contreras RA, Marisol P, Hans K, Zamora P, Zúñiga GE. UV-B shock induces photoprotective flavonoids but not antioxidant activity in Antarctic Colobanthus quitensis (Kunth) Bart. Environmental and Experimental Botany. 2019;**159**:179-190. DOI: 10.1016/j.envexpbot.2018.12.022

[42] Zhao X, Zheng W, Qu T, Zhong Y, Xu J, Jiang Y, et al. Dual roles of reactive oxygen species in intertidal macroalgae *Ulva prolifera* under ultraviolet-B radiation. Environmental and Experimental Botany. 2021;**189**:104534. DOI: 10.1016/j.envexpbot.2021.104534

[43] Kolackova M, Chaloupsky P, Cernei N, Klejdus B, Huska D, Adam V. Lycorine and UV-C stimulate phenolic secondary metabolites production and miRNA expression in Chlamydomonas reinhardtii. Journal of Hazardous Materials. 2020;**391**:122088. DOI: 10.1016/j.jhazmat.2020.122088

[44] Rodrigues-Corrêa KCDS, Honda MDH, Borthakur D, Fett-Neto AG. Mimosine accumulation in *Leucaena leucocephala* in response to stress signaling molecules and acute UV exposure. Plant Physiology and Biochemistry. 2019;**135**:432-440. DOI: 10.1016/j.plaphy.2018.11.018

[45] Li Y, Liu S, Shawky E, Tao M, Liu A, Sulaiman K, et al. SWATH-based quantitative proteomic analysis of Morus alba L. leaves after exposure to ultraviolet-B radiation and incubation in the dark. Journal of Photochemistry and Photobiology. B. 2022;**230**:112443. DOI: 10.1016/j.jphotobiol.2022.112443

[46] Zhu W, Zheng W, Hu X, Xu X, Zhang L, Tian J. Variations of metabolites and proteome in *Lonicera japonica* Thunb. Buds and flowers under UV radiation. Biochimica et Biophysica Acta Proteins and Proteomica. 2017;**1865**(4):404-413. DOI: 10.1016/j.bbapap.2017.01.004

[47] Pandey A, Jaiswal D, Agrawal SB. Ultraviolet-B mediated biochemical and metabolic responses of a medicinal plant *Adhatoda vasica* Nees. at different growth stages. Journal of Photochemistry and Photobiology B: Biology. 2021;**216**:112142. DOI: 10.1016/j.jphotobiol.2021.112142

[48] Müller V, Albert A, Winkler J, Lankes C, Noga G, Hunsche M. Ecologically relevant UV-B dose combined with high PAR intensity distinctly affect plant growth and accumulation of secondary metabolites in leaves of *Centella asiatica* L. Urban. Journal of Photochemistry and Photobiology B: Biology. 2013;**127**:161-169. DOI: 10.1016/j.jphotobiol.2013.08.014

[49] Mariz-Ponte N, Mendes RJ, Sario S, Ferreira de Oliveira JMP, Melo P, Santos C. Tomato plants use non-enzymatic antioxidant pathways to

cope with moderate UV-A/B irradiation: A contribution to the use of UV-A/B in horticulture. Journal of Plant Physiology. 2018;**221**:32-42. DOI: 10.1016/j.jplph.2017.11.013

[50] Pizarro-Oteíza S, Salazar F. Effect of UV-LED irradiation processing on pectolytic activity and quality in tomato (*Solanum lycopersicum*) juice. Innovative Food Science and Emerging Technologies. 2022;**80**:103097. DOI: 10.1016/j.ifset.2022.103097

[51] Mariz-Ponte N, Martins S, Goncalves A, Correia CM, Ribeiro C, Dias MC, et al. The potential use of the UV-A and UV-B to improve tomato quality and preference for consumers. Scientia Horticulturae. 2019;**246**:777-784. DOI: 10.1016/j.scienta.2018.11.058

[52] Bano C, Nimisha A, Sunaina N, Singh B. UV-B radiation escalate allelopathic effect of benzoic acid on Solanum lycopersicum L. Scientia Horticulturae. 2017;**220**:199-205. DOI: 10.1016/j.scienta.2017.03.052

[53] Yin Y, Tian X, Yang J, Yang Z, Tao J, Fang W. Melatonin mediates isoflavone accumulation in germinated soybeans (*Glycine max* L.) under ultraviolet-B stress. Plant Physiology Biochemistry. 2022;**175**:23-32. DOI: 10.1016/j.plaphy.2022.02.001

[54] Raipuria RK, Kataria S, Watts A, Jain M. Magneto-priming promotes nitric oxide via nitric oxide synthase to ameliorate the UV-B stress during germination of soybean seedlings. Journal of Photochemistry and Photobiology B: Biology. 2021;**220**:112211. DOI: 10.1016/j.jphotobiol.2021.112211

[55] Ma M, Zhang H, Jiao y, Yang M, Tang J, Wang P, Yang R, Gu Z. Response of nutritional and functional composition, anti-nutritional factors and antioxidant activity in germinated soybean under UV-B radiation. Food Science and Technology. 2020;**118**:108709. DOI: 10.1016/j.lwt.2019.108709

[56] Ma M, Wang P, Yang R, Gu Z. Effects of UV-B radiation on the isoflavone accumulation and physiological-biochemical changes of soybean during germination: Physiological-biochemical change of germinated soybean induced by UV-B. Food Chemistry. 2018;**250**:259-267. DOI: 10.1016/j.foodchem.2018.01.051

[57] Mata-Ramírez D, Román S, Serna-Saldívar O, Antunes-Ricardo M. Enhancement of anti-inflammatory and antioxidant metabolites in soybean (*Glycine max*) calluses subjected to selenium or UV-light stresses. Scientia Horticulturae. 2019;**257**:108669. DOI: 10.1016/j.scienta.2019.108669

[58] Ma M, Xu W, Wang P, Gu Z, Zhang H, Yang R. UV-B- triggered H2O2 production mediates isoflavones synthesis in germinated soybean. Food Chemistry: X. 2022;**14**:100331. DOI: 10.1016/j.fochx.2022.100331

[59] Kovács V, Gondor OKS, Majláth G, Janda I, Pál T. UV-B radiation modifies the acclimation processes to drought or cadmium in wheat. Environmental and Experimental Botany. 2014;**100**:122-131. DOI: 10.1016/j.envexpbot.2013.12.019

[60] Chen Z, Ma Y, Yang R, Gu Z, Wang P. Effects of exogenous Ca2+ on phenolic accumulation and physiological changes in germinated wheat (Triticum aestivum L.) under UV-B radiation. Food Chemistry. 2019;**288**:368-376. DOI: 10.1016/j.foodchem.2019.02.131

[61] Huang H, Ge Z, Limwachiranon J, Li L, Li W, Luo Z. UV-C treatment

affects browning and starch metabolism of minimally processed lily bulb. Postharvest Biology and Technology. 2017;**128**:105-111. DOI: 10.1016/j.postharvbio.2017.02.010

[62] Martínez-Sánchez A, Guirao-Martínez J, Martínez JA, Lozano-Pastor P, Aguayo E. Inducing fungal resistance of spinach treated with preharvest hormetic doses of UV-C. LWT-Food Science and Technology. 2019;**113**:108302. DOI: 10.1016/j.lwt.2019.108302

[63] Wang M, Leng C, Zhu Y, Wang P, Gu Z, Yang R. UV-B treatment enhances phenolic acids accumulation and antioxidant capacity of barley seedlings. LWT LWT-Food Science and Technology. 2022;**153**:112445. DOI: 10.1016/j.lwt.2021.112445

[64] Nazari M, Fatemeh Z. Ultraviolet-B induced changes in *Mentha aquatica* (a medicinal plant) at early and late vegetative growth stages: Investigations at molecular and genetic levels. Industrial Crops and Products. 2020;**154**:112618. DOI: 10.1016/j.indcrop.2020.112618

[65] Surjadinata BB, Jacobo-Velázquez DA, Cisneros-Zevallos L. Physiological role of reactive oxygen species, ethylene, and jasmonic acid on UV light-induced phenolic biosynthesis in wounded carrot tissue. Postharvest Biology and Technology. 2021;**172**:111388. DOI: 10.1016/j.postharvbio.2020.111388

[66] Li L, Li C, Sun J, Xin M, Yi P, He X, et al. Synergistic effects of ultraviolet light irradiation and high-oxygen modified atmosphere packaging on physiological quality, microbial growth and lignification metabolism of fresh-cut carrots. Postharvest Biology and Technology. 2021;**173**:111365. DOI: 10.1016/j.postharvbio.2020.111365

[67] Heidrun E-K, Heller W, Sonnenbichler J, Zetl I, Schäfer W, Ernst D, et al. Oxidative stress and plant secondary metabolism: 6″-O-malonylapiin in parsley. Phytochemistry. 1993;**34**(3):687-691. DOI: 10.1016/0031-9422(93)85340-W

[68] Duarte-Sierra A, Nadeau F, Angers P, Michaud D, Arul J. UV-C hormesis in broccoli florets: Preservation, phyto-compounds and gene expression. Postharvest Biology and Technology. 2019;**157**:110965. DOI: 10.1016/j.postharvbio.2019.110965

[69] (a)Zhang K, Chen L, Wei M, Qiao H, Zhang S, Li Z, et al. Metabolomic profile combined with transcriptomic analysis reveals the value of UV-C in improving the utilization of waste grape berries. Food Chemistry. 2021;**363**:130288. DOI: 10.1016/j.foodchem.2021.130288

[70] Csepregi K, Kőrösi L, Teszlák P, Hideg E. Postharvest UV-A and UV-B treatments may cause a transient decrease in grape berry skin flavonol-glycoside contents and total antioxidant capacities. Phytochemistry Letters. 2019;**31**:63-68. DOI: 10.1016/j.phytol.2019.03.010

[71] Li M, Li X, Han C, Ji N, Jin P, Zheng Y. UV-C treatment maintains quality and enhances antioxidant capacity of fresh-cut strawberries. Postharvest Biology and Technology. 2019;**156**:110945. DOI: 10.1016/j.postharvbio.2019.110945

[72] Luengo-Escobar A, Magnum de Oliveira SF, Acevedo P, Nunes-Nesi A, Alberdi M, Reyes-Díaz M. Different levels of UV-B resistance in Vaccinium corymbosum cultivars reveal distinct backgrounds of phenylpropanoid metabolites. Plant Physiology and Biochemistry. 2017;**118**:541-550. DOI: 10.1016/j.plaphy.2017.07.021

[73] Inostroza-Blancheteau C, Acevedo P, Loyola R, Arce-Johnson P,

Alberdi M, Reyes-Díaz M. Short-term UV-B radiation affects photosynthetic performance and antioxidant gene expression in highbush blueberry leaves. Plant Physiology and Biochemistry. 2016;**107**:301-309. DOI: 10.1016/j.plaphy.2016.06.019

[74] Sun T, Ouyang H, Sun P, Zhang W, Wang Y, Cheng S, et al. Postharvest UV-C irradiation inhibits blackhead disease by inducing disease resistance and reducing mycotoxin production in 'Korla' fragrant pear (*Pyrus sinkiangensis*). International Journal of Food Microbiology. 2022;**362**:109485. DOI: 10.1016/j.ijfoodmicro.2021.109485

[75] (b)Zhang W, Jiang H, Cao J, Jiang W. UV-C treatment controls brown rot in postharvest nectarine by regulating ROS metabolism and anthocyanin synthesis. Postharvest Biology and Technology. 2021;**180**:111613. DOI: 10.1016/j.postharvbio.2021.111613

[76] Gil M, Pontin M, Berli F, Bottini R, Piccoli P. Metabolism of terpenes in the response of grape (*Vitis vinifera* L.) leaf tissues to UV-B radiation. Phytochemistry. 2012;**77**:89-98. DOI: 10.1016/j.phytochem.2011.12.011

[77] Interdonato R, Rosa M, Nieva CB, González JA, Hilal M, Prado FF. Effects of low UV-B doses on the accumulation of UV-B absorbing compounds and total phenolics and carbohydrate metabolism in the peel of harvested lemons. Environmental and Experimental Botany. 2011;**70**(2-3):204-211. DOI: 10.1016/j.envexpbot.2010.09.006

[78] Dias MC, Pinto DCGA, Freitas H, Santos C, Silva AMS. The antioxidant system in Olea europaea to enhanced UV-B radiation also depends on flavonoids and secoiridoids. Phytochemistry. 2020;**170**:112199. DOI: 10.1016/j.phytochem.2019.112199

[79] Santin M, Lucini L, Castagna A, Rocchetti G, Hauser MT, Ranieri A. Comparative "phenol-omics" and gene expression analyses in peach (*Prunus persica*) skin in response to different postharvest UV-B treatments. Plant Physiology and Biochemistry. 2019;**135**:511-519. DOI: 10.1016/j.plaphy.2018.11.009

[80] Parihar P, Singh R, Singh A, Prasad SM. Role of oxylipin on luffa seedlings exposed to NaCl and UV-B stresses: An insight into mechanism. Plant Physiology and Biochemistry. 2021;**167**:691-704. DOI: 10.1016/j.plaphy.2021.08.032

[81] Mishra P, Prasad SM. Low dose UV-B radiation-induced mild oxidative stress impact on physiological and nutritional competence of Spirulina (Arthrospira) species. Plant Stress. 2021;**2**:100039. DOI: 10.1016/j.stress.2021.100039

[82] Lambrecht CD, da Silveira MM, Kröning DP, Ziegler V, de Oliveira M, Ferreira CD. Chemical, physical, and sensory changes in rice subjected to UV-C radiation and its acceptability to rice weevil *Sitophilus oryzae* (L.) (Coleoptera: Curculionidae) and humans. Journal of Stored Products Research. 2021;**90**:101760. DOI: 10.1016/j.jspr.2020.101760

[83] Michailidis M, Karagiannis E, Polychroniadou C, Tanou G, Karamanoli K, Molassiotis A. Metabolic features underlying the response of sweet cherry fruit to postharvest UV-C irradiation. Plant Physiology and Biochemistry. 2019;**144**:49-57. DOI: 10.1016/j.plaphy.2019.09.030

[84] Bernhard GH, Neale RE, Barnes PW, Neale PJ, Zepp RG, Wilson SR... White CC. Environmental effects of stratospheric ozone depletion, UV radiation and interactions with

climate change: UNEP environmental effects assessment panel, update 2019. Photochemical & Photobiological Sciences. 2020;**19**:542-584. DOI: 10.1039/d0pp90011g

[85] Ulm R, Baumann A, Oravecz A, Mate Z, Adam E, Oakeley EJ, et al. Genome-wide analysis of gene expression reveals function of the bZIP transcription factor HY5 in the UV-B response of Arabidopsis. Proceedings of the National Academy of Sciences. 2004;**101**:1397-1402. DOI: 10.1073/pnas.0308044100

[86] Ulm R, Nagy F. Signalling and gene regulation in response to ultraviolet light. Current Opinion in Plant Biology. 2005;**8**:477-482. DOI: 10.1016/j.pbi.2005.07.004

[87] Casati P, Walbot V. Rapid transcriptome responses of maize (*Zea mays*) to UV-B in irradiated and shielded tissues. Genome Biology. 2004;**5**:R16. DOI: 10.1186/gb-2004-5-3-r16

[88] Britt AB. Repair of DNA damage induced by solar UV. Photosynthesis Research. 2004;**81**:105-112. DOI: 10.1023/B:PRES.0000035035.12340.58

[89] Jenkins GI, Long JC, Wade HK, Shenton MR, Bibikova TN. UV and blue light signalling: Pathways regulating chalcone synthase gene expression in Arabidopsis. New Phytologist. 2001;**151**:121-131. DOI: 10.1046/j.1469-8137.2001.00151.x

[90] Liu Y, Roof S, Ye Z, Barry C, van Tuinen A, Vrebalov J, et al. Manipulation of light signal transduction as a means of modifying fruit nutritional quality in tomato. Proceedings of the National Academy of Sciences. 2004;**101**:9897-9902. DOI: 10.1073/pnas.0400935101

[91] Osterlund MT, Hardtke CS, Deng XW. Targeted destabilization of HY5 during light-regulated development of Arabidopsis. Nature. 2000;**405**(6785):462-466. DOI: 10.1038/35013076

[92] Quail PH. Photosensory perception and signalling in plant cells: New paradigms? Current Opinion in Cell Biology. 2002;**14**(2):180-188. DOI: 10.1016/0959-437x(94)90131-l

[93] Yi C, Deng XW. COP1—From plant photomorphogenesis to mammalian tumorigenesis. Trends in Cell Biology. 2005;**15**:618-625. DOI: 10.1016/j.tcb.2005.09.007

[94] Lockhart J. How ELONGATED HYPOCOTYL5 helps protect Plants from UV-B rays. The Plant Cell. 2014;**26**(10):3826-3826. DOI: 10.1105/tpc.114.133363

[95] Oravekz A, Baumann A, Máté Z, Brzezinska A, Molinier J, Oakeley EJ, et al. CONSTITUTIVELY PHOTOMORPHOGENIC is required for the UV-B response in Arabidopsis. The Plant Cell. 2006;**18**:1975-1990. DOI: 10.1105/tpc.105.040097

[96] Favory JJ, Stec A, Gruber H, Rizzini L, Oravekz A, Funk M, et al. Interaction of COP1 and UVR8 regulates UV-B induced photomorphogenesis and stress acclimation in Arabidopsis. The EMBO Journal. 2009;**28**:591-601. DOI: 10.1038/emboj.2009.4

[97] von Arnim AG, Deng XW. Light inactivation of Arabidopsis photomorphogenic repressor COP1 involves a cell-specific regulation of its nucleocytoplasmic partitioning. Cell. 1994;**79**:1035-1045. DOI: 10.1016/0092-8674(94)90034-5±

[98] Wang H, Ma LG, Li JM, Zhao HY, Deng XW. Direct interaction of

Arabidopsis cryptochromes with COP1 in light control development. Science. 2001;**294**:154-158. DOI: 10.1126/science.1063630

[99] Seo HS, Watanabe E, Tokutomi S, Nagatani A, Chua NH. Photoreceptor ubiquitination by COP1 E3 ligase desensitizes phytochrome A signaling. Genes and Development. 2004;**18**:617-622. DOI: 10.1101/gad.1187804

[100] Sheerin DJ, Menon C, Zur Oven-Krockhaus S, Enderle B, Zhu L, Johnen P, et al. Light-activated phytochrome A and B interact with members of the SPA family to promote photomorphogenesis in Arabidopsis by reorganizing the COP1/SPA complex. The Plant Cell. 2015;**27**:189-201. DOI: 10.1105/tpc.114.134775

[101] Christie JM, Arvai AS, Baxter KJ, Heilmann M, Pratt AJ, O'Hara A, et al. Plant UVR8 photoreceptor senses UV-B by tryptophan-mediated disruption of cross-dimer salt bridges. Science. 2012;**335**:1492-1496. DOI: 10.1126/science.1218091

[102] Wu D, Hu Q, Yan Z, Chen W, Yan C, Huang X, et al. Structural basis of ultraviolet-B perception by UVR8. Nature. 2012;**484**:214-219. DOI: 10.1038/nature10931

[103] O'Hara A, Jenkins GI. In vivo function of tryptophans in the Arabidopsis UV-B photoreceptorUVR8. The Plant Cell. 2012;**24**:3755-3766. DOI: 10.1105/tpc.112.101451

[104] Ulm R, Jenkins GI. Q&A: How do plants sense and respond to UV-B radiation? BMC Biology. 2015;**13**:45. DOI: 10.1186/s12915-015-0156-y

[105] Jenkins GI. Structure and function of the UV-B photoreceptor UVR8. Current Opinion in Structural Biology. 2014;**29**:52-57. DOI: 10.1016/j.sbi.2014.09.004

[106] Fernandez MB, Tossi V, Lamattina L, Cassia R. A comprehensive phylogeny reveals functional conservation of the UV-B photoreceptor UVR8 from green algae to higher plants. Frontiers in Plant Science. 2016;**7**:1698. DOI: 10.3389/fpls.2016.01698

[107] Kaiserli E, Jenkins GI. UV-B promotes rapid nuclear translocation of the Arabidopsis UV-B specific signaling component UVR8 and activates its function in the nucleus. The Plant Cell. 2007;**19**:2662-2673. DOI: 10.1105/tpc.107.053330

[108] Heijde M, Ulm R. Reversion of the Arabidopsis UV-B photoreceptor UVR8 to the homodimeric ground state. Proceedings of the National Academy of Sciences. 2013;**110**:1113-1118. DOI: 10.1073/pnas.1214237110

[109] Gruber H, Heijde M, Heller W, Albert A, Seidlitz H, Ulm R. Negative feedback regulation of UV-B-induced photomorphogenesis and stress acclimation in *Arabidopsis*. Proceedings of the National Academy of Sciences. 2010;**107**(46):20132-20137. DOI: 10.1073/pnas.0914532107

Section 2

Physiology: Plant Response to Stressors

Chapter 3

Role of Plant Hormones in Mitigating Abiotic Stress

Nazima Rasool

Abstract

Agricultural productivity world over is threatened by abiotic stress, intensifying food security issues. The plant hormones play a significant role in mitigating abiotic stresses, including drought stress, salinity stress, heat stress, and heavy metal stress, faced by the plants. Considerable research has been conducted to understand hormone-mediated abiotic stress responses in plants and the underlying biosynthetic and regulatory pathways. Deciphering these pathways would allow their manipulation in the laboratory and possible extension to the field. In the present chapter, an overview of the role plant hormones play in mitigating abiotic stress, the underlying mechanisms of their action, and the cross-talk between their signaling pathways to mitigate abiotic stress is presented.

Keywords: abiotic stress, plant hormones, stress response, stress mitigation, plant productivity

1. Introduction

Plant hormones or phytohormones are biochemicals required for the normal growth and development of plants [1–3]. Plant hormones include auxins (IAA), gibberellins (GAs) cytokinins (CK), abscisic acid (ABA), ethylene (ET), besides jasmonates (JA), salicylic acid (SA), brassinosteroids (BR), strigolactones (SL), and nitric oxide (NO). Apart from their role in plant growth and development, hormones also mediate response to biotic (disease, pathogens, herbivores, etc.) and abiotic (drought, heat, salinity, heavy metals, etc.) stress [3–6]. Hormones act at the site of their biosynthesis or some distance away from it [3, 6–8]. Hormone biosynthesis, distribution, and patterns of their signal transduction change under stress conditions [8, 9]. Ethylene and ABA play remarkable roles in regulating the abiotic stress response [8]. The exogenous supply of phytohormones also increases stress endurance in plants [10, 11]. Abiotic stress factors rarely occur individually, and many stresses produce the same effects at the cellular level with an overlap in the expression pattern of stress response genes [12]. In the current chapter, the hormone-mediated response of plants to the abiotic stress, including drought, heat, and salinity, is discussed.

 IntechOpen

1.1 What is abiotic stress?

Ecological factors favor plant growth at optimum levels and constitute stress at sub- or supra-optimal levels. Abiotic stress reduces crop productivity by about 50% (**Table 1**) [8, 29]. High temperatures lead to 20% decrease in the yield, low temperatures 7%, salinity 10%, drought 9% and other forms of stress cause 4% yield loss [30]. In grain crops, grain size, number, and dry weight are influenced by abiotic stress, especially if present during the reproductive phase [30]. Various aspects of plant growth as affected by abiotic stress are presented in **Figure 1**. Crop productivity may be reduced by 2.5–16% by a 1°C rise in seasonal temperature in tropical and subtropical regions [31]. The stress response depends on the genetic constitution and adaptive response of a plant [32].

1.2 Drought

Drought has been defined as "a period of abnormally dry weather sufficiently prolonged for the lack of water to cause a serious hydrologic imbalance in the affected area" [33]. Drought is one of the dominant factors diminishing crop productivity [34, 35]. Drought has been called as "one of the world's extreme weather-related natural hazards" [35, 36]. It threatens the sustainability of agricultural systems around the world [37]. From 1994 to 2013, it represented 5% of

Crop	Abiotic stress factor	Loss in productivity	References
Wheat	Drought	27.5%	Zhang et al., [13]
	Temperature	29–44%	Djanaguiraman et al., [14]
	Salinity	45%	Ali et al., [15]
Rice	Drought	25.4%	Zhang et al., [13]
	Temperature	3.2%*	Zhao et al., [16]
	Salinity	30–50%	Eynard et al., [17]
Maize	Drought	5–15%	Campos et al., [18]
	Temperature	7.4%*	Zhao et al., [16]
	Salinity	34%	Cucci et al., [19]
Chickpea	Drought	45–69%	Nayyar et al., [20]
	Temperature	39%	Devasirvatham et al., [21]
	Salinity	8–10%	Zawude and Shanko, [22]
Soybean	Drought	46–71%	Samarah et al., [23]
	Temperature	42–64%	Jumrani & Bhatia, [24]
	Salinity	66–86%	Bustingorri & Lavado [25]
Sunflower	Drought	50%	Hussain et al., [26]
	Temperature	*6%	Rondanini et al., [27]
	Salinity	50%	El-Kader et al., [28]

*****_Estimates for 1°C rise in temperature._

Table 1.
Loss of productivity in the major staple crops due to abiotic stress.

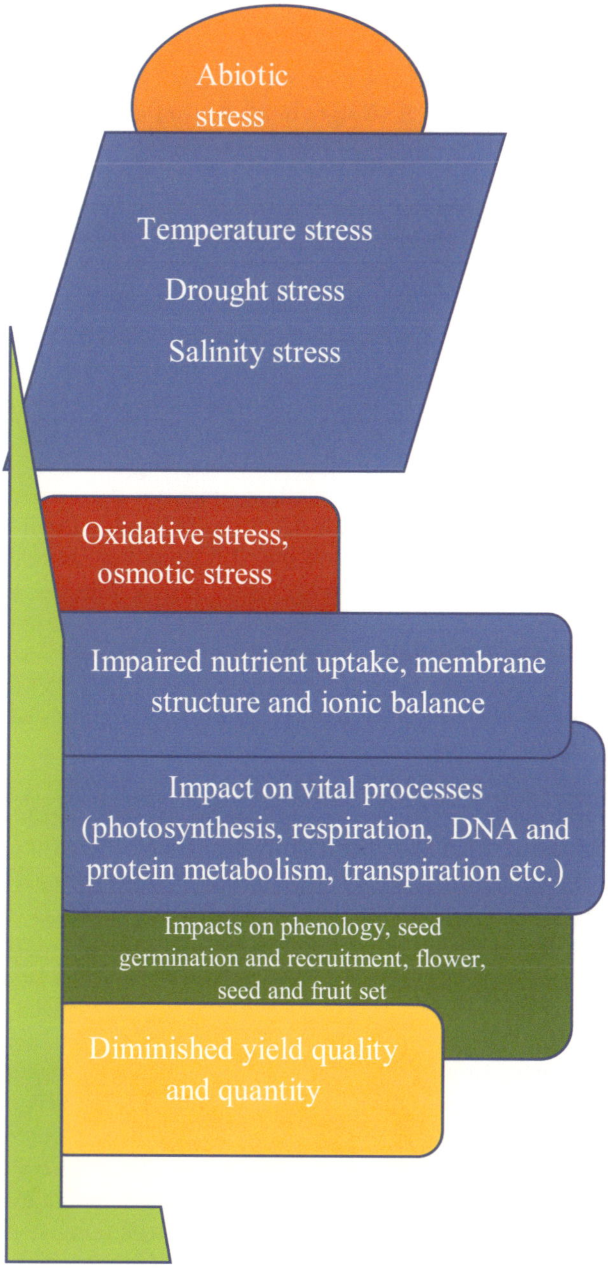

Figure 1.
Schematic presentation of impact of abiotic stress on growth & development of plants.

all natural disasters and affected one billion people [35, 38]. All aspects of plant growth, including photosynthesis, protein synthesis, water relations, cell turgidity, membrane integrity, and nutrient uptake, are affected by drought [8, 39, 40]. It causes oxidative stress and damages the biological molecules, including DNA, proteins, and photosynthetic pigments. [8, 35, 41–45]. Plants synthesize a whole range of molecules as protection against drought stress, for example, proline, glycine betaine, soluble sugars (mannitol, sorbitol, and trehalose), polyamines, and proteins [37, 46].

ABA levels increase in plants under drought stress [47, 48], inducing the expression of ABA-dependent genes [6, 49]. ABA signaling leads to the closure of stomata, reducing transpiration [48]. Expression levels of *ZEP* (*Zeaxanthin Epoxidase*) gene, AAO3 (*Arabidopsis Aldehyde Oxidase*) gene, NCED3 (*Nine-Cis-Epoxycarotenoid Dioxygenase*) gene, and the *MCSU* (*Molybdenum Cofactor Sulfurase*) *Gene* are increased upon osmotic stress [50]. Overexpression of *NCED3* improves water use efficiency and its mutation causes drought susceptibility [49, 51, 52]. ABA is transported into the guard cells through passive diffusion via members of ABC (ABCG25 and ABCG40) and nitrate (AIT1/NRT1.2 and NPF4.6) transporter families. *ABCG25* is an ABA exporter with tissue-specific expression induced by ABA and drought stress [49, 53]. *ABCG40*, AIT1/NRT1.2, and NPF4.6 import ABA into the guard cells. ABA also generates ROS, which leads to increased cytosolic Ca^{2+} levels and stomatal closure [54–56].

About 14 ABA receptor proteins mediate ABA signaling. Pyrabactin Resistance 1 (PYR1) and PYR1-like (PYL) regulatory elements undergo a conformational change after ABA binding and inactivate the clade A Serine/Threonine Protein Phosphatase 2C (PP2C) [48, 57, 58]. This in turn triggers the ABA signaling cascade by phosphorylation of serine/threonine kinases [48, 59]. Transcription of ABA-responsive genes is upregulated by binding of ABRE (ABA-Responsive Elements) to the ABRE-Binding Proteins (AREBs) or ABRE-Binding Factors (ABFs) [48, 60]. ABFs are activated by their ABA-mediated phosphorylation [48, 61]. *AREB1/ABF2*, *AREB2/ABF4,* and *ABF3* are induced by abiotic stress, including dehydration and high salinity [48]. Transcription factors belonging to MYC, MYB, and NAC protein families are also known to work in an ABA-dependent manner [48, 62, 63]. Stress response improves in plants overexpressing RD26 (Responsive to Desiccation 26), a stress-inducible NAC transcription factor [48, 63]. Dehydration-Responsive Element (DRE)-Binding Protein (DREB) transcription factors are regulated by ABA-dependent pathways under osmotic stress [48, 64]. The binding of AREB1, AREB2, and ABF3 to the DREB2A promoter results in the activation of *DREB2A* in an ABA-dependent manner [48, 65].

In *Arabidopsis thaliana,* many genes involved in ABA biosynthesis and signaling have been characterized [6, 66]. When *A. thaliana* is exposed to drought or salt stress, expression of the *ABA3/LOS5* gene increases considerably [67]. Constitutive or drought-induced expression of this gene has been reported to increase rice yield [6, 68]. *AtNCED3* plays an important role in drought tolerance [6]. Higher expression of *SgNCED1* in transgenic tobacco plants with this gene from *Stylosanthes guianensis* had improved drought and salinity tolerance and higher (51–77%) ABA content [6, 69]. In tomato plants, overexpression of *LeNCED1* constitutively resulted in the accumulation of ABA [6, 70]. Drought-inducible *rd29A* promoter-driven gene construct in *Brassica napus* increased yield under mild drought conditions [6, 71]. The wild form of this gene codes for the b-subunit of farnesyltransferase, which is involved in ABA-dependent signal transduction [72]. Exogenous application of ABA enhances the activities of GT, CAT, APX, and SOD [31, 37]. ABA priming increases the relative water content in drought-stressed wheat cultivars [73].

ABA is negatively regulated by cytokinin receptor HKs, AHK2 (*Arabidopsis* histidine kinase 2), AHK3, and AHK4 mutations in these genes increase drought tolerance [49, 74, 75]. CK, being an ABA antagonist, is decreased in conditions of drought stress; however, CK has also been reported to increase proline levels, inhibit senescence, and promote survival under drought conditions [6, 76]. Exogenously applied 6-benzylaminopurine increased the photosynthetic rate and stimulated protective enzymes in the maize seedlings [77]. BRs increase drought tolerance in many plants when applied exogenously [6, 78]. However, some reports also suggest that endogenous BRs or their perception are not involved in the water stress response [6, 79]. Auxins regulate ABA [37, 80]; Indole-3-acetic acid (IAA)-amido synthetase encoding gene *TLD1/OsGH3.13* increases expression of LEA (Late Embryogenesis Abundant) genes increasing drought tolerance in rice seedlings [6, 81]. Several studies indicate mitigating effects of SA on drought, salinity, and high-temperature stress [82, 83]. SA has been reported to increase catalase activity in wheat under drought stress [37, 84]. In *Portulaca oleracea*, SA improved photosynthetic pigments, secondary metabolites, and gas exchange [37, 85]. SA application increased water use efficiency, photosynthesis, and activity of antioxidant enzymes and also prevented cell damage under drought [86]. Ethylene activates DREB transcription factors [87]. Under mild drought stress shoot dry weight of six cultivars of wheat ranging from sensitive to tolerant was higher in the tolerant ones, which was related to higher ethylene content [37, 88]. *Etol1* mutants of rice that accumulate more ethylene than *OsETOL1* tolerate drought better.

1.3 Temperature

Temperature affects the distribution, phenology, and physiology of plants [89]. Temperature is increasing under dry as well as wet conditions in the changing global climate scenario [89, 90]. For 2081 – 2100, the IPCC has predicted average temperatures higher by 1.0°C to 1.8°C under very low, 2.1°C to 3.5°C under intermediate and 3.3°C to 5.7°C under very high GHG emission scenarios in comparison to 1850-1900 [91]. The crop productivity decreases by 6% for one degree rise in temperature beyond the optimum [8, 92]. Temperature stress causes accumulation of ROS, denaturation, misfolding, and aggregation of proteins, changes the membrane structure affecting permeability and raft distribution, besides its impact on leaf area, leaf retention, stomatal conductance, water potential, rate of transpiration, etc. [89–91, 93]. Photosynthetic capacity may be diminished or permanently damaged due to heat stress [91, 94].

Plants produce transcription factors, heat signaling proteins, and molecular chaperones to prevent protein misfolding and aggregation after heat shock (HS) [95, 96]. In response to HS, the endogenous ABA levels increase transiently increasing the antioxidant capacity [47, 97, 98], for example, by inducing RBOH-NADPH oxidases. Out of 10 different *RBOH* genes identified in *Arabidopsis,* only *AtRBOHD* is upregulated in response to heat stress [91, 99]. The mutants for this gene show low germination and seedling survival at higher temperatures [91, 97, 100]. ABA biosynthesis inhibitors and ABA signaling mutants have impaired heat stress tolerance [91, 97]. Both the heat shock proteins and their transcription factors are regulated by ABA. Expression levels of *ABA1/ZEP* and *NCED2/5/9* increase in *Arabidopsis* at 32°C increasing the ABA levels. Cucumbers and red-skinned grapes show higher ABA levels at 35°C [98, 101]. Drought priming in *Festuca arundinacea* and *Arabidopsis* increases their heat tolerance. *Arabidopsis* ABA biosynthesis mutants or plants treated with ABA biosynthesis inhibitors lack the drought priming effect [91, 102]. ABA treatment increases the expression of tall fescue heat stress transcription factor A2c (*FaHSFA2c*). ABA may also modulate carbohydrate and energy status to strengthen the heat stress response [103].

Auxins play an important role in thermomorphogenesis [91, 98, 104]. Auxin biosynthesis genes *TAA1, CYP79B2,* and *YUCCA8* are upregulated at higher temperatures; thermomorphogenesis response is abolished in *shy2–2* (short hypocotyl) mutation affecting the auxin-responsive *IAA3* gene [98, 105, 106]. Auxin-mediated thermomorphogenesis is regulated through phytochrome interacting factors (PIFs) and bHLH transcriptional regulators. PIFs also upregulate auxin biosynthesis; HDA9 (histone deacetylase 9), a chromatin-modifying enzyme, facilitates the binding of PIF4 to the promoter of YUCCA8 [91, 98, 106, 107]. *pif4* mutants have very low levels of enzymes of the YUCCA family, aminotransferase, and cytochrome P450s, which are involved in the heat stress response [91, 105, 106]. In *pif4* plants, ectopic expression of *PIF4* under an epidermis-specific promoter restores hypocotyl elongation induced by heat stress [91, 98, 108]. PIF4 and PIF7 loss of function mutants lose their heat stress-induced thermomorphogenesis. Thermomorphogenesis also requires HSP90. Thermomorphogenesis also involves brassinosteroids through phyB-PIF4 [98, 109, 110]; the temperature-sensitivity of hypocotyl elongation is inhibited by the application of PPZ (propiconazole), a BR biosynthesis inhibitor [109, 111].

BRs increase the production of HSPs [91, 112] and regulate the heat-induced accumulation of proton–pumping ATPase and aquaporins [91, 113], besides inducing the expression and activity of ROS scavenging enzymes under heat stress [91, 114]. In tomatoes, BR treatment increases the expression of RBOH1 and apoplast H_2O_2 levels [115]. Interestingly, H_2O_2 activates MPK2, which in turn enhances RBOH1 expression [91, 116]. Heat stress causes the accumulation of BZR1 (Brassinazole-resistant 1), an important transcription factor in BR signaling, in the nucleus [110].

Ethylene is another hormone involved in heat stress tolerance. EIN2 and ER1 mutants have poor survival rates under heat stress [95, 97]. *Arabidopsis* plants overexpressing ERF1 are more tolerant to heat stress than the control plants; ERF-1 overexpressing plants have higher transcript levels of *HsfA3* and *HSP70*. Studies have indicated increased synthesis of ET under heat stress [98, 117]. However, in *Arabidopsis*, *ein2–1* mutants exhibit greater tolerance to heat stress [91, 118]. ET is involved in CO_2-induced heat stress responses in tomatoes [11] and increased thermotolerance in rice [91, 95].

CKs play an important role in heat stress responses in plants [91, 119, 120]. They increase the activities of APX, SOD, and GP and also upregulate genes responsible for photosynthesis and carbohydrate metabolism under heat stress [91, 121]. CK oxidase/dehydrogenase inhibitors improve heat stress tolerance [91, 122]. Heat stress tolerance is also increased in plants with ectopic expression of isopentenyl transferase (ipt) from *Agrobacterium tumefaciens* [91, 123]. In *Arabidopsis*, rice, and passion fruit, external CK application decreased the negative effects of heat stress [124, 125].

In *Medicago sativa*, plant height, photosynthetic efficiency, and plant biomass were improved by pre-treatment with SA [126]. SA promotes the activities of CAT, SOD, and POX, which improve photosynthetic efficiency, ROS scavenging, and HSP21 levels [83, 95]. Under heat stress, it protects photosystem II and maintains high Rubisco activity [95, 127]. SA application in tomatoes under heat stress decreased oxidative damage and significantly improved gas exchange, proline content, and water use efficiency [83]. ET and JA accumulate in *Arabidopsis* after heat stress [97]. Plants with constitutive expression of PR1 (cpr5–1) have higher heat stress tolerance [95, 118]. Exogenous JA application reduced the negative effects of heat stress [95, 118]. External application of strigolactone to SL biosynthesis mutants restores seed thermo-inhibition [98, 128].

GA biosynthesis and accumulation increase under elevated temperatures in Carrizo citrange seedlings, wheat, and soybean hypocotyl [98, 129]. PIF4 upregulates the *GA20ox1* gene [130]. PIF4 transcription factors Class I TCP14 and TCP15 (Teosinte Branched 1, Cycloidea, and PCF) play an important role during this process; *TCP14* and *TCP15* mutants have reduced temperature sensitivity [130]. Seeds at

32°C have lower expression of *GA20ox1, GA20ox2, GA20ox3, GA3ox1, and GA3ox2* in comparison to those kept at 28–29°C [98, 131].

SA biosynthesis is suppressed at higher temperatures in tobacco after TMV infection and in *Arabidopsis* after Pst 350 DC3000 infection [132]. High temperatures suppress the expression of *Isochorismate Synthase 1* (*ICS1*), and *ics1* do not show temperature sensitivity to infection [132]. JA biosynthesis genes are upregulated by moderately higher (29–30°C) temperatures after wounding or Pst DC3000 infection in *Arabidopsis* [132, 133]. High temperatures have a tissue-specific effect on JA systemic transport in plants [122].

1.4 Salinity

Soils with electric conductivity higher than 4 dS/m at 25°C are classified as saline [134]. Salinity has affected more than 800 million hectares of land globally, decreasing potential agricultural land by 1–2% per year [8, 135]. More than 50% of the land in developing countries, particularly that falling in the arid region, is affected by salinity, causing yield losses to the tune of 40%, for example, in the case of wheat [8, 136]. It decreases the quantity as well as the quality of the produce [8, 137]. Salinity impairs water uptake and causes ion toxicity, osmotic stress, nutrient deficiency, and oxidative stress [138]. Salt stress causes physiological drought impairing protein and photosynthesis [8, 137]. Changes in the intracellular Ca^{2+} levels, excess Na^+, and ROS accumulation are the signals that trigger the salt stress response [139].

Ethylene is the major hormone in the salt stress response [138]. The levels of ET as well as its precursor ACC (1-aminocyclopropane-1-carboxylate) increase under salt stress [138, 140]. While salt tolerance can be increased by the application of ET or its precursor ACC [141, 142], inhibition of ET synthesis or signaling may increase salt sensitivity [138]. Ethylene signaling involves five ethylene receptors [ETR1 (Ethylene Response 1), ERS1 (Ethylene Response Sensor 1), ETR2, EIN4 (Ethylene Insensitive 4), and ERS2], a protein kinase, CTR1 (Constitutive Triple Response 1—a negative regulator) and a key positive regulator EIN2, which signals primary transcription factors EIN3, EIL1 (Ethylene Insensitive Like 1) and EIL2 and many downstream ethylene response factors. Osmotic stress, induced by many abiotic stresses, including salinity, suppresses the expression of *ETR1* [138, 143]. During short- and long-term salt stress, ethylene receptor genes (*ETR1, ETR2*, and *EIN4*), signaling genes (*CTR1, EIN3, ERF1*, and *ERF2*), and MAPK cascade genes (*MEKK1-MKK2-MPK4/6*), are upregulated in cotton [144]. Many *ERF* (Ethylene-Responsive Element Binding Factor) genes *ESE1–ESE3* are induced by ethylene and salt stress. Accumulation and transcriptional activity of EIN3 and EBF1/EBF2 degradation are promoted under salt stress [144]. The levels of 1-aminocyclopropane-1-carboxylic acid synthases (ACSs) increase significantly under salinity stress [138, 144, 145]. ACC pretreatment increases salt stress tolerance in *Arabidopsis* seedlings [138, 141, 142, 146]. Salinity induces ACS1 transcription in tobacco [138, 147]. Salt stress given to salt-acclimated and non-acclimated plants upregulated four ACSs [138, 148]. In the post-transcriptional regulation of ACSs, stress-induced MAPK cascades phosphorylate CSs, preventing their 26S proteasome-mediated degradation [138, 149]. The effect of salt acclimation is diminished by the loss of function of MAPK6 [148]. Stabilization of ACSs apparently needs MPK6 to maintain high ethylene levels [138, 148]. ACSs are also stabilized by CDPKs (Calcium-Dependent Protein Kinases) in tomatoes [138].

ACC content and activity of 1-aminocyclopropane-1-carboxylic acid oxidase (ACO) is increased under salt stress in *Cicer arietinum* roots [138].

200 mM NaCl induces the expression of *ETOL1* in rice [138, 150]. *ETO1's* loss of function promotes ethylene production in *Arabidopsis*. Root-to-shoot delivery of Na^+ is restricted in the absence of ETO1, which also increases RBOHF-dependent ROS accumulation in root stele tissue. Loss of ETO1 also increases K^+ levels by increasing K^+-transporter *HAK5* transcripts [138]. *Arabidopsis etr1* loss-of-function mutants have increased salt tolerance [141, 142, 151]. ET sensitivity decreases and salt sensitivity increases in tobacco and *Arabidopsis* on overexpression of *NTHK1* [138, 147]. Loss of function of *CTR1* increases salinity tolerance [142, 145]. *Arabidopsis* loss-of-function *EIN2* mutants are salt sensitive; overexpression of the C-terminus of *EIN2* in *ein2–5* mutants decreases salt sensitivity [138, 141, 142 152]. *ein3eil1* double mutants of *Arabidopsis* are highly sensitive to salinity. Also, *ein3*–1 mutants are highly salt-sensitive whereas plants overexpressing *EIN3* are salt tolerant [138, 142, 145, 152].

An array of stress-responsive genes is regulated by ABA [153]. ABA coordinates with ET in mediating salt stress. On exposure to salt, many genes involved in ABA biosynthesis, including ZEP, AAO, and MCSU, are stimulated through Ca_2^+ –dependent phosphorylation events and their downstream signaling pathways [153, 154]. Increased ABA levels have been reported in many plants, including *Oryza sativa* [155], *Brassica* [156], *Phaseolus vulgaris* [157], and *Zea mays* [158]. Higher ABA levels help accumulate proteins for osmotic adjustment and also cause stomatal closure. High accumulation of ABA due to ectopic expression of drought-responsive *Genes OsDSM2* (*Drought-Hypersensitive Mutant 2*) and *OsCam1–1* (*Oryza Sativa Calmodulin 1–1*) in rice increases salt stress tolerance [153]. Salt stress as well as ABA treatment upregulates several MAPKs [153], and plants with higher expression levels of MAPKs have higher salt stress tolerance [153, 159]. ABA-regulated Ca^{2+}-dependent kinases and SnRks phosphorylate ABA-related transcription factors, affect gene expression, and modulate salt stress [12, 153, 159]. Promoters of stress-responsive genes contain many regulatory sequences (DRE/CRT, ABRE, MYC recognition sequence (MYCRS), and MYB recognition sequence (MYBRS)). Activation of salt stress-responsive genes is stimulated by ABA-dependent transcription factors ABFs, MYCs, and MYBs, which directly bind to these sequences on the promoters [153]. Promoters of all LEA genes have ABRE motifs that bind ABF [153]. ABFs and DREB2 regulate the drought-inducible Dihydroorotate Dehydrog*enase1* gene, which is important in salt and drought stress responses [153]. Since ABA and ET enter into crosstalk during the stress response, tolerance to salt, osmotic, and heat stresses is alleviated by a mutation in ACS7 [138].

The information available on the mechanism of salt stress response via auxins is scarce [153]. *YUCCA3*, a gene involved in auxin biosynthesis, causes hypersensitivity to salt stress, leading to increased auxin production [160]. Auxin accumulation and redistribution in response to salt stress change the root architecture [161]. Salt stress in tomatoes decreases auxin levels by 75% [162]. The reduced growth under salt stress is a manifestation of altered levels of IAA biosynthesis and its distribution [8]. The salinity stress in wheat decreases CKs biosynthesis [8]. In *Arabidopsis,* wild-type plants were not as tolerant to salt stress as CK-deficient mutants [75]. Mutants with decreased CK levels had higher expression of the HKT1–1 gene, which encodes a Na^+ transporter [75]. Salt tolerance was reduced in *Arabidopsis* plants overexpressing *IPT8* genes [163]. However, the positive role of CK in salinity stress has also been reported. Applying cytokinin oxidase inhibitor (INCYDE) to salt-stressed tomato plants improved flower production and photosynthesis [164].

2. Conclusions

Plant hormones play an important role in the growth and development of plants and also represent an important line of defense against abiotic stress. Hormones change the pattern of growth to enable the plants to withstand stress. The plant stress response involves many hormones, their downstream response factors, associated gene networks, and transcription factors. The crosstalk between hormones and their synergistic or antagonistic interactions play central role in phytohormone-mediated abiotic stress tolerance [165]. Understanding the molecular level interaction between elements of different pathways controlling stress response is critical to allow their manipulation to improve stress tolerance. This is important, as the diversity, duration, and intensity of abiotic stresses are increasing in the changing global climate scenario. Plant hormones are an important target for better management of abiotic stress, especially, in view of the limited success of conventional breeding techniques in dealing with it. Phytohormone pathways and the intermediaries therein can go a long way in the production of climate-resilient crops.

New technologies to bioengineer plants have proven useful in achieving this end; examples include soybean [166], maize [167], rice [168], and potato [169]. Techniques including transcriptome analysis, next-generation sequencing analysis, transgenic plants, genome editing, etc. are being used to identify the hormone-mediated regulatory mechanisms of the plant stress response. Transcriptome analysis using microarrays, a survey of transcriptome profiles, and levels of microRNAs in plants under stress using RNA-seq have helped understand the mechanism of stress tolerance in plants [170]. With genome editing technology, genomes can now be modified in a site-specific manner using specifically designed endonucleases like zinc finger nucleases (ZFN) or TAL effector nucleases (TALEN; [49, 171]) and the CRISPR/CAS system [49, 172].

In a nutshell, new pathways are already emerging. However, the complex interactions between the hormones and their ability to regulate a wide array of plant developmental and physiological processes complicate teasing out the effect of an individual hormone. Lack of information about the tissue-specific stress response and genetic plasticity as well as the extreme complexity of thresholds for different stress responses makes mechanistic understanding of abiotic stress tolerance difficult [173]. In order to better understand the hormone mediated abiotic stress response, the future research should focus on identifying the antagonistic and synergistic interactions between various hormones and the critical regulatory junctures in the hormone crosstalk.

Author details

Nazima Rasool
Department of Botany, University of Kashmir, Srinagar, Jammu and Kashmir, India

*Address all correspondence to: rasoolnazima@gmail.com

References

[1] Jiang K, Asami T. Chemical regulators of plant hormones and their applications in basic research and agriculture. Bioscience, biotechnology, biochemistry. 2018;**82**(8):1265-1300

[2] Wani SH, Kumar V, Shriram V, Sah SK. Phytohormones and their metabolic engineering for abiotic stress tolerance in crop plants. The Crop Journal. 2016;**4**(3):162-176

[3] Rasool N. Plant hormones: Role in alleviating biotic stress. Plant Hormones: Recent Advances, New Perspectives and Applications. 2022;**25**:17

[4] Ku YS, Sintaha M, Cheung MY, Lam HM. Plant hormone signaling crosstalks between biotic and abiotic stress responses. International Journal of Molecular Sciences. 2018;**19**(10):3206

[5] Rhaman MS, Imran S, Rauf F, Khatun M, et al. Seed priming with phytohormones: An effective approach for the mitigation of abiotic stress. Plants. 2020;**10**(1):37

[6] Peleg Z, Blumwald E. Hormone balance and abiotic stress tolerance in crop plants. Current opinion in plant biology. 2011;**14**(3):290-295

[7] Chantre Nongpiur R, Lata Singla-Pareek S, Pareek A. Genomics approaches for improving salinity stress tolerance in crop plants. Current Genomics. 2016;**17**(4):343-357

[8] Talaat NB. Abiotic stresses-induced physiological alteration in wheat. In: Wheat Production in Changing Environments. Singapore: Springer; 2019. pp. 1-30

[9] Eyidogan F, Oz MT, Yucel M, Oktem HA. Signal transduction of phytohormones under abiotic stresses. In: Phytohormones and Abiotic Stress Tolerance in Plants. Berlin, Heidelberg: Springer; 2012. pp. 1-48

[10] Nguyen TQ, Sesin V, Kisiala A, Emery RN. Phytohormonal roles in plant responses to heavy metal stress: Implications for using macrophytes in phytoremediation of aquatic ecosystems. Environment Toxicology Chemistry. 2021;**40**(1):7-22

[11] Sytar O, Kumari P, Yadav S, Brestic M, Rastogi A. Phytohormone priming: Regulator for heavy metal stress in plants. Journal of Plant Growth Regulation. 2019;**38**(2):739-752

[12] Tuteja N. Mechanisms of high salinity tolerance in plants. Methods in Enzymology. 2007;**428**:419-438

[13] Zhang J, Zhang S, Cheng M, Jiang H, et al. Effect of drought on agronomic traits of rice and wheat: A meta-analysis. International Journal of Environemt Research Public Health. 2018;**15**(5):839

[14] Djanaguiraman M, Narayanan S, Erdayani E, Prasad PV. Effects of high temperature stress during anthesis and grain filling periods on photosynthesis, lipids and grain yield in wheat. BMC Plant Biology. 2020;**20**(1):1-2

[15] Ali A, Basra SM, Ahmad R, Wahid A. Optimizing silicon application to improve salinity tolerance in wheat. Soil Environment. 2009;**28**(2):136-144

[16] Zhao C, Liu B, Piao S, et al. Temperature increase reduces global yields of major crops in four independent estimates. Proceedings of the National Academy of Sciences. 2017;**114**(35):9326-9331

[17] Eynard A, Lal R, Wiebe K. Crop response in salt-affected soils. Journal of Sustainable Agriculture. 2005;**27**(1):5-0

[18] Campos H, Cooper M, Habben JE, Edmeades GO, Schussler JR. Improving drought tolerance in maize: A view from industry. Field Crops Research. 2004;**90**(1):19-34

[19] Cucci G, Lacolla G, Boari F, Mastro MA, Cantore V. Effect of water salinity and irrigation regime on maize (Zea mays L.) cultivated on clay loam soil and irrigated by furrow in southern Italy. Agricultural Water Management. 2019;**222**:118-124

[20] Nayyar H, Singh S, Kaur S, Kumar S, Upadhyaya HD. Differential sensitivity of macrocarpa and microcarpa types of Cicer arietinum L to water stress: Association of contrasting stress response with oxidative injury. Journal of Integrative Plants Biology. 2006;**48**(11):1318-1329

[21] Devasirvatham V, Gaur PM, Raju TN, Trethowan RM, Tan DK. Field response of Cicer arietinum L. to high temperature. Field Crops Research. 2015;**172**:59-71

[22] Shanko D, Jateni G, Debela A. Effects of salinity on chickpea landraces during germination stage. Biochemistry & Molecular Biology Journal. 2017;**3**:214-219

[23] Samarah NH, Mullen RE, Cianzio SR, Scott P. Dehydrin-like proteins in soybean seeds in response to drought stress during seed filling. Crop Science. 2006;**46**(5):2141-2150

[24] Jumrani K, Bhatia VS. Impact of combined stress of high temperature and water deficit on growth and seed yield of soybean. Physiology and Molecular Biology of Plants. 2018;**24**(1):37-50

[25] Bustingorri C, Lavado RS. Soybean growth under stable versus peak salinity. Scientia Agricola. 2011;**68**:102-108

[26] Hussain M, Farooq S, Hasan W, Ul-Allah S, Tanveer M, Farooq M, et al. Drought stress in sunflower: Physiological effects and its management through breeding and agronomic alternatives. Agricultural water management. 31 Mar 2018;**201**:152-166

[27] Rondanini D, Mantese A, Savin R, Hall AJ. Responses of sunflower yield and grain quality to alternating day/night high temperature regimes during grain filling: Effect of timing, duration intensity of exposure to stress. Field Crops Research. 2006;**96**(1):48-62

[28] Abd El-Kader AA, Mohamedin AA, Ahmed MK. Growth and yield of sunflower as affected by different salt affected soils. International Journal of Agriculture and Biology. 2006;**8**:583-587

[29] Vandenbroucke KO, Metzlaff MI. Abiotic stress tolerant crops: Genes, pathways and bottlenecks. Sustainable Food Production. 2013:1-7

[30] Kajla M, Yadav VK, Khokhar J, et al. Increase in wheat production through management of abiotic stresses: A review. Journal of Applied and Natural Science. 2015;**7**(2):1070-1080

[31] Li S, Liu J, Liu H, Qiu R, Gao Y, Duan A. Role of hydraulic signal and ABA in decrease of leaf stomatal and mesophyll conductance in soil drought-stressed tomato. Frontiers in plant science. 2021;**12**:653186

[32] Pereira A. Plant abiotic stress challenges from the changing environment. Frontiers in plant science. 2016;7:1123

[33] Jane AB, George DH, Damon PC. Chapter 3 - Hazards, In: Jane A. Bullock, George D. Haddow, Damon P. Coppola editors. Introduction to Homeland Security (Sixth Edition), Butterworth-Heinemann; 2021, Pages 81-140, ISBN 9780128171370

[34] Vicente-Serrano SM, Gouveia C, Camarero JJ, et al. Response of vegetation to drought time-scales across global land biomes. Proceedings of the National Academy of Sciences. 2013;**110**(1):52-57

[35] Haro-Monteagudo D, Daccache A, Knox J. Exploring utility of drought indicators to assess climate risks to agricultural productivity in humid climate. Hydraulic Research. 2018;**49**(2):539-551

[36] Wilhite DA, editor. Drought as a natural hazard: Concepts and definitions. In: Drought: A Global Assessment. London, UK: Routledge; 2000. pp. 3-18

[37] Wahab A, Abdi G, Saleem MH, et al. Plants' physio-biochemical and phyto-hormonal responses to alleviate adverse effects of drought stress: A comprehensive review. Plants. 2022;**11**(13):1620

[38] UNISDR CRED. The Human Cost of Natural Disasters: A Global Perspective. Geneva: Centre for Research on Epidemiology of Disasters (CRED). 2015

[39] Todorova D, Talaat NB, Katerova Z, et al. Polyamines and brassinosteroids in drought stress responses and tolerance in plants. Water stress and crop plants: a sustainable approach. 2016;**2**:608-627

[40] Liu H, Bruce DR, Sissons M, Able AJ, et al. Genotype-dependent changes in the phenolic content of durum under water-deficit stress. Cereal Chemistry. 2018;**95**(1):59-78

[41] Perveen S, Hussain SA. Methionine-induced changes in growth, glycinebetaine, ascorbic acid, total soluble proteins and anthocyanin contents of two Zea mays L. varieties under salt stress. JAPS: Journal of Animal & Plant Sciences. 2021;**31**(1):131-142

[42] Sofy MR, Aboseidah AA, Heneidak SA, et al. ACC deaminase containing endophytic bacteria ameliorate salt stress in Pisum sativum through reduced oxidative damage induction of antioxidative defense systems. Environmental Science and Pollution Research. 2021;**28**(30):40971-40991

[43] McDowell NG, Sapes G, Pivovaroff A, Adams HD, et al. Mechanisms of woody-plant mortality under rising drought, CO2 and vapour pressure deficit. Nature Reviews Earth & Environment. 2022;**3**(5):294-308

[44] Pepe M, Crescente MF, Varone L. Effect of water stress on physiological and morphological leaf traits: A comparison among the three widely-spread invasive alien species Ailanthus altissima, Phytolacca americana, and Robinia pseudoacacia. Plants. 2022;**11**(7):899

[45] Zandi P, Schnug E. Reactive oxygen species, antioxidant responses and implications from a microbial modulation perspective. Biology. 2022;**11**(2):155

[46] Ozturk M, Turkyilmaz Unal BP, et al. Osmoregulation and its actions during the drought stress in plants. Physiologia Plantarum. 2021;**172**(2):1321-1335

[47] Cutler SR, Rodriguez PL, Finkelstein RR, Abrams SR. Abscisic acid: Emergence of a core signaling network. Annual Review of Plant Biology. 2010;**4**:61

[48] Shukla S, Zhao C, Shukla D. Dewetting controls plant hormone perception and initiation of drought resistance signaling. Structure. 2019;**27**(4):692-702

[49] Osakabe Y, Osakabe K, Shinozaki K, Tran LS. Response of plants to water stress. Frontiers in Plant Science. 2014;**5**:86

[50] Verma V, Ravindran P, Kumar PP. Plant hormone-mediated regulation of stress responses. BMC Plant Biology. 2016;16(1):1-0.

[51] Tung SA, Smeeton R, White CA, Black CR, et al. Over-expression of LeNCED1 in tomato with the rbcS3C promoter allows recovery of lines that accumulate high levels of ABA exhibit severe phenotypes. Plant Cell & Environment. 2008;**31**(7):968-981

[52] Behnam BA, Iuchi SA, Fujita MI, Fujita YA, et al. Characterization of the promoter region of an Arabidopsis gene for 9-cis-epoxycarotenoid dioxygenase involved in dehydration-inducible transcription. DNA research. 2013;**20**(4):315-324

[53] Kang J, Hwang JU, Lee M, Kim YY, Assmann SM, Martinoia E, et al. PDR-type ABC transporter mediates cellular uptake of the phytohormone abscisic acid. Proceedings of the National Academy of sciences. 2010;**107**(5):2355-2360

[54] Pei ZM, Kuchitsu K, Ward JM, Schwarz M, Schroeder JI. Differential abscisic acid regulation of guard cell slow anion channels in Arabidopsis wild-type and abi1 and abi2 mutants. The Plant Cell. 1997;**9**(3):409-423

[55] Kwak JM, Mori IC, Pei ZM, Leonhardt N, Torres MA, et al. NADPH oxidase AtrbohD and AtrbohF genes function in ROS-dependent ABA signaling in Arabidopsis. The EMBO Journal. 2003;**22**(11):2623-2633

[56] Negi J, Matsuda O, Nagasawa T, Oba Y, et al. CO2 regulator SLAC1 its homologues are essential anion homeostasis plant cells. Nature. 2008;**452**(7186):483-486

[57] Miyazono KI, Miyakawa T, Sawano Y, Kubota K, et al. Structural basis of ABA signaling. Nature. 2009;**462**(7273):609-614

[58] Park SY, Fung P, Nishimura N, Jensen DR, et al. Abscisic acid inhibits type 2C protein phosphatases via PYR/PYL family START proteins. Science. 2009;**324**(5930):1068-1071

[59] Soon FF, Ng LM, Zhou XE, West GM, Kovach A, et al. Molecular mimicry regulates ABA signaling by SnRK2 kinases and PP2C phosphatases. Science. 2012;**335**(6064):85-88

[60] Fujita Y, Fujita M, Satoh R, Maruyama K, et al. AREB1 is a transcription activator of novel ABRE-dependent ABA signaling that enhances drought stress tolerance in Arabidopsis. The Plant Cell. 2005;**17**(12):3470-3488

[61] Kagaya Y, Hobo T, Murata M, et al. ABA–induced transcription is mediated by phosphorylation ABA response element binding factor. The Plant Cell. 2002;**14**(12):3177-3189

[62] Abe H, Urao T, Ito T, Seki M, et al. Arabidopsis AtMYC2 (bHLH) and AtMYB2 (MYB) function as transcriptional activators in ABA signaling. The Plant Cell. 2003;**15**(1):63-78

[63] Fujita M, Fujita Y, Maruyama K, et al. Dehydration-induced NAC protein, RD26, involved novel ABA-dependent stress-signaling pathway. Plant Journal. 2004;**39**(6):863-876

[64] Lata C, Prasad M. Role of DREBs in regulation of abiotic stress responses in plants. Journal of experimental botany. 2011;**62**(14):4731-4748

[65] Kim JS, Mizoi J, Yoshida T, Fujita Y, et al. An ABRE promoter sequence is involved in osmotic stress-responsive expression of the DREB2A gene, which encodes a transcription factor regulating drought-inducible genes in Arabidopsis. Plant and Cell Physiology. 2011;**52**(12):2136-2146

[66] Xue-Xuan X, Hong-Bo S, Yuan-Yuan M, et al. Biotechnological

implications from ABA roles in cold stress and leaf senescence as important signal for improving plant survival under abiotic-stressed conditions. Critical Reviews in Biotechnology. 2010;**30**(3):222-230

[67] Xiong L, Ishitani M, Lee H, Zhu JK. The Arabidopsis LOS5/ABA3 locus encodes a molybdenum cofactor sulfurase and modulates cold stress–and osmotic stress–responsive gene expression. The Plant Cell. 2001;**13**(9):2063-2083

[68] Xiao BZ, Chen X, Xiang CB, Tang N, Zhang QF, Xiong LZ. Evaluation of seven function-known candidate genes for their effects on improving drought resistance of transgenic rice under field conditions. Molecular Plant. 2009;**2**(1):73-83

[69] Zhang Y, Yang J, Lu S, Cai J, Guo Z. Overexpressing SgNCED1 in tobacco increases ABA level, antioxidant enzyme activities, and stress tolerance. Journal of Plant Growth Regulation. 2008;**27**(2):151-158

[70] Thompson AJ, Andrews J, Mulholland BJ, et al. Overproduction of abscisic acid in tomato increases transpiration efficiency and root hydraulic conductivity and influences leaf expansion. Plant Physiology. 2007;**143**(4):1905-1917

[71] Wang Y, Ying J, Kuzma M, Chalifoux M, et al. Molecular tailoring of farnesylation for plant drought tolerance and yield protection. The Plant Journal. 2005;**43**(3):413-424

[72] Pei ZM, Ghassemian M, Kwak CM, et al. Role of farnesyltransferase in ABA regulation of guard cell anion channels and plant water loss. Science. 1998;**282**(5387):287-290

[73] Khosravi-nejad F, Khavari-nejad RA, Moradi F, Najafi F. Cytokinin and abscisic acid alleviate drought stress through changing organic acids profile, ion immolation, and fatty acid profile to improve yield of wheat cultivars. Physiology and Molecular Biology of Plants. 2022;**24**:1-1

[74] Tran LS, Urao T, Qin F, Maruyama K, Kakimoto T, et al. Functional analysis of AHK1/ATHK1 and cytokinin receptor histidine kinases in response to abscisic acid, drought, and salt stress in Arabidopsis. Proceedings of National Academic and Sciences. 2007;**104**(51):20623-20628

[75] Nishiyama R, Watanabe Y, Fujita Y, et al. Analysis of cytokinin mutants and regulation of cytokinin metabolic genes reveals important regulatory roles of cytokinins in drought, salt and ABA responses, and ABA biosynthesis. The Plant Cell. 2011;**23**(6):2169-2183

[76] Alvarez S, Marsh EL, Schroeder SG, Schachtman DP. Metabolomic and proteomic changes in the xylem sap of maize under drought. Plant, Cell & Environment. 2008;**31**(3):325-340

[77] Yuan Z, Wang C, Li S, Li X, Tai F. Effects of different plant hormones or PEG seed soaking on maize resistance to drought stress. Canadian Journal of Plant Science. 2014;**94**(8):1491-1499

[78] Divi UK, Krishna P. Overexpression of the brassinosteroid biosynthetic gene AtDWF4 in Arabidopsis seeds overcomes abscisic acid-induced inhibition of germination and increases cold tolerance in transgenic seedlings. Journal of Plant Growth Regulation. 2010;**29**(4):385-393

[79] Jager CE, Symons GM, Ross JJ, Reid JB. Do brassinosteroids mediate the water stress response? Physiologia Plantarum. 2008;**133**(2):417-425

[80] Farhangi-Abriz S, Torabian S. Biochar increased plant growth-promoting hormones helped to alleviates

salt stress in common bean seedlings. Journal of Plant Growth Regulation. 2018;**37**(2):591-601

[81] Zhang SW, Li CH, Cao J, Zhang YC, et al. Altered architecture and enhanced drought tolerance in rice via the down-regulation of indole-3-acetic acid by TLD1/OsGH3. 13 activation. Plant Physiology. 2009;**151**(4):1889-1901

[82] Abdelaal KA, Attia KA, Alamery SF, El-Afry MM, et al. Exogenous application of proline and salicylic acid can mitigate the injurious impacts of drought stress on barley plants associated with physiological histological characters. Sustainability. 2020;**12**(5):1736

[83] Jahan MS, Wang Y, Shu S, Zhong M, et al. Exogenous salicylic acid increases the heat tolerance in tomato (Solanum lycopersicum L) by enhancing photosynthesis efficiency and improving antioxidant defense system through scavenging of reactive oxygen species. Scientia Horticulturae. 2019;**247**:421-429

[84] Maghsoudi K, Emam Y, Ashraf M, Arvin MJ. Alleviation of field water stress in wheat cultivars by using silicon and salicylic acid applied separately or in combination. Crop and Pasture Science. 2019;**70**(1):36-43

[85] Munsif F, Shah T, Arif M, Jehangir M, et al. Combined effect of salicylic acid and potassium mitigates drought stress through the modulation of physio-biochemical attributes and key antioxidants in wheat. Saudi Journal of Biological Sciences. 2022;**29**(6):103294

[86] Lobato AK, Barbosa MA, Alsahli AA, Lima EJ, Silva BR. Exogenous salicylic acid alleviates the negative impacts on production components, biomass and gas exchange in tomato plants under water deficit improving redox status and anatomical responses. Physiologia Plantarum. 2021;**172**(2):869-884

[87] Bashar KK. Hormone dependent survival mechanisms of plants during post-waterlogging stress. Plant Signaling & Behavior. 2018;**13**(10):e1529522

[88] Chandwani S, Amaresan N. Role of ACC deaminase producing bacteria for abiotic stress management and sustainable agriculture production. Environment Science Pollution Research. Apr 2022;**29**(16):22843-22859

[89] Mathur S, Raikalal P, Jajoo A. Physiological responses of wheat to environmental stresses. In: Wheat Production in Changing Environments. Singapore: Springer; 2019. pp. 31-61

[90] Hao Z, AghaKouchak A, Phillips TJ. Changes in concurrent monthly precipitation and temperature extremes. Environmental Research Letters. 2013;**8**(3):034014

[91] IPCC: Summary for Policymakers. In: Climate Change 2021: The Physical Science Basis. Contribution of Working Group I to the Sixth Assessment Report of the Intergovernmental Panel on Climate Change. Masson DV, Zhai P, Pirani A, Connors SL, Péan C, Berger S. et al. editiors. Cambridge, United Kingdom and New York, NY, USA: Cambridge University Press; 2021. pp. 3–32

[92] Gourdji SM, Sibley AM, Lobell DB. Global crop exposure to critical high temperatures in reproductive period: Trends future projections. Environment Research Letter. 2013;**8**(2):024041

[93] Lippmann R, Babben S, Menger A, Delker C, Quint M. Development of wild and cultivated plants under global warming conditions. Current Biology. 2019;**29**(24):R1326-R1338

[94] Li N, Euring D, Cha JY, Lin Z, Lu M, et al. Plant hormone-mediated

regulation of heat tolerance in response to global climate change. Fron. Plant Sci. 2021;**11**:627969

[95] Hu S, Ding Y, Zhu C. Sensitivity and responses of chloroplasts to heat stress in plants. Frontiers in Plant Science. 2020;**11**:375

[96] Wu YS, Yang CY. Ethylene-mediated signaling confers thermotolerance and regulates transcript levels of heat shock factors in rice seedlings under heat stress. Botanical studies. 2019;**60**(1):1-2

[97] Guihur A, Rebeaud ME, Goloubinoff P. How do plants feel the heat and survive? Trends in Biochemical Sciences. 2022

[98] Larkindale J, Hall JD, Knight MR, Vierling E. Heat stress phenotypes of Arabidopsis mutants implicate multiple signaling pathways in the acquisition of thermotolerance. Plant Physiology. 2005;**138**(2):882-897

[99] Castroverde CD, Dina D. Temperature regulation of plant hormone signaling during stress and development. Journal of Experimental Botany. 2021;**72**(21):7436-7458

[100] Kaya H, Takeda S, Kobayashi MJ, Kimura S, et al. Comparative analysis of the reactive oxygen species-producing enzymatic activity of Arabidopsis NADPH oxidases. Plant Journal. 2019;**98**(2):291-300

[101] Silva-Correia J, Freitas S, Tavares RM, et al. Phenotypic analysis of Arabidopsis heat stress response during germination early seedling development. Plant Methods. 2014;**10**(1):1-1

[102] Gao-Takai M, Katayama-Ikegami A, et al. A low temperature promotes anthocyanin biosynthesis but does not accelerate endogenous ABA accumulation in red-skinned grapes. Plant Science. 2019;**283**:165-176

[103] Zhang X, Wang X, Zhuang L, Gao Y, Huang B. ABA mediation of drought priming-enhanced heat tolerance in Festuca arundinacea and Arabidopsis. Physiologia Plantarum. 2019;**167**(4):488-501

[104] Rezaul IM, Baohua F, Tingting C, Weimeng F, et al. Abscisic acid prevents pollen abortion under high-temperature stress by mediating sugar metabolism in rice spikelets. Physiologia Plantarum. 2019;**165**(3):644-663

[105] Küpers JJ, Oskam L, Pierik R. Photoreceptors regulate plant developmental plasticity through auxin. Plants. 2020;**9**(8):940

[106] Franklin KA, Lee SH, Patel D, Kumar SV, et al. Phytochrome-interacting factor 4 regulates auxin biosynthesis high temperature. Proceedings of the National Academy of Sciences. 2011;**108**(50):20231-20235

[107] Sun J, Qi L, Li Y, Chu J, Li C. PIF4–mediated activation of YUCCA8 expression integrates temperature into the auxin pathway in regulating Arabidopsis hypocotyl growth. PLoS genetics. 2012;**8**(3):e1002594

[108] Van Der Woude LC, Perrella G, Snoek BL, et al. HISTONE DEACETYLASE 9 stimulates auxin-dependent thermomorphogenesis in Arabidopsis thaliana by mediating H2A. Z depletion. Proceedings of National Academic Sciences. 2019;**116**(50):25343-25354

[109] Kim EJ. Russinova E Brassinosteroid signalling. Current Biology. 2020;**30**:R294-R298

[110] Oh E, Zhu JY, Wang ZY. Interaction between BZR1 and PIF4 integrates brassinosteroid and environmental

responses. Nature Cell Biology. 2012;**14**(8):802-809

[111] Ibañez C, Delker C, Martinez C, Bürstenbinder K, et al. Brassinosteroids dominate hormonal regulation of plant thermomorphogenesis via BZR1. Current Biology. 2018;**28**(2):303-310

[112] Bellstaedt J, Trenner J, Lippmann R, et al. A mobile auxin signal connects temperature sensing in cotyledons with growth responses in hypocotyls. Plant Physiology. 2019;**180**(2):757-766

[113] Dhaubhadel S, Browning KS, Gallie DR, Krishna P. Brassinosteroid functions to protect the translational machinery and heat-shock protein synthesis following thermal stress. The Plant Journal. 2002;**29**(6):681-691

[114] Sadura I, Libik-Konieczny M, Jurczyk B, Gruszka D, Janeczko A. Plasma membrane ATPase and the aquaporin HvPIP1 in barley brassinosteroid mutants acclimated to high and low temperature. Journal of Plant Physiology. 2020;**244**:153090

[115] Nie WF, Wang MM, Xia XJ, Zhou YH, et al. Silencing of tomato RBOH1 and MPK2 abolishes BR-induced H_2O_2 generation and stress tolerance. Plant, Cell Environment. 2013;**36**(4):789-803

[116] Yin Y, Qin K, Song X, Zhang Q, Zhou Y, Xia X, et al. BZR1 transcription factor regulates heat stress tolerance through FERONIA receptor-like kinase-mediated reactive oxygen species signaling in tomato. Plant and Cell Physiology. 2018;**59**(11):2239-2254

[117] Zhou J, Xia XJ, Zhou YH, Shi K, Chen Z, Yu JQ. RBOH1-dependent H_2O_2 production and subsequent activation of MPK1/2 play an important role in acclimation-induced cross-tolerance in tomato. Journal of Experimental Botany. 2014;**65**(2):595-607

[118] Fei Q, Wei S, Zhou Z, Gao H, Li X. Adaptation of root growth to increased ambient temperature requires auxin and ethylene coordination in Arabidopsis. Plant Cell Reports. 2017;**36**(9):1507-1518

[119] Clarke SM, Cristescu SM, Miersch O, et al. Jasmonates act with salicylic acid to confer basal thermotolerance in Arabidopsis thaliana. The New Phytologist. 2009;**182**(1):175-187

[120] Cortleven A, Leuendorf JE, Frank M, et al. Cytokinin action in response to abiotic and biotic stresses in plants. Plant, Cell & Environment. 2019;**42**(3):998-1018

[121] Dobrá J, Černý M, Štorchová H, Dobrev P, et al. The impact of heat stress targeting on the hormonal and transcriptomic response in Arabidopsis. Plant Science. 2015;**231**:52-61

[122] Prerostova S, Dobrev PI, Kramna B, Gaudinova A, Knirsch V, Spichal L, et al. Heat acclimation and inhibition of cytokinin degradation positively affect heat stress tolerance of Arabidopsis. Frontiers in Plant Science. 2020;**18**(11):87

[123] Skalák J, Černý M, Jedelský P, Dobrá J, et al. Stimulation of ipt overexpression as a tool to elucidate the role of cytokinins in high temperature responses of Arabidopsis thaliana. Journal of Experimental Botany. 2016;**67**(9):2861-2873

[124] Sobol S, Chayut N, Nave N, Kafle D, et al. Genetic variation in yield under hot ambient telmperatures spotlights a role for cytokinin in protection of developing floral primordia. Plant, Cell & Environment. 2014;**37**(3):643-657

[125] Wu C, Cui K, Wang W, Li Q, Fahad S, Hu Q, et al. Heat-induced cytokinin transportation and degradation are associated with reduced

panicle cytokinin expression and fewer spikelets per panicle in rice. Frontiers in Plant Science. 2017;**8**:371

[126] Wassie M, Zhang W, Zhang Q, Ji K, Cao L, Chen L. Exogenous SA ameliorates heat stress-induced damages and improves growth and photosynthetic efficiency in alfalfa (Medicago sativa L.). Ecotoxicology and Environmental Safety. 2020;**191**:110206

[127] Wang LJ, Fan L, Loescher W, et al. Salicylic acid alleviates decreases in photosynthesis under heat stress and accelerates recovery in grapevine leaves. BMC Plant Biology. 2010 10(1):1-0.

[128] Toh S, Kamiya Y, Kawakami N, Nambara E, et al. Thermoinhibition uncovers a role for SL in Arabidopsis seed germination. Plant Cell Physiology. 2012;**53**(1):107-117

[129] Bawa G, Feng L, Chen G, Chen H, et al. Gibberellins and auxin regulate soybean hypocotyl elongation under low light and high-temperature interaction. Physiologia Plantarum. 2020;**170**(3):345-356

[130] Ferrero LV, Viola IL, Ariel FD, Gonzalez DH. Class I TCP transcription factors target the gibberellin biosynthesis gene GA20ox1 and the growth-promoting genes HBI1 and PRE6 during thermomorphogenic growth in Arabidopsis. Plant and Cell Physiology. 2019;**60**(8):1633-1645

[131] Toh S, Imamura A, Watanabe A, Nakabayashi K, et al. High temperature-induced ABA biosynthesis and its role in the inhibition of gibberellin action in Arabidopsis seeds. Plant Physiology. 2008;**146**(3):1368-1385

[132] Huot B, Castroverde CD, Velásquez AC, et al. Dual impact of elevated temperature on plant defence and bacterial virulence in Arabidopsis. Nat. com. 2017;**8**(1):1-2

[133] Havko NE, Kapali G, Das MR, Howe GA. Stimulation of insect herbivory by elevated temperature outweighs protection by jasmonate pathway. Plants. 2020;**9**(2):172

[134] Richards LA. Diagnosis and improvement of saline and alkali soils. Agriculture Handbook. 1968;**60**:210-220

[135] Farooq M, Gogoi N, Hussain M, Barthakur S, et al. Effects, tolerance mechanisms management of salt stress in grain legumes. Plant Physiology and Biochemistry. 2017;**118**:199-217

[136] Singh RP, Jha P, Jha PN. Bio-inoculation of plant growth-promoting rhizobacterium Enterobacter cloacae ZNP-3 increased resistance against salt and temperature stresses in wheat plant (Triticum aestivum L.). Journal of Plant Growth Regulation. 2017;**36**(3):783-798

[137] Jusovic M, Velitchkova MY, Misheva SP, Börner A, Apostolova EL, Dobrikova AG. Photosynthetic responses of a wheat mutant (Rht-B1c) with altered DELLA proteins to salt stress. Journal of Plant Growth Regulation. 2018;**37**(2):645-656

[138] Tao JJ, Chen HW, Ma B, Zhang WK, Chen SY, Zhang JS. The role of ethylene in plants under salinity stress. Frontiers in Plant Science. 2015;**6**:1059

[139] Zhao C, Jiang W, Zayed O, Liu X, Tang K, Nie W, et al. The LRXs-RALFs-FER module controls plant growth and salt stress responses by modulating multiple plant hormones. National Science Review. 2021;**8**(1):nwaa149

[140] Morgan PW, Drew MC. Ethylene and plant responses to stress. Physiologia Plantarum. 1997;**100**(3):620-630

[141] Cao WH, Liu J, He XJ, Mu RL, et al. Modulation of ethylene responses affects plant salt-stress responses. Plant Physiology. 2007;**143**(2):707-719

[142] Peng J, Li Z, Wen X, Li W, Shi H, Yang L, et al. Salt-induced stabilization of EIN3/EIL1 confers salinity tolerance by deterring ROS accumulation in Arabidopsis. PLoS Genetics. 2014;**10**(10):e1004664

[143] Zhao XC, Schaller GE. Effect of salt and osmotic stress upon expression of ethylene receptor ETR1 in Arabidopsis thaliana. Febs Letters. 2004;**562**(1-3):189-192

[144] Peng Z, He S, Gong W, Sun J, Pan Z, Xu F, et al. Comprehensive analysis of differentially expressed genes and transcriptional regulation induced by salt stress in two contrasting cotton genotypes. BMC Genomics. 2014;**15**(1):1-28

[145] Achard P, Cheng H, De Grauwe L, et al. Integration of plant responses to environmentally activated phytohormonal signals. Science. 2006;**311**(5757):91-94

[146] Li G, Meng X, Wang R, Mao G, Han L, Liu Y, et al. Dual-level regulation of ACC synthase activity by MPK3/MPK6 cascade and its downstream WRKY transcription factor during ethylene induction in Arabidopsis. PLoS Genetics. 2012;**8**(6):e1002767

[147] Cao WH, Liu J, Zhou QY, Cao YR, et al. Expression of tobacco ethylene receptor NTHK1 alters plant responses to salt stress. Plant Cell Environment. 2006;**29**(7):1210-1219

[148] Shen X, Wang Z, Song X, Xu J, Jiang C, Zhao Y, et al. Transcriptomic profiling revealed an important role of cell wall remodeling and ethylene signaling pathway during salt acclimation in Arabidopsis. Plant molecular biology. 2014;**86**(3):303-317

[149] Liu Y, Zhang S. Phosphorylation of 1-aminocyclopropane-1-carboxylic acid synthase by MPK6, a stress-responsive mitogen-activated protein kinase, induces ethylene biosynthesis in Arabidopsis. The Plant Cell. 2004;**16**(12):3386-3399

[150] Du H, Wu N, Cui F, You L, Li X, Xiong L. A homolog of ETHYLENE OVERPRODUCER, OsETOL1, differentially modulates drought and submergence tolerance in rice. The Plant Journal. 2014;**78**(5):834-849

[151] Zhou HL, Cao WH, Cao YR, Liu J, Hao YJ, Zhang JS, et al. Roles of ethylene receptor NTHK1 domains in plant growth, stress response and protein phosphorylation. FEBS letters. 2006;**580**(5):1239-1250

[152] Lei G, Shen M, Li ZG, Zhang B, et al. EIN2 regulates salt stress response and interacts with a MA3 domain-containing protein ECIP1 in Arabidopsis. Plant, Cell & Environment. 2011;**34**(10):1678-1692

[153] Ryu H, Cho YG. Plant hormones in salt stress tolerance. Journal of Plant Biology. 2015;**58**(3):147-155

[154] Saeng-ngam S, Takpirom W, Buaboocha T, et al. The role of the OsCam1-1 salt stress sensor in ABA accumulation salt tolerance in rice. Journal of Plant Biology. 2012;**55**(3):198-208

[155] Moons A, Bauw G, Prinsen E, Van Montagu M, Van Der Straeten D. Molecular and physiological responses to abscisic acid and salts in roots of salt-sensitive and salt-tolerant Indica rice varieties. Plant Physiology. 1995;**107**(1):177-186

[156] He T, Cramer GR. ABA concentrations are correlated with leaf area reductions in two salt-stressed rapid-cycling brassica species. Plant and Soil. 1996;**179**(1):25-33

[157] Cabot C, Sibole JV, Barceló J, Poschenrieder C. Abscisic acid decreases leaf Na+ exclusion in salt-treated Phaseolus vulgaris L. Journal of Plant Growth Regulation. 2009;**28**(2):187-192

[158] Cramer GR, Quarrie SA. Abscisic acid is correlated with the leaf growth inhibition of four genotypes of maize differing in their response to salinity. Functional plant biology. 2002;**29**(1):111-115

[159] Rao KP, Richa TA, Kumar KU, et al. In silico analysis reveals 75 members of mitogen-activated protein kinase kinase kinase gene family in rice. DNA research. 2010;**17**(3):139-153

[160] Jung JH, Park CM. Auxin modulation of salt stress signaling in Arabidopsis seed germination. Plant signaling & behavior. 2011;**6**(8): 1198-1200

[161] Wang L, Wang Z, Xu Y, Joo SH, et al. OsGSR1 is involved in crosstalk between gibberellins and brassinosteroids in rice. The Plant Journal. 2009;**57**(3):498-510

[162] Kazan K. Auxin and the integration of environmental signals into plant root development. Annals of botany. 2013;**112**(9):1655-1665

[163] Wang X, Dinler BS, Vignjevic M, Jacobsen S, Wollenweber B. Physiological and proteome studies of responses to heat stress during grain filling in contrasting wheat cultivars. Plant Science. 2015;**230**:33-50

[164] Aremu AO, Masondo NA, Sunmonu TO, et al. A novel inhibitor of cytokinin degradation (INCYDE) influences the biochemical parameters and photosynthetic apparatus in NaCl-stressed tomato plants. Planta. 2014;**240**(4):877-889

[165] Rivero RM, Gimeno J, Van Deynze A, Walia H, Blumwald E. Enhanced cytokinin synthesis in tobacco plants expressing PSARK: IPT prevents degradation of photosynthetic protein complexes during drought. Plant & Cell Physiology. 2010;**51**(11):1929-1941

[166] Li, Y.J. Zhang, J.C. Zhang, J. etal. Expression of an Arabidopsis molybdenum factor sulphurase gene in soybean enhances drought tolerance and increases yield under field conditions Plant Biotechnology Journal, 11 (2013), pp. 747-758

[167] Lu Y, Li Y, Zhang J, Xiao Y, et al. Overexpression of Arabidopsis molybdenum cofactor sulfurase gene confers drought tolerance in maize. PLoS One. 2013;**8**:e52126

[168] Zhang JJ, Li WJ, Zhang SN, et al. The putative auxin efflux carrier OsPIN3t is involved in drought stress response drought tolerance. The Plant Journal. 2012;**72**:805-816

[169] Kim IJD, Baek HC, Park HJ, et al. Overexpression of Arabidopsis YUCCA6 in potato results in high-auxin developmental phenotypes and enhanced resistance to water deficit Mol. Plant. 2013;**6**:337-349

[170] Cai ZQ, Gao Q. Comparative physiological and biochemical mechanisms of salt tolerance five contrasting highland quinoa cultivars. BMC plant biology. 2020;**20**(1):1-5

[171] Shukla VK, Doyon Y, Miller JC, DeKelver RC, et al. Precise genome modification in the crop species Zea mays using zinc-finger nucleases. Nature. 2009;**459**(7245):437-441

[172] Nekrasov V, Staskawicz B, Weigel D, Jones JD, Kamoun S. Targeted mutagenesis in the model plant Nicotiana benthamiana using Cas9 RNA-guided endonuclease. Nature Biotechnology. 2013;**31**(8):691-693

[173] Kohli A, Sreenivasulu N, Lakshmanan P, Kumar PP. The phytohormone crosstalk paradigm takes center stage in understanding how plants respond to abiotic stresses. Plant Cell Reports. 2013 32(7):945-957.

Chapter 4

Reorganization of the Endomembrane System and Protein Transport Pathways under Abiotic Stress

Miguel Sampaio, João Neves, Tatiana Cardoso, José Pissarra, Susana Pereira and Cláudia Pereira

Abstract

Stress compromises protein trafficking in plants, which often results in modifications to the endomembrane system and trafficking pathways. Proteins travel in unexpected ways during stress, and cell compartments alter their appearance, activity, and content to cope with the difficulties that stress brings. We will piece together material on the issue in this chapter, emphasizing how the endomembrane system processes such changes and how it reacts to a dynamic environment. The intricate dynamics of protein transport pathways and how they maintain cellular homeostasis under challenging circumstances is illustrated.

Keywords: abiotic stress, endomembranes, protein trafficking, vacuolar routes, endoplasmic reticulum

1. Introduction

Diverse environmental stresses frequently trigger signals and pathways that lead to cellular responses, such as increased antioxidant expression, solute accumulation, altered protein transport, and endomembrane remodeling [1–4]. In fact, nowadays, crop failures caused by climate change and human action pose the biggest hazard to human and environmental health through food safety declining [5]. Trying to face this everchanging environment, plants have developed the capacity to adapt to and benefit from changes in their surroundings, activating stress defense mechanisms [6]. The processes behind the stress response are only partially understood, and alterations in the transcriptome are still the outcome of a complex chain of circumstances. One of the most important mechanisms, especially concerning interorganellar connections, occurs at the endomembrane level [7, 8], from which new markers for the assisted selection of stress-resistant crop types can be found. Since the plants' successful adaptation likely relies on balanced interactions and synergistic effects among ordinarily unrelated proteins, defining each participant's precise roles in the game is a crucial aspect of plant genetic improvement [9]. Recent experimental

 IntechOpen

evidence [10] points to a variety of protein classes (including aquaporins, soluble N-ethylmaleimide-sensitive factor attachment protein receptors (SNAREs), ATPase pumps, or channels) that regulate particular membrane transport events, resulting in significant cell reorganization events in challenging environmental conditions. As an example, the AKT1/KC1, a shaker-like potassium channel, was selectively accumulated on small vacuoles [11] and is sufficient to confer stress tolerance when overexpressed. Several research groups discovered intriguing connections between stress tolerance and previously unrecognized membrane rearrangements. However, the relationship between the architecture of membranous structures and their ability to withstand stress has only recently gained the attention of researchers. Numerous research items have supported the notion that endomembrane trafficking is closely related to stress signaling pathways; nevertheless, these studies lack a better understanding of the underlying mechanisms. In the last several years, there has been a notable advancement in our understanding of the mechanisms behind protein sorting. Due to their significance in maintaining the homeostasis of plant cells, particular attention has been paid to the study of proteins that are directed toward the vacuole and the inherent sorting mechanisms. Regarding this matter, recent results imply that alternative routes may challenge the orthodox concept of protein transport to the vacuole [12–14]. These alternative routes are regarded as one of the plant's adaptations to challenging circumstances. As a result, it is believed that certain conditions may cause the vacuolar trafficking pathways to change to better serve the demands of the plant. Alongside the vacuole, the endoplasmic reticulum, as the entrance to the endomembrane trafficking routes, also plays an important role in the folding, quality control, and sorting of newly produced proteins [15–17]. Additionally, as the link between the actin cytoskeleton and the endomembrane system is essential to maintaining many aspects of plant cell function and development, the cell cytoskeleton also plays a significant role in the response and adaptation to stress [18]. This chapter aims at describing the more recent findings on the effects of abiotic stress in the endomembrane system, alterations in vacuolar trafficking routes, and the importance of the cell cytoskeleton in these processes. Also, examples of proteins and endomembrane effectors with altered expression/localization were depicted from the available literature that can represent a collection of putative markers for abiotic stress studies (**Table 1**).

2. Endoplasmic reticulum and stress

A network of tubules and cisternae that extends across the entire cell and links with several other organelles, the endoplasmic reticulum (ER), is crucial for maintaining cellular homeostasis as well as for detecting and disseminating external signals [7]. The ER is one of the main organelles that mediate the stress response in both plants and animals [15–17]. Protein misfolding and accumulation following adverse environmental conditions can lead to ER stress [19–21]. In response, the cell activates various mechanisms to maintain the homeostasis of the ER, such as the expression of genes encoding chaperones and other proteins with the ability to fold proteins, degradation linked to the ER, or a reduction in the amount of protein translation loaded into the ER [19, 22]. As an example, unfolded proteins can bind to BIP proteins (binding proteins), which activate bZIP transcription factors like bZIP17/bZIP28 that are transported to the Golgi to be cleaved (**Figure 1**) [17, 19]. To regain ER equilibrium, this transport will upregulate genes related to the ER stress pathway [17]. The

	Protein	Stress-related response	Refs
ER-related	bZIP28	Involved in the activation of heat stress response genes	[17, 19]
	bZIP17	Participates in the activation of salt stress response genes	[20, 21]
	IRE1	Responsible for the splicing of bZIP60 mRNA, required for the activation of genes involved in the ER stress reaction; regulates the stress transcriptome by degrading several mRNAs	[22–25]
	NPR1	Suppresses the transcriptional role of bZIP28 and bZIP60 in ER stress responses triggered during pathogen attack	[26, 27]
	ATG8	Following ER stress, many ER components are delivered for degradation *via* autophagy, forming ER-derived autophagic bodies	[28–30]
Vacuole-related	CBL–CIPK	Important role in the detoxification of $Mg2^{+}$ in the vacuole during salt stress conditions	[31]
	VPEs	Hydrolytic enzymes, such as proteases and antimicrobial compounds, are released to the cytosolic environment, or extracellularly, to fight pathogen attacks.	[32–34]
Cytoskeleton-related	CesA	Osmotic stress induces endocytosis of cellulose synthase complex and their interaction with cortical microtubules	[35, 36]
	CSI1-dependent SmaCCs/MASCs	During endocytosis, CSI1-dependent SmaCCs/MASCs are formed, allowing a quick regulation of cellulose synthesis under abiotic stress	[37]
	NET1A	Reacts to extracellular signals, such as stress related to pathogen infection	[38]
Vacuolar trafficking	RMR1; VSR1; SYP51; VTI12; VTI11; VSR2	Genes involved in the PSV sorting are positively regulated in plants under abiotic stress, while genes involved in the LV sorting downregulated	[4]
	VSR1	Important for the regulation of abscisic acid (ABA) biosynthesis, a signaling molecule in several stress conditions	[39]
	RabG3e	Arabidopsis plants overexpressing AtRabG3e showed increased tolerance to salt and osmotic stress along with a reduction in the accumulation of reactive oxygen species	[40]
	VAMP7C	Suppression of the v-SNARE AtVAMP7C had a positive impact in improving plant salt tolerance by inhibiting the fusion of H_2O_2-containing vesicles with the vacuole	[41]
Unconventional vacuolar routes	PSIB	Overexpression of PSIB in *Arabidopsis thaliana* correlates with salt and osmotic stress conditions, in some cases improving plant fitness	[42]
	Cysteine Proteinases	Cysteine proteinases accumulate in long ER bodies, whose fusion with the PSV may be triggered by stress	[43]
	PR1 PDF1.2	ER bodies filled with defense proteins are formed and eventually fuse with the plasma membrane or with the vacuole in a Golgi-independent manner	[44]

Table 1.
Endomembrane-associated proteins with responses to adverse abiotic conditions.

upregulation of genes implicated in stress response, such as bZIP28, which activates heat stress response genes [17, 23] and bZIP17, which activates salt stress responses [20, 24], may also be mediated *via* this transport. As so, it is easy to see that ER stress responses are frequently triggered in scenarios of heat and salt stress [17, 23, 24]. However, other significant proteins, such as the ER-resident transmembrane protein inositol-requiring enzyme-1 (IRE1), are implicated in the unfolded protein responses (UPR) that react to unfavorable environmental conditions (**Figure 1**) [17, 25]. The heat stress response is said to be mediated by this protein. It is necessary for the activation of the genes involved in the ER's stress response because bZIP60 mRNA, which is spliced by the heat-activated enzyme IRE1 [26], is present. By destroying several mRNAs, this protein also controls the stress transcriptome [27, 28]. Other UPRs are triggered in this sort of stress in addition to the previously reported mechanism, but their overexpression is irrelevant, suggesting that salt stress can merely increase the misfolding of a new group of proteins [17, 24, 29, 30]. Another transcriptional component that affects plant UPR has recently been discovered. It has been shown that the nonexpressor of PR1 gene 1 (NPR1) inhibits the transcriptional function of bZIP28 and bZIP60 in ER stress responses (**Figure 1**). NPR1 is a critical redox-regulated master regulator of salicylic acid (SA)-dependent responses to pathogens. NPR1 is translocated to the nucleus and physically interacts with bZIP28 and bZIP60, acting as an antagonist of such UPR proteins to maximize their cytoprotective function in the UPR (**Figure 1**). This occurs when ER stress causes the cytosolic redox potential to decrease. A negative feedback loop that is crucial for regulating energy consumption and preserving basal cellular homeostasis during ER stress signaling may be

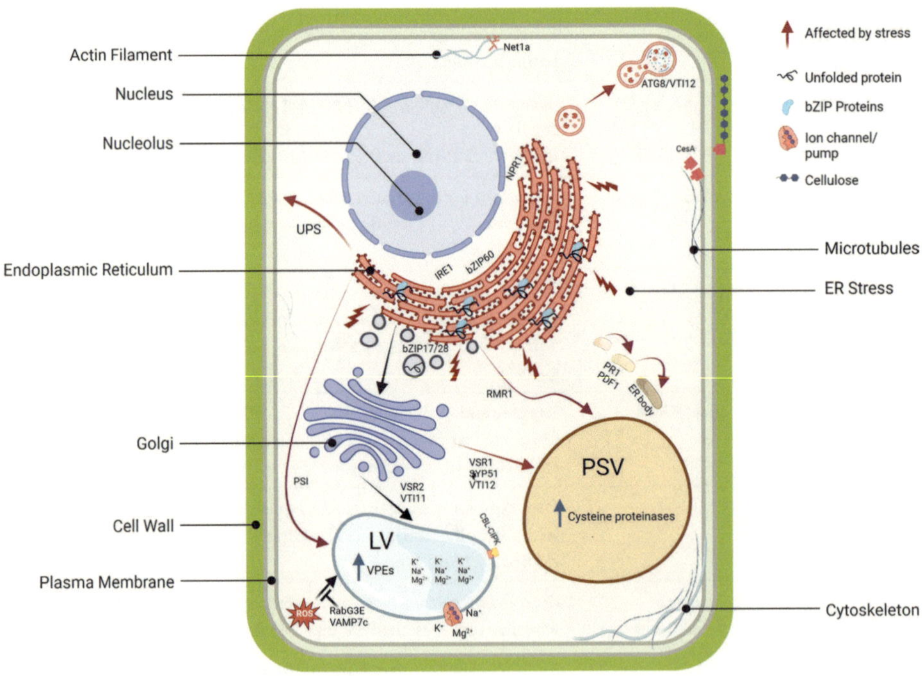

Figure 1.
Diagram showing the modifications in endomembrane trafficking and related protein effectors under abiotic stress conditions. LV—Lytic Vacuole; PSI—Plant Specific Insert; PSV—Protein Storage Vacuole; ROS—Reactive Oxygen Species; UPS—Unconventional Protein Secretion; VPEs—Vacuolar Processing Enzymes. Image created with BioRender.com, accessed on 21 November 2022.

promoted by NPR1 roles in plant UPR monitoring [31]. The intercellular mobility of bZIP60, which promotes systemic UPR signaling, has been shown to govern a non-cell-autonomous component in addition to cell-intrinsic UPR signaling. Evidence suggests that the sbZIP60 protein can move between cells and activate a target gene's promoter, thus promoting UPR gene expression in cells far from the region of ER stress. Such findings imply that ER stress systemic signaling may represent a mode of anticipation of a potentially imminent ER stress, as the cells of tissues that have not yet been subjected to ER stress are prepared by triggering the accumulation of ER stress attenuating protein transcripts [32].

Under stress, the ER's other compartmentalization mechanisms, such as autophagy, are also engaged in addition to ER stress responses. The sequestration of cytosolic components by a newly generated, double-membrane vesicle known as an autophagosome, which is subsequently directed to the plant vacuole, is known as macroautophagy [33, 34]. Notably, it was also claimed that selective autophagy delivered vacuolar resident proteins to this organelle *via* specific trafficking channels. The degradative aspect that autophagy is typically associated with contrasts with its role in the biogenesis-mediating process. It was demonstrated that triggering ER stress in Arabidopsis causes the transfer of ER components, like the ribosome-decorated ER membrane, to vacuoles *via* autophagy [34], supporting the idea that autophagy may be involved in the trafficking of storage proteins. Additionally, it is now understood that both yeast and mammalian cells use autophagy to transfer ER components for destruction during ER stress [35–37]. Plants share many similarities with other eukaryotes in this regard, and the accumulation of Atg8-positive bodies that co-localized with the ER marker GFP-HDEL was found after ER stress. In the vacuoles of ER-stressed plants, the presence of autophagic structures, including ER membranes, was also discovered by electron microscopy [34]. IRE1b has been linked to this kind of response, according to a different study conducted by Bao and colleagues [38]. This pathway, however, is not related to BZIP60, but instead to the regulated IRE1-dependent decay of messenger RNA (RIDD), in which IRE1 degrades the mRNAs of factors encoded by genes that prevent the activation of autophagy processes in response to ER stress [39], such as BGLU21, a member of the β-glucosidase family and one of the main elements of ER bodies.

The ER is in a unique position to identify extracellular stimuli and coordinate the cellular response to adverse and demanding situations in the cell because it is the origin of the endomembrane system. Its central network-like structure, which permeates the entire cell, enables it to interact with other organelles at several points, demonstrating the high complexity of the ER mechanisms that are crucial to preserving the functionality of cellular homeostasis and signaling cascades.

3. The vacuole as a major player in cell homeostasis

Vacuoles perform physical and metabolic tasks, can occupy up to 80% of the volume of a cell, and are crucial for cellular responses to abiotic and biotic stimuli as well as to general cell homeostasis [40, 41]. These organelles often house water, nutrients, ions, and secondary metabolites, but they can also act as a deposition location for waste materials, excess solutes, and toxic cell remnants [42–45]. They also play a role in programmed cell death [46]. The protein storage vacuole (PSV) and the lytic vacuole (LV) are the two main forms of vacuoles found in plant cells. Proteins predominate in storage tissues (such as cotyledons, endosperm, and tubers) and vegetative tissues (bark, leaves, and pods) of adult plants, and they often accumulate in the PSVs because of their higher pH and lower hydrolytic activity when compared to the LVs [47, 48]. LVs, on the other hand, are mostly present in vegetative tissues and are employed for storing and depositing undesirable substances. This form of vacuole controls the breakdown of a wide range of macromolecules and other chemicals because of its high hydrolytic activity and acidic pH [49, 50]. Initially, it was not expected to find both forms of vacuoles in the same cell; however, research done in root tip cells of barley and pea seedlings proved this was not the case [51, 52]. In addition, a study employing the model plant Arabidopsis thaliana found that, rather than being created from scratch, the LV is embedded in the PSV during germination [53]. Two distinct types of vacuoles suggest that plants have unique trafficking processes and pathways for various proteins. Additionally, it has been suggested that the coexistence of LVs and PSVs in a single cell may function as a plant flexibility mechanism in response to shifting environmental conditions [54–57].

According to a recent study by Neves and colleagues, Arabidopsis plants exposed to abiotic stress exhibit differential expression of genes involved in vacuolar trafficking, with the pathway to the PSV becoming enhanced [4]. In fact, under abiotic stress, plants are able to control their growth and development by changing cellular and morphological mechanisms, and cellular responses/adaptations to stress may affect the distribution and sorting of particular proteins and molecules. Additionally, numerous studies highlight the crucial function of the vacuole as a defense mechanism against abiotic stress. In fact, the vacuole appears to respond to stress through various processes, including the build-up of hazardous products and the maintenance of cell-turgor pressure. According to a study using suspension-cultured mangrove (*Bruguiera sexangula*) cells, when cells are exposed to salt stress, their vacuolar volume quickly increases at the expense of their cytoplasm volume in order to maintain turgor pressure, most likely due to an increase in the concentration of Na^+ in the vacuole [58]. Another study employing the Arabidopsis thaliana plant demonstrates the significance of the vacuole during oxidative stress. In fact, the vacuole developed large concentrations of GSSG (oxidized glutathione) as a defense mechanism against a too positive shift in the cytosolic glutathione redox potential [59]. In addition, the vacuole plays a role in systems that counteract environmental stress, such as lowering the cytoplasmic toxicity of high ion concentrations to prevent cell death. According to a study by Tang and colleagues [60], the excessive Mg^{2+} vacuolar sequestration that plants use to survive Mg^{2+} stress is a novel function of the Calcineurin B-like (CLB) interacting protein kinases' (CIPK) (CBL-CIPK) signaling network. A generic mechanism underpinning the detoxification of additional ions, such as Na^+, may be represented by the reported Mg^{2+} partitioning process in the vacuole controlled by the CBL-CIPK pathway (**Figure 1**). Contrary to abiotic stress, where the vacuole's integrity is crucial for maintaining the cell's homeostasis, pathogen infections necessitate the breakdown of the vacuole and the release of its contents (for a review on the subject, see [61]). The vacuole stores vast amounts of hydrolytic enzymes, such as proteases and antimicrobial substances as an innate defense mechanism, that are subsequently released under pathogen attack in a procedure that is not fully understood [62, 63]. The release of vacuolar contents has been attributed to two distinct mechanisms, including hypersensitive reaction and programmed cell death (PCD) [64]. The disruption of the tonoplast and involvement of vacuolar processing enzymes (VPEs) in one case, and the fusion of the tonoplast and plasma membrane (PM) in the other, were observed (**Figure 1**). Vacuolar contents are released as a consequence in both situations.

Alterations in vacuolar morphology, such as changes in vacuolar trafficking, are a crucial aspect of cell homeostasis under stress and also help maintain plant homeostasis. The actin cytoskeleton and SNARE proteins, which control these adaptations, allow the vacuolar network to be structurally reorganized while preserving its dynamics [65, 66].

4. The dynamic cytoskeleton concept

The notion of cytoskeleton has been transformed from a static, supporting structure to a dynamic mechanism in energetic balance that fine-tunes its time and space resolutions to adjust its functions to driving changes and stress reactions [67]. In plant cells, intracellular transport is primarily driven by myosin motors and actin filament bundles. Modifications in Golgi body motility show that changes in the pace of actin remodeling also have an impact on its functionality [68]. Depolymerization of actin inhibits both ER remodeling and Golgi movement, highlighting the significance of the actin cytoskeleton [69, 70]. Four members of the Myosin XI family (xi-k, xi-1, xi-2, and xi-i) were subjected to mutant knock-out studies, which revealed the importance of these proteins for normal cellular and whole-organism development as well as Golgi body dynamics [71]. However, microtubules are believed to be crucial at specific times in the formation of plant cells [72]. Given that stress is a condition that the cell finds to be quite difficult, it is necessary to test the idea that the cytoskeleton network will also have to adapt because its contact with membranes is essential for the cell's ability to self-organize. Reviewing the complexity of organelle movement within the plant secretory pathway, Brandizzi and Wasteneys [72] cast doubt on the actin-centric view of the motility of secretory organelles. They analyzed past studies and recent discoveries that support the critical function of microtubules in plant cell development, positioning of Golgi stacks, involvement in cellulose synthesis, and polar auxin transport.

The research of Ambrose and collaborators [73], which used hybrid and *in vivo* bimolecular fluorescence complementation techniques, was a turning point in understanding the relationship between endomembrane trafficking and microtubules. They found that the microtubule-associated protein CLASP interacts with the retromer, facilitating the association between TGN/early endosomes and cortical microtubules through interaction with sorting nexin1 (SNX1). The retromer protein complex, which SNX1 is a part of, recycles the plasma membrane auxin efflux carrier PIN2, hence regulating auxin transport.

Further investigations demonstrated the anchoring of compartments transporting cellulose synthase complexes to microtubules, confirming the importance of microtubules in organelle location and function. Cellulose is created at the plasma membrane by multi-enzyme complexes, in contrast to some cell wall polysaccharides made by glycosyl transferases and altered by Golgi-located enzymes [72]. This multi-enzyme complex must be delivered to the appropriate places at the PM. The trans-Golgi network (TGN) compartment is used to secrete cellulose synthase (CESA) complexes (CSCs) to the plasma membrane [74]. Through research on the intracellular trafficking of cellulose synthase complexes, small CesA-containing compartments (SmaCCs) [75] and microtubule-associated cellulose synthase compartments (MASCs) [76] were identified. Osmotic stress or the reduction of cellulose synthesis causes the endocytosis processes of the cellulose synthase complex, which causes the concentration of organelles containing CESA and their intense interaction with cortical microtubules [75, 76]. The SmaCCs associated with CSC transport may constitute a specialized secretory route involved in cell wall production, according to a theory where microtubule-associated compartments constitute functional secretory vesicles when plants are under cellular stress (**Figure 1**) [74]. This is a result of the fact that, before releasing CSC to the PM, these organelles reduce osmotic stress [75]. The SmaCC/MASC-mediated fast recovery of CSCs after stress relief depends on the protein cellulose synthase interactive 1 (CSI1), which is connected to cortical microtubules and involved in the interaction between CSCs and these structures [77].

SmaCCs/MASCs are also formed as a result of AP2M, a part of clathrin-mediated endocytosis. Lei and colleagues [77] suggest a concept in which CSI1-dependent SmaCCs/MASCs are produced during endocytosis, enabling rapid modulation of cellulose synthesis in response to abiotic stress. All of these methods help to decipher a spatiotemporal model of trafficking processes in cell wall deposition under both stress-free and demanding circumstances. Actin-binding proteins from the NET super-family [18] are recruited to various membrane compartments *via* a C-terminal region and directly interact with F-actin. These proteins include NET1A, which labels the plasma membrane, NET4A, which labels the tonoplast, and NET3B, which labels the endoplasmic reticulum. The fact that NET1A is among the actin-associated endoplasmic reticulum-plasma membrane contact site (EPCSs) proteins and that they react to extracellular signals like stress brought on by pathogen infection is further evidence for this claim [78]. The protein complex composed by the membrane-anchored protein VAP27 (At3g60600) and the actin-binding protein NET3C (At2g47920), which has an affinity for microtubules, is indicated to define the contact points between the plasma membrane and the cortical endoplasmic reticulum network [79]. In conclusion, the coordination of endomembrane trafficking requires the precise control of endomembrane carriage in space and time, incorporating both actin- and microtubule-based processes.

Additionally, vesicle shuttles (also known as transport vesicles) are the primary means of moving cargo molecules across compartments, and the cytoskeleton plays a function in making this process easier [80]. When the plant is exposed to harsh conditions, the relevance of this "shuttle transport" may take on more substantial outlines in the context of cellular rearrangement.

5. Vacuolar transport under stress

A sophisticated network of receptors and vesicles controls the movement of proteins into the vacuole. Because of this, proteins can be sorted differently, arriving at various locations depending on the receptors and vesicles employed [81, 82]. The vacuolar sorting receptors (VSRs), which are in charge of cargo binding and release as well as traffic regulation from and to the prevacuolar compartment (PVC) [14, 83], are involved in the transport of soluble cargoes by the conventional pathway. In addition to these receptors, proteins with the receptor homology region-transmembrane domain-RING-H2 (RMR) have been found to be involved in the flow to the PSV. These receptors, however, cannot be regenerated again [81, 84, 85]. The type of vesicles is another distinguishing element for the eventual location of the vacuolar proteins. Clathrin-coated vesicles (CCVs), which are located in the trans-Golgi Network (TGN) and are engaged in post-Golgi transport, are in charge of transporting proteins to the LV [81, 84, 86]. Dense vesicles (DVs), which are larger carriers compared to CCVs, fuse with PVCs and go to the PSV [14, 87–89]. It is evident that it is a flexible and well-coordinated network when all the information on protein trafficking to the vacuole is considered collectively [90]. Therefore, it is not surprising that this delicate balance can be disrupted in response to abiotic stress in order for the cell and, eventually, the plant, to meet their demands and survive.

Few studies have focused on this topic, and the changes in vacuolar trafficking that occur as a result of stress in cells have not yet been fully defined. Nevertheless, a few singular observations and reports are noteworthy because they might pave the way for further focused investigation. In recent work, Neves and colleagues [4] examined the expression of multiple endomembrane system effectors to assess how various abiotic stresses affect the endomembrane system in *A. thaliana*. The authors demonstrate that during abiotic stress, the PSV sorting genes AtRMR1, AtVSR1, AtSYP51, and AtVTI12 are positively regulated, whereas the LV sorting genes AtVTI11 and AtVSR2 are negatively regulated. The authors' theory, which is based on these observations, is that under abiotic stress circumstances, the PSV route would be strengthened at the expense of the LV pathway. Despite being very preliminary, this research identifies several crucial genes that are involved in the vacuolar route that may help understand how the cell responds to challenging circumstances. One example is the v-SNAREs VTI12 and its homolog VTI11, which work in several vesicle transport routes and mediate the transport to various vacuolar types [88]. VTI12, however, performs different activities, including helping autophagic vesicle binding and fusion [91]. Along with SYP61 and SYP41, it is a protein complex component located at the TGN. SYP61 has been linked to osmotic stress reactions [92], and it is hypothesized that it may also be a part of stress-responsive transport pathways, as that SYP121 at the plasma membrane has been linked to [10]. As a member of the same complex as SYP61, VTI12 might possibly take part in this mechanism. In fact, it has been demonstrated that VTI12 expression is 20–30 times higher in Arabidopsis plants grown under abiotic stress than in control circumstances [4], which is suggestive of a probable function in cells' adaptation or stress response. Additionally, the VSRs implicated in the trafficking of the PSV appear to react to stress. A unique role for AtVSR1 in osmotic stress tolerance and the control of abscisic acid (ABA) production, which is a key regulator of the signaling pathways generated by osmotic stressors, was recently proposed by Wang and collaborators [93]. With the aid of a vsr mutant, the authors demonstrated that vacuolar trafficking, which is mediated by VSR1, was essential for ABA production and osmotic stress tolerance. A different study found that Arabidopsis plants overexpressing AtRabG3e were more tolerant to salt and osmotic stress and produced fewer reactive oxygen species [94]. AtRabG3e engages in membrane fusion between the PVC and the vacuole, highlighting the importance of this pathway in stress response (**Figure 1**). The Rab GTPases are a broad family of proteins that regulate vesicle targeting and specificity [95]. In addition to the traditional pathway, the endocytic route to the vacuole has also been linked to plant tolerance to salt stress. This was demonstrated in a study by Leshem and colleagues [96], who found that suppressing the v-SNARE AtVAMP7C, which is necessary for endosomal vesicle fusion with the tonoplast, had a favorable effect on enhancing plant salt tolerance. Overall, the SNARE proteins are essential for protein trafficking to the vacuole, which is critical for both responses to stress and adaptations to it (See [97] for a review on SNAREs in plant stress responses). As with Adaptor protein 3 (AP-3) and the adaptor complex that interacts with VTI12 in the TGN, it is also important to investigate the role of other post-Golgi pathways [98]. This system interacts with the traditional pathway in a way that appears to affect how plants respond to stress circumstances while enabling a quicker supply of vital proteins for the vacuole's biogenesis. In addition,

DVs-mediated transport, which still has to be studied, is a good substitute for conventional transport in these challenging circumstances.

6. Taking a shortcut to the vacuole

Studies have described proteins and vacuolar signals that do not follow the mainstream route to the vacuole. The Golgi apparatus is required for some alternative sorting routes, like AP-3 and dense vesicle sorting, although other pathways also seem to be Golgi-independent [14]. Stress may activate these alternative sorting routes to better meet the plant's unique needs at the cellular level, but the relationship between stress and unorthodox sorting routes is largely unknown. In fact, autophagy-related processes, which can be triggered by a variety of environmental perturbations, seem to be connected to direct ER-to-vacuole pathways. A different pathway from the ER to the vacuole has been described for a variety of proteins or vacuolar sorting determinants in recent years [99–101]. Cardosin, a Plant Specific Insert (PSI), stands out among them because other similar domains lack this capability [13].

It is thought that additional unidentified, unconventional routes operate identically to the PSI-mediated vacuolar transport when plants are under stress, providing plants the option to sort proteins by the conventional approach or by a direct ER-to-vacuole transfer. In fact, a recent exploratory study [102] showed that Arabidopsis thaliana overexpression of PSIB correlates with conditions of salt and osmotic stress, occasionally improving plant fitness. A distinct family of proteins known as cysteine proteases also appears to be connected to salt stress. In both seedlings (as demonstrated in Vigna mungo [103] and Ricinus communis [104]) and the epidermis of vegetative tissues (Arabidopsis thaliana [105]), these proteins accumulate in lengthy ER bodies that eventually merge with the vacuole. Recent evidence suggests that direct ER body fusion with the vacuole might well be induced by stress, which sheds fresh light on the relevance of this kind of transport. The breakdown of storage proteins during plant growth is brought on by these proteins and the vacuolar processing enzymes.

The formation of ER bodies filled with defense proteins like pathogenesis-related 1 (PR1) or plant defensin 1.2 (PDF1.2) in response to pathogen attacks has been described in a similar way (for a review, see [61]) (**Figure 1**). These ER bodies then fuse with the plasma membrane or the vacuole in a way that is Golgi-independent. Additionally, autophagy markers are regularly seen in the ER and vacuole membranes [93], and stress frequently induces autophagic compartments [3, 106]. It is yet unclear how autophagy in and of itself can aid in vacuolar sorting, and additional mechanisms or regulators undoubtedly need to be engaged. An intriguing example of unusual trafficking involves the exocyst pathway, which plays a role in autophagy and plant defense, and anthocyanins that are imported to the vacuole during cycles of stress and famine [107, 108]. After reviewing all the available instances, it is critical to research the direct ER-to-vacuole transfer in stressed plants. In fact, defining atypical sorting routes along with stress responses would offer fresh perspectives on the scant knowledge that has previously been known. Given that it speeds up and increases the dynamic of protein transport to the vacuole, the Golgi bypass may significantly impact stress responses. As a matter of fact, a number of unconventional pathways are triggered by modifications in the cell environment rather than being constitutive.

7. Conclusions

For many years, studies and discussions on the effects of stress on plants have dominated the headlines. However, because of the discussion's main emphasis on the physiology and antioxidant system of plants, essential cell activities are frequently overlooked. However, given that a significant number of genes and proteins are *de novo* generated in response to stress and must be transported to their correct locations, this is a crucial problem to investigate. Understanding trafficking processes and proteins linked with transport is crucial in this situation. The major "sensor" for stress is thought to be the ER, where the stress responses begin and from which proteins and signals are either transported to other parts of the cell or destroyed. Given its various dimensions and functions, the vacuole is also crucial to this process. As a result, one of the key mechanisms in plant defense and cellular homeostasis is the transit of vesicles between the ER and the vacuole. According to a recent study, in which it was demonstrated that the Golgi is hypertrophied and associated with high vesiculation in plants under stress using Transmission Electron Microscopy [4], the high amount of proteins and molecules newly produced will likely cause saturation of the Golgi trafficking pathways. In this case, the ER is directly connected to the vacuole, which is a speedier path and can be thought of as an escape from the gridlock that started between the ER, Golgi, and prevacuolar compartments. In fact, it appears that stress or other challenging conditions are connected to these atypical pathways to the vacuole or the plasma membrane. We still have a long way to go before we fully understand the mechanisms underlying these pathways and how they are regulated, but the first steps are being made, and in the near future, we anticipate having a clearer picture of the process and a better understanding of the mechanisms underlying plant tolerance and adaptation to stress.

Acknowledgements

This research was supported by national funds through FCT, within the scope of UIDB/05748/2020 and UIDP/05748/2020. M.S. is the recipient of a PhD fellowship (Ref.: SFRH/UIDB/151042/2021) granted by GreenUPorto (Sustainable Agrifood Production Research Centre/Inov4Agro, with financial support from FCT.

Conflict of interest

The authors declare no conflict of interest.

Author details

Miguel Sampaio[1], João Neves[2], Tatiana Cardoso[2], José Pissarra[1], Susana Pereira[1*] and Cláudia Pereira[1*]

1 Department of Biology, GreenUPorto - Sustainable Agrifood Production Research Centre/Inov4Agro, Faculty of Sciences of the University of Porto, Porto, Portugal

2 Faculty of Sciences of the University of Porto, Department of Biology, Porto, Portugal

*Address all correspondence to: mspereir@fc.up.pt and cpereira@fc.up.pt

References

[1] Cushman JC, Bohnert HJ. Genomic approaches to plant stress tolerance. Current Opinion in Plant Biology. 2000;**3**:117-124. DOI: 10.1016/S1369-5266(99)00052-7

[2] Moellering ER, Benning C. Galactoglycerolipid metabolism under stress: A time for remodeling. Trends in Plant Science. 2011;**16**:98-107. DOI: 10.1016/j.tplants.2010.11.004

[3] Bassham DC, Laporte M, Marty F, Moriyasu Y, Ohsumi Y, Olsen LJ, et al. Autophagy in development and stress responses of plants. Autophagy. 2006;**2**:2-11. DOI: 10.4161/auto.2092

[4] Neves J, Sampaio M, Séneca A, Pereira S, Pissarra J, Pereira C. Abiotic stress triggers the expression of genes involved in protein storage vacuole and exocyst-mediated routes. International Journal of Molecular Sciences. 2021;**22**: 1-20. DOI: 10.3390/ijms221910644

[5] Mousavi-Derazmahalleh M, Bayer PE, Hane JK, Valliyodan B, Nguyen HT, Nelson MN, et al. Adapting legume crops to climate change using genomic approaches. Plant, Cell & Environment. 2019;**42**:6-19

[6] Kalinowska K, Isono E. All roads lead to the vacuole - Autophagic transport as part of the endomembrane trafficking network in plants. Journal of Experimental Botany. 2018;**69**:1313-1324

[7] Liu L, Li J. Communications between the endoplasmic reticulum and other organelles during abiotic stress response in plants. Frontiers in Plant Science. 2019;**10**:749. DOI: 10.3389/fpls.2019.00749

[8] Rosquete MR, Drakakaki G. Plant TGN in the stress response: A compartmentalized overview. Current Opinion in Plant Biology. 2018;**46**:122-129. DOI: 10.1016/j.pbi.2018.09.003

[9] Chevalier AS, Chaumont F. Trafficking of plant plasma membrane aquaporins: Multiple regulation levels and complex sorting signals. Plant & Cell Physiology. 2015;**56**:819-829

[10] Hachez C, Laloux T, Reinhardt H, Cavez D, Degand H, Grefen C, et al. Arabidopsis SNAREs SYP61 and SYP121 coordinate the trafficking of plasma membrane aquaporin PIP2;7 to modulate the cell membrane water permeability. Plant Cell. 2014;**26**:3132-3147. DOI: 10.1105/tpc.114.127159

[11] Ahmad I, Devonshire J, Mohamed R, Schultze M, Maathuis FJM. Overexpression of the potassium channel TPKb in small vacuoles confers osmotic and drought tolerance to rice. The New Phytologist. 2016;**209**:1040-1048. DOI: 10.1111/nph.13708

[12] Pereira C, Pereira S, Satiat-Jeunemaitre B, Pissarra J. Cardosin A contains two vacuolar sorting signals using different vacuolar routes in tobacco epidermal cells. The Plant Journal. 2013;**76**:87-100. DOI: 10.1111/tpj.12274

[13] Vieira V, Peixoto B, Costa M, Pereira S, Pissarra J, Pereira C. N-linked glycosylation modulates Golgi-independent vacuolar sorting mediated by the plant specific insert. Plants. 2019;**8**:1-21. DOI: 10.3390/plants8090312

[14] Di Sansebastiano G-P, Barozzi F, Piro G, Denecke J, De C, Lousa M, et al. Trafficking routes to the plant vacuole: Connecting alternative and classical pathways. Journal of Experimental

Botany. 2017;**69**:79-90. DOI: 10.1093/jxb/erx376

[15] Ellgaard L, Helenius A. Quality control in the endoplasmic reticulum. Nature Reviews. Molecular Cell Biology. 2003;**4**:181-191. DOI: 10.1038/nrm1052

[16] Schröder M, Kaufman RJ. The mammalian unfolded protein response. Annual Review of Biochemistry. 2005;**74**:739-789. DOI: 10.1146/annurev.biochem.73.011303.074134

[17] Park CJ, Park JM. Endoplasmic reticulum plays a critical role in integrating signals generated by both biotic and abiotic stress in plants. Frontiers in Plant Science. 2019;**10**:399. DOI: 10.3389/fpls.2019.00399

[18] Wang P, Hussey PJ. Interactions between plant endomembrane systems and the actin cytoskeleton. Frontiers in Plant Science. 2015;**6**:1-8. DOI: 10.3389/fpls.2015.00422

[19] Zhu J-K. Abiotic stress signaling and responses in plants. Cell. 2016;**167**:313-324. DOI: 10.1016/j.cell.2016.08.029. Abiotic

[20] Wang X, Xu M, Gao C, Zeng Y, Cui Y, Shen W, et al. The roles of endomembrane trafficking in plant abiotic stress responses. Journal of Integrative Plant Biology. 2020;**62**:55-69. DOI: 10.1111/jipb.12895

[21] Howell SH. Endoplasmic reticulum stress responses in plants. Annual Review of Plant Biology. 2013;**64**:477-499. DOI: 10.1146/annurev-arplant-050312-120053

[22] Walter P, Ron D. The unfolded protein response: From stress pathway to homeostatic regulation. Science (80-.). 2011;**334**:1081-1086. DOI: 10.1126/science.1209038

[23] Gao H, Brandizzi F, Benning C, Larkin RM. A membrane-tethered transcription factor defines a branch of the heat stress response in Arabidopsis thaliana. Proceedings of the National Academy of Sciences of the United States of America. 2008;**105**:16397-16403. DOI: 10.1073/pnas.0808463105

[24] Liu JX, Srivastava R, Che P, Howell SH. Salt stress responses in Arabidopsis utilize a signal transduction pathway related to endoplasmic reticulum stress signaling. The Plant Journal. 2007;**51**:897-909. DOI: 10.1111/j.1365-313X.2007.03195.x

[25] Koizumi N, Martinez IM, Kimata Y, Kohno K, Sano H, Chrispeels MJ. Molecular characterization of two Arabidopsis Ire1 homologs, endoplasmic reticulum-located transmembrane protein kinases. Plant Physiology. 2001;**127**:949-962. DOI: 10.1104/pp.010636

[26] Deng Y, Humbert S, Liu JX, Srivastava R, Rothstein SJ, Howell SH. Heat induces the splicing by IRE1 of a mRNA encoding a transcription factor involved in the unfolded protein response in Arabidopsis. Proceedings of the National Academy of Sciences of the United States of America. 2011;**108**:7247-7252. DOI: 10.1073/pnas.1102117108

[27] Mishiba KI, Nagashima Y, Suzukia E, Hayashi N, Ogata Y, Shimada Y, et al. Defects in IRE1 enhance cell death and fail to degrade mRNAs encoding secretory pathway proteins in the Arabidopsis unfolded protein response. Proceedings of the National Academy of Sciences of the United States of America. 2013;**110**:5713-5718. DOI: 10.1073/pnas.1219047110

[28] Maurel M, Chevet E, Tavernier J, Gerlo S. Getting RIDD of RNA: IRE1 in cell fate regulation. Trends in

Biochemical Sciences. 2014;**39**:245-254. DOI: 10.1016/j.tibs.2014.02.008

[29] Alvim FC, Carolino SMB, Cascardo JCM, Nunes CC, Martinez CA, Otoni WC, et al. Enhanced accumulation of BiP in transgenic plants confers tolerance to water stress. Plant Physiology. 2001;**126**:1042-1054. DOI: 10.1104/pp.126.3.1042

[30] Valente MAS, Faria JAQA, Soares-Ramos JRL, Reis PAB, Pinheiro GL, Piovesan ND, et al. The ER luminal binding protein (BiP) mediates an increase in drought tolerance in soybean and delays drought-induced leaf senescence in soybean and tobacco. Journal of Experimental Botany. 2009;**60**:533-546. DOI: 10.1093/jxb/ern296

[31] Lai YS, Renna L, Yarema J, Ruberti C, He SY, Brandizzi F. Salicylic acid-independent role of NPR1 is required for protection from proteotoxic stress in the plant endoplasmic reticulum. Proceedings of the National Academy of Sciences of the United States of America. 2018;**115**:E5203-E5212. DOI: 10.1073/pnas.1802254115

[32] Lai YS, Stefano G, Zemelis-Durfee S, Ruberti C, Gibbons L, Brandizzi F. Systemic signaling contributes to the unfolded protein response of the plant endoplasmic reticulum. Nature Communications. 2018;**9**:1-11. DOI: 10.1038/s41467-018-06289-9

[33] Li F, Vierstra RD. Autophagy: A multifaceted intracellular system for bulk and selective recycling. Trends in Plant Science. 2012;**17**:526-537. DOI: 10.1016/j.tplants.2012.05.006

[34] Liu Y, Bassham DC. Autophagy: Pathways for self-eating in plant cells. Annual Review of Plant Biology. 2012;**63**:215-237. DOI: 10.1146/annurev-arplant-042811-105441

[35] Bernales S, McDonald KL, Walter P. Autophagy counterbalances endoplasmic reticulum expansion during the unfolded protein response. PLoS Biology. 2006;**4**:2311-2324. DOI: 10.1371/journal.pbio.0040423

[36] Ogata M, Hino S, Saito A, Morikawa K, Kondo S, Kanemoto S, et al. Autophagy is activated for cell survival after endoplasmic reticulumstress. Molecular and Cellular Biology. 2006;**26**:9220-9231. DOI: 10.1128/mcb.01453-06

[37] Deegan S, Saveljeva S, Gorman AM, Samali A. Stress-induced self-cannibalism: On the regulation of autophagy by endoplasmic reticulum stress. Cellular and Molecular Life Sciences. 2013;**70**:2425-2441. DOI: 10.1007/s00018-012-1173-4

[38] Bao Y, Pu Y, Yu X, Gregory BD, Srivastava R, Howell SH, et al. IRE1B degrades RNAs encoding proteins that interfere with the induction of autophagy by ER stress in Arabidopsis thaliana. Autophagy. 2018;**14**:1562-1573. DOI: 10.1080/15548627.2018.1462426

[39] Hollien J, Lin JH, Li H, Stevens N, Walter P, Weissman JS. Regulated Ire1-dependent decay of messenger RNAs in mammalian cells. The Journal of Cell Biology. 2009;**186**:323-331. DOI: 10.1083/jcb.200903014

[40] Marty F. Plant vacuoles. Plant Cell. 1999;**11**:587-599. DOI: 10.1016/j.cub.2014.11.056

[41] Martinoia E, Maeshima M, Neuhaus HE. Vacuolar transporters and their essential role in plant metabolism. Journal of Experimental Botany. 2007;**58**:83-102. DOI: 10.1093/jxb/erl183

[42] Bethke PC, Jones RL. Vacuoles and prevacuolar compartments. Current Opinion in Plant Biology. 2000;**3**:469-475. DOI: 10.1016/S1369-5266(00)00115-1

[43] Jiang L, Phillips TE, Rogers SW, Rogers JC. Biogenesis of the protein storage vacuole crystalloid. The Journal of Cell Biology. 2000;**150**:755-770

[44] Martinoia E, Massonneau A, Frangne N. Transport processes of solutes across the vacuolar membrane of higher plants. Plant & Cell Physiology. 2000;**41**:1175-1186. DOI: 10.1093/pcp/pcd059

[45] Hatsugai N, Kuroyanagi M, Yamada K, Meshi T, Tsuda S, Kondo M, et al. A plant vacuolar protease, VPE, mediates, virus-induced hypersensitive cell death. Science (80-.). 2004;**305**:855-858. DOI: 10.1126/science.1099859

[46] Hara-Nishimura I, Hatsugai N. The role of vacuole in plant cell death. Cell Death and Differentiation. 2011;**18**:1298-1304. DOI: 10.1038/cdd.2011.70

[47] Zouhar J, Muñoz A, Rojo E. Functional specialization within the vacuolar sorting receptor family: VSR1, VSR3 and VSR4 sort vacuolar storage cargo in seeds and vegetative tissues. The Plant Journal. 2010;**64**:577-588. DOI: 10.1111/j.1365-313X.2010.04349.x

[48] Müntz K. Protein dynamics and proteolysis in plant vacuoles. Journal of Experimental Botany. 2007;**58**:2391-2407. DOI: 10.1093/jxb/erm089

[49] Frigerio L, Hinz G, Robinson DG. Multiple vacuoles in plant cells: Rule or exception? Traffic. 2008;**9**:1564-1570. DOI: 10.1111/j.1600-0854.2008.00776.x

[50] Rogers JC. Multiple vacuoles in plant cells. Plant Physiology. 2008;**146**:1024-1025. DOI: 10.1104/pp.107.900248

[51] Paris N, Stanley CM, Jones RL, Rogers JC. Plant cells contain two functionally distinct vacuolar compartments. Cell. 1996;**85**:563-572. DOI: 10.1016/S0092-8674(00)81256-8

[52] Olbrich A, Hillmer S, Hinz G, Oliviusson P, Robinson DG. Newly formed vacuoles in root meristems of barley and pea seedlings have characteristics of both protein storage and lytic vacuoles. Plant Physiology. 2007;**145**:1383-1394. DOI: 10.1104/pp.107.108985

[53] Martínez DE, Costa ML, Gomez FM, Otegui MS, Guiamet JJ. "Senescence-associated vacuoles" are involved in the degradation of chloroplast proteins in tobacco leaves. The Plant Journal. 2008;**56**:196-206. DOI: 10.1111/j.1365-313X.2008.03585.x

[54] Neuhaus J-M, Rogers JC. Sorting of proteins to vacuoles in plant cells. In: Protein Trafficking in Plant Cells. Springer Netherlands: Dordrecht; 1998. pp. 127-144

[55] Jürgens G. Membrane trafficking in plants. Annual Review of Cell and Developmental Biology. 2004;**20**:481-504. DOI: 10.1146/annurev.cellbio.20.082503.103057

[56] Vitale A, Hinz G. Sorting of proteins to storage vacuoles: how many mechanisms? Trends in Plant Science. 2005;**10**:316-323. DOI: 10.1016/j.tplants.2005.05.001

[57] Pereira C, Pereira S, Pissarra J. Delivering of proteins to the plant vacuole-an update. International Journal of Molecular Sciences. 2014;**15**:1-20. DOI: 10.3390/ijms15057611

[58] Mimura T, Kura-Hotta M, Tsujimura T, Ohnishi M, Miura M, Okazaki Y, et al. Rapid increase of vacuolar volume in response to salt stress. Planta. 2003;**216**:397-402. DOI: 10.1007/s00425-002-0878-2

[59] Queval G, Jaillard D, Zechmann B, Noctor G. Increased intracellular H2O2 availability preferentially drives glutathione accumulation in vacuoles and chloroplasts. Plant, Cell and Environment. 2011;**34**:21-32. DOI: 10.1111/j.1365-3040.2010.02222.x

[60] Tang RJ, Zhao FG, Garcia VJ, Kleist TJ, Yang L, Zhang HX, et al. Tonoplast CBL-CIPK calcium signaling network regulates magnesium homeostasis in Arabidopsis. Proceedings of the National Academy of Sciences of the United States of America. 2015;**112**:3134-3139. DOI: 10.1073/pnas.1420944112

[61] Ruano G, Scheuring D. Plant cells under attack: Unconventional. Plants. 2020;**2020**:1-18

[62] Heard W, Sklenář J, Tomé DFA, Robatzek S, Jones AME. Identification of regulatory and cargo proteins of endosomal and secretory pathways in arabidopsis thaliana by proteomic dissection. Molecular & Cellular Proteomics. 2015;**14**:1796-1813. DOI: 10.1074/mcp.M115.050286

[63] Carter C, Pan S, Zouhar J, Avila EL, Girke T, Raikhel NV. The vegetative vacuole proteome of Arabidopsis thaliana reveals predicted and unexpected proteins. Plant Cell. 2004;**16**:3285-3303. DOI: 10.1105/tpc.104.027078

[64] Hatsugai N, Hara-Nishimura I. Two vacuole-mediated defense strategies in plants. Plant Signaling & Behavior. 2010;**2010**:568-1570. DOI: 10.4161/psb.5.12.13319

[65] Scheuring D, Löfke C, Krüger F, Kittelmann M, Eisa A, Hughes L, et al. Actin-dependent vacuolar occupancy of the cell determines auxin-induced growth repression. Proceedings of the National Academy of Sciences of the United States of America. 2016;**113**:452-457. DOI: 10.1073/pnas.1517445113

[66] Löfke C, Dünser K, Scheuring D, Kleine-Vehn J. Auxin regulates SNARE-dependent vacuolar morphology restricting cell size. eLife. 2015;**2015**:1-16. DOI: 10.7554/eLife.05868

[67] Nick P. Microtubules, signalling and abiotic stress. The Plant Journal. 2013;**75**:309-323. DOI: 10.1111/tpj.12102

[68] Tolmie F, Poulet A, McKenna J, Sassmann S, Graumann K, Deeks M, et al. The cell wall of Arabidopsis thaliana influences actin network dynamics. Journal of Experimental Botany. 2017;**68**:4517-4527. DOI: 10.1093/jxb/erx269

[69] Boevink P, Oparka K, Cruz SS, Martin B, Betteridge A, Hawes C. Stacks on tracks: The plant Golgi apparatus traffics on an actin/ER network †. The Plant Journal. 1998;**15**:441-447. DOI: 10.1046/j.1365-313X.1998.00208.x

[70] Sparkes I, Runions J, Hawes C, Griffing L. Movement and remodeling of the endoplasmic reticulum in nondividing cells of tobacco leaves. Plant Cell. 2009;**21**:3937-3949. DOI: 10.1105/tpc.109.072249

[71] Peremyslov VV, Morgun EA, Kurth EG, Makarova KS, Koonin EV, Dolja VV. Identification of myosin XI receptors in arabidopsis defines a distinct class of transport vesicles. Plant Cell. 2013;**25**:3022-3038. DOI: 10.1105/tpc.113.113704

[72] Brandizzi F, Wasteneys GO. Dependent endomembrane organization

in plant cells : An emerging role for microtubules. Cytoskeleton. 2013;**2013**:339-349. DOI: 10.1111/tpj.12227

[73] Ambrose C, Ruan Y, Gardiner J, Tamblyn LM, Catching A, Kirik V, et al. CLASP interacts with sorting Nexin 1 to link microtubules and auxin transport via PIN2 recycling in *Arabidopsis thaliana*. Developmental Cell. 2013;**24**:649-659. DOI: 10.1016/j.devcel.2013.02.007

[74] Vellosillo T, Dinneny JR, Somerville CR, Ehrhardt DW. TRANVIA (TVA) facilitates cellulose synthase trafficking and delivery to the plasma membrane. Proceedings of the National Academy of Sciences. 2021;**118**:1-9. DOI: 10.1073/PNAS.2021790118

[75] Gutierrez R, Lindeboom JJ, Paredez AR, Emons AMC, Ehrhardt DW. Arabidopsis cortical microtubules position cellulose synthase delivery to the plasma membrane and interact with cellulose synthase trafficking compartments. Nature Cell Biology. 2009;**11**:797-806. DOI: 10.1038/ncb1886

[76] Crowell EF, Bischoff V, Desprez T, Rolland A, Stierhof YD, Schumacher K, et al. Pausing of golgi bodies on microtubules regulates secretion of cellulose synthase complexes in Arabidopsis. Plant Cell. 2009;**21**:1141-1154. DOI: 10.1105/tpc.108.065334

[77] Lei L, Singh A, Bashline L, Li S, Yingling YG, Gua Y. Cellulose synthase interactive1 Is required for fast recycling of cellulose synthase complexes to the plasma membrane in arabidopsis. Plant Cell. 2015;**27**:2926-2940. DOI: 10.1105/tpc.15.00442

[78] Wang P, Hawkins TJ, Hussey PJ. Connecting membranes to the actin cytoskeleton. Current Opinion in Plant Biology. 2017;**40**:71-76. DOI: 10.1016/j.pbi.2017.07.008

[79] Wang P, Hawkins TJ, Richardson C, Cummins I, Deeks MJ, Sparkes I, et al. The plant cytoskeleton, NET3C, and VAP27 mediate the link between the plasma membrane and endoplasmic reticulum. Current Biology. 2014;**24**:1397-1405. DOI: 10.1016/j.cub.2014.05.003

[80] Boutte Y, Vernhettes S, Satiat-Jeunemaitre B. Involvement of the cytoskeleton in the secretory pathway and plasma membrane organisation of higher plant cells Involvement of the cytoskeleton in the secretory pathway and plasma membrane organisation of higher plant cells. Cell Biology International. 2007;**31**:649-654. DOI: 10.1016/j.cellbi.2007.01.006

[81] Zhang X, Li H, Lu H, Hwang I. The trafficking machinery of lytic and protein storage vacuoles: How much is shared and how much is distinct? Journal of Experimental Botany. 2021;**72**:3504-3512. DOI: 10.1093/jxb/erab067

[82] Pereira C, Di Sansebastiano G. Pietro Mechanisms of membrane traffic in plant cells. Plant Physiology and Biochemistry. 2021;**169**:102-111. DOI: 10.1016/j.plaphy.2021.11.003

[83] Marcos Lousa C, Gershlick DC, Denecke J. Mechanisms and concepts paving the way towards a complete transport cycle of plant vacuolar sorting receptors. Plant Cell. 2012;**24**:1714-1732. DOI: 10.1105/tpc.112.095679

[84] Kim H, Kang H, Jang M, Chang JH, Miao Y, Jiang L, et al. Homomeric interaction of AtVSR1 is essential for its function as a vacuolar sorting receptor. Plant Physiology. 2010;**154**:134-148. DOI: 10.1104/pp.110.159814

[85] Miao Y, Yan PK, Kim H, Hwang I, Jiang L. Localization of green fluorescent protein fusions with the seven Arabidopsis vacuolar sorting receptors to prevacuolar compartments in tobacco BY-2 cells. Plant Physiology. 2006;**142**:945-962. DOI: 10.1104/pp.106.083618

[86] Park M, Song K, Reichardt I, Kim H, Mayer U, Stierhof YD, et al. Arabidopsis μ-adaptin subunit AP1M of adaptor protein complex 1 mediates late secretory and vacuolar traffic and is required for growth. Proceedings of the National Academy of Sciences of the United States of America. 2013;**110**:10318-10323. DOI: 10.1073/pnas.1300460110

[87] Hillmer S, Movafeghi A, Robinson DG, Hinz G. Vacuolar storage proteins are sorted in the cis-cisternae of the pea cotyledon Golgi apparatus. The Journal of Cell Biology. 2001;**152**:41-50. DOI: 10.1083/jcb.152.1.41

[88] Hinz G, Colanesi S, Hillmer S, Rogers JC, Robinson DG. Localization of vacuolar transport receptors and cargo proteins in the Golgi apparatus of developing Arabidopsis embryos. Traffic. 2007;**8**:1452-1464. DOI: 10.1111/j.1600-0854.2007.00625.x

[89] Hinz G, Hillmer S, Bäumer M, Hohl I. Vacuolar storage proteins and the putative vacuolar sorting receptor BP-80 exit the Golgi apparatus of developing pea cotyledons in different transport vesicles. Plant Cell. 1999;**11**:1509-1524. DOI: 10.1105/tpc.11.8.1509

[90] Jiang L, Erickson AH, Rogers JC. Multivesicular bodies: A mechanism to package lytic and storage functions in one organelle? Trends in Cell Biology. 2002;**12**:362-367. DOI: 10.1016/S0962-8924(02)02322-X

[91] Surpin M, Zheng H, Morita MT, Saito C, Avila E, Blakeslee JJ, et al. The VTI family of SNARE proteins is necessary for plant viability and mediates different protein transport pathways. Plant Cell. 2003;**15**:2885-2899. DOI: 10.1105/tpc.016121

[92] Zhu J-K. Salt and drought stress signal transduction in plants. Annual Review of Plant Biology. 2002;**53**:247-273. DOI: 10.1146/annurev.arplant.53.091401.143329

[93] Wang ZY, Gehring C, Zhu J, Li FM, Zhu JK, Xiong L. The arabidopsis vacuolar sorting receptor1 is required for osmotic stress-induced abscisic acid biosynthesis. Plant Physiology. 2015;**167**:137-152. DOI: 10.1104/pp.114.249268

[94] Mazel A, Leshem Y, Tiwari BS, Levine A. Induction of salt and osmotic stress tolerance by overexpression of an intracellular vesicle trafficking protein AtRab7 (AtRabG3e). Plant Physiology. 2004;**134**:118-128. DOI: 10.1104/pp.103.025379

[95] Rutherford S, Moore I. The Arabidopsis Rab GTPase family: Another enigma variation. Current Opinion in Plant Biology. 2002;**5**:518-528. DOI: 10.1016/S1369-5266(02)00307-2

[96] Leshem Y, Melamed-Book N, Cagnac O, Ronen G, Nishri Y, Solomon M, et al. Suppression of Arabidopsis vesicle-SNARE expression inhibited fusion of H2O2-containing vesicles with tonoplast and increased salt tolerance. Proceedings of the National Academy of Sciences of the United States of America. 2006;**103**:18008-18013. DOI: 10.1073/pnas.0604421103

[97] Kwon C, Lee JH, Yun HS. Snares in plant biotic and abiotic stress responses. Molecules and Cells. 2020;**43**:501-508. DOI: 10.14348/molcells.2020.0007

[98] Feraru E, Paciorek T, Feraru MI, Zwiewka M, de Groodt R, de Rycke R, et al. The AP-3 β adaptin mediates the biogenesis and function of lytic vacuoles in Arabidopsis. Plant Cell. 2010;**22**:2812-2824. DOI: 10.1105/tpc.110.075424

[99] Stigliano E, Faraco M, Neuhaus J-MM, Montefusco A, Dalessandro G, Piro G, et al. Pietro Two glycosylated vacuolar GFPs are new markers for ER-to-vacuole sorting. Plant Physiology and Biochemistry. 2013;**73**:337-343. DOI: 10.1016/j.plaphy.2013.10.010

[100] Occhialini A, Gouzerh G, Sansebastiano D, Pietro G, Neuhaus JM. Dimerization of the vacuolar receptors AtRMR1 and -2 from Arabidopsis thaliana contributes to their localization in the trans-Golgi network. International Journal of Molecular Science. 2016;**2016**:17. DOI: 10.3390/ijms17101661

[101] Pompa A, De Marchis F, Pallotta MT, Benitez-Alfonso Y, Jones A, Schipper K, et al. Unconventional transport routes of soluble and membrane proteins and their role in developmental biology. International Journal of Molecular Science. 2017;**2017**:18. DOI: 10.3390/ijms18040703

[102] Moura I, Pereira S, Séneca A, Pissarra J, Pereira C. Overexpression of plant specific Insert from Cardosin B (PSI B) in Arabidopsis correlates with cell responses to stresses. Biological Life Science Forum. 2021;**1**:1-8. DOI: doi.org/10.3390

[103] Toyooka K, Okamoto T, Minamikawa T. Mass transport of proform of a KDEL-tailed cysteine proteinase (SH-EP) to protein storage vacuoles by endoplasmic reticulum-derived vesicle is involved in protein mobilization in germinating seeds. The Journal of Cell Biology. 2000;**148**:453-463. DOI: 10.1083/jcb.148.3.453

[104] Schmid M, Simpson DJ, Sarioglu H, Lottspeich F, Gietl C. The ricinosomes of senescing plant tissue bud from the endoplasmic reticulum. Proceedings of the National Academy of Sciences of the United States of America. 2001;**98**:5353-5358. DOI: 10.1073/pnas.061038298

[105] Hayashi Y, Yamada K, Shimada T, Matsushima R, Nishizawa NK, Nishimura M, et al. A proteinase-storing body that prepares for cell death or stresses in the epidermal cells of arabidopsis. Plant & Cell Physiology. 2001;**42**:894-899. DOI: 10.1093/pcp/pce144

[106] Xiong Y, Contento AL, Nguyen PQ, Bassham DC. Degradation of oxidized proteins by autophagy during oxidative stress in arabidopsis. Plant Physiology. 2007;**143**:291-299. DOI: 10.1104/pp.106.092106

[107] Kulich I, Žárský V. Autophagy-related direct membrane import from ER/Cytoplasm into the vacuole or apoplast: A hidden gateway also for secondary metabolites and phytohormones? International Journal of Molecular Sciences. 2014;**15**:7462-7474. DOI: 10.3390/ijms15057462

[108] Žárský V, Kulich I, Fendrych M, Pečenková T. Exocyst complexes multiple functions in plant cells secretory pathways. Current Opinion in Plant Biology. 2013;**16**:726-733. DOI: 10.1016/j.pbi.2013.10.013

Chapter 5

Photosynthetic Response and Adaptation of Plants in Perspective of Global Climate Change

Mohammad Javad Ahmadi-Lahijani and Saeed Moori

Abstract

The intense agricultural and human being activities, especially after the industrialization era, have increased the CO_2 concentration, which led to changes in the global climate. Climate change and its consequences, that is, elevated CO_2, water stress, and extreme temperatures, have induced many biotic and abiotic stresses and have caused alterations in plant physiology, leading to a reduced photosynthetic capacity of plants. Photosynthesis is the most crucial biochemical process in plants that determines the final dry matter production and productivity of plants. The efficiency and status of the photosynthetic apparatus can be measured by the measurement of chlorophyll fluorescence. Measurements of chlorophyll fluorescence are easy, non-destructive, and quick, and it reflects changes in the general bioenergy status of a plant. Studies have indicated that abiotic stresses emerging from climate changes cause changes in the biological processes of plants and damage the internal structure of photosynthesis and control of the cellular process. Chlorophyll fluorescence, meanwhile, is an effective parameter and an indicator of photosynthetic status and its mechanisms under stressful conditions. Therefore, the photosynthetic changes and adaptation and the role of chlorophyll fluorescence in determining its status under climate change are discussed in this chapter.

Keywords: abiotic stress, chlorophyll fluorescence, drought, elevated CO_2, extreme temperatures, leaf physiology

1. Introduction

Food production is required to be increased by ~70% to feed the global population of 9 billion by 2050 [1], since the food demand, especially in developing countries, will be immensely enhanced. During the last 160,000 years, the concentration of atmospheric carbon dioxide has been varying between 170 and 300 μmol mol^{-1}. But with the beginning of the industrial revolution in Western Europe (between 1750 and 1800), the concentration of CO_2 increased from 280 to 385 μmol mol^{-1} [2]. According to predictions, with the rapid increase in world population, consumption of fossil fuels, industrial development, and deforestation, the concentration of carbon dioxide, which is ~400 μmol mol^{-1}, will reach 700 micromoles by the end of this century [3].

Climate change and global warming have been one of the most controversial issues in the recent decade. Intense agricultural and industrial activities since the industrial revolution have hastened the process of global warming. The chemistry of the climate has been changed by agricultural and human being activities and consequently, many abiotic and biotic stresses have emerged and negatively affected plants' physiology and biochemistry. Crops resistant to environmental stresses should be the focus of agricultural plant development under the increased global temperatures and climate changes.

Due to continuously increasing the greenhouse gases, such as CO_2, in the atmosphere, climate change is happening rapidly. Climate change by increasing temperatures and reducing precipitations imposes abiotic stress exposure in many areas. Abiotic stresses, such as drought, salinity, cold, heat, UV radiation, and heavy metals, are the major limitations in agricultural products and adversely influence plant growth. It is estimated that abiotic stresses reduce crop yield by approximately 50% [4]. Drought, salinity, and extreme temperatures are among the most dreadful abiotic stresses in modern agriculture.

One of the most vital processes of plants that are affected by global climate change is photosynthesis. Photosynthesis is a vital biochemical process in plants that supplies the carbon and energy required for the biosynthesis of organic compounds and controls plant growth and development [5]. Photosynthesis is particularly sensitive to environmental constraints [6]. The environmental stresses adversely affect the photosynthetic capacity of plants. The increasing global population and climate change over the coming decades require enhanced photosynthetic efficiency to ensure food security. Thus, an understanding of the photosynthetic response and optimization under future climate uncertainties will be required for an improvement in crop production to meet future food requirements.

Chlorophyll fluorescence is one of the effective, non-destructive, and quick methods for evaluating the photochemical status of the plant photosynthetic system. Chlorophyll fluorescence is a useful parameter for the measurement of environmental

Plant species	Environmental conditions	Parameters	Reference
Potato (*Solanum tuberosum*)	Elevated CO_2	g_m, T_r, g_s ↓ A_n, C_i, R_D ↑	[10, 11]
Tomato (*Solanum lycopersicum* L.)	Elevated CO_2	A_n, V_{cmax}, J_{max}, f_v/f_m, ETR, $NADP^+$/NADPH ↑ NPQ, RL ↓	[12]
Fagus sylvatica	Elevated CO_2	A_n, R_D ↑ g_s, V_{cmax} ↓	[13]
Yucca (*Y. brevifolia and Y. schidigera*)	Elevated CO_2	A_n, f_v/f_m, Φ_{PSII} ↑ g_s ↓	[14]
Cotton (*Gossypium hirsutum* L.)	Elevated CO_2	F_o', F_m', Φ_{CO2}, ↑ fv'/fm', qP, ETR, Φ_{PSII}, Φ_{PSII}/Φ_{CO2}, ETR/A_n, ↓	[15]
Grape (*Vitis vinifera* L.)	Elevated CO_2	qP, Φ_{PSII}, ETR↑ f_v/f_m, NPQ ↓	[16]
Oak (*Picea abies*) and (*Quercus petraea*)	Elevated CO_2	A_n, g_s, T_r, WUE ↑	[17]

Plant species	Environmental conditions	Parameters	Reference
Pea (*Pisum sativum* L.)	High temperatures	A_n, g_s ↓	[18]
Wheat (*Triticum aestivum*)	High temperatures	WUE ↓	[19]
Barley (*Hordeum vulgare* L.)	High temperatures	f_v/f_m, Φ_{PSII} ↓	[20]
Tomato (*S. lycopersicum* L.)	High temperatures	ETR ↓	[21]
Alfalfa (*Medicago sativa*)	High temperatures	Chl ↓ F_o, F_m ↑	[22]
Tomato (*S. lycopersicum* L.)	High temperatures	A_n, Vc_{max}, J_{max}, f_v/f_m, ETR, $NADP^+$/ NADPH ↓ NPQ ↑	[12]
Lentil (*Lens culinaris*)	Low temperatures	*fv'/fm'*, *fq'/F*$_m$' ↓	[23]
Salvia leriifolia Benth, *Visia faba*	Low temperatures	*fv'/fm'* ↓	[24–26]
Faba bean (*Vicia faba* L.)	Low temperatures	g_m, A_n, T_r, g_s, C_i, C_i:C_a ↓	[27]
Chickpea (*Cicer arietinum* L.)	Low temperatures	*fv'/fm'*, *fq'/F*$_m$' ↓	[28, 29]
Barley (*H. vulgare* L.)	Low temperatures	Φ_{PSII}, ETR ↓ NPQ ↑	[30]
Oats (*Avena sativa*)	Low temperatures	f_v/f_m ↓	[31]
Barley (*H. vulgare* L.)	Drought	Chl, F_o, f_v/f_o, f_v/f_m, ETR ↓	[32]
Maize (*Zea mays* L.)	Drought	Rubisco ↓	[33]
Black-eyed pea (*Vigna unguiculata*)	Drought	A_n, *fv'/fm'* ↓	[34]
Barley (*H. vulgare* L.)	Drought	NPQ ↑	[35]
Castor bean (Ricinus communis)	Drought	A_n, C_i ↓	[36]
Wheat (T. aestivum)	Drought	g_m, A_n, T_r, g_s ↓	[37]
Oak (P. abies) and (Q. petraea)	Drought	A_n, g_s, T_r, WUE, V_C, J ↓	[17]
Sweet corn (Z. mays L.)	Drought	f_v/f_m ↓	[38]

Increase (↑), decrease (↓).

Table 1.
Effect of climate changes induced stresses on photosynthetic and chlorophyll fluorescence parameters.

stress effects on photosynthetic apparatus and an effective indicator of photosynthesis limiting factors. The photochemical efficiency of photosystem II (PSII) is strongly influenced by the climate change consequences such as elevated CO_2, extreme temperatures, and water stress, and a reduction in leaf relative water content and the accumulation of carbohydrates in leaves decreases the quantum efficiency of PSII [7].

More food must be produced by global agriculture to sustain a growing human population in the twenty-first century [8]. Producing more food, however, is threatened by the climate change constraints that limit plant productivity [9]. Under natural conditions, plants are exposed to many adverse environmental stresses that disrupt the photosynthetic apparatus, causing a decrease in plant productivity and overall yield. In the present chapter, the impacts of changing climatic conditions on photosynthesis, with an emphasis on the main consequences of climate change, that is, elevated CO_2, extreme temperatures, and drought are discussed (**Table 1**).

2. Climate change consequences and photosynthetic response

2.1 Elevated CO_2

Carbon dioxide, like other important factors, such as light, water, and nutrients, is one of the determinant factors in plant production. Carbon dioxide is the key substrate for photosynthesis and the source of carbon for plants; however, high, or low CO_2 concentration diversely affects plant growth and productivity [39]. Carbon dioxide stimulates photosynthesis, inhibits photorespiration, and increases the efficiency of water and nitrogen use, which leads to more biomass production and changes in plant composition. Increasing CO_2 concentration by preventing photorespiration in C_3 plants increases the efficiency of photosynthesis because, in the current CO_2 concentration the carboxylation capacity of Ribulose-1,5-bisphosphate carboxylase/oxygenase (Rubisco) does not reach the saturation limit (Drake et al., 1997). The increase in growth and yield of crop species due to doubling the CO_2 concentration was primarily due to the faster photosynthetic rate and secondarily due to less photorespiration [40].

Photosynthesis of C_3 plants is not completely saturated at the current CO_2 concentration. Increasing CO_2 concentration stimulates the rate of photosynthesis and has a positive effect on the growth and performance of plants [41]. Idso and Idso [42] believe that by doubling the current CO_2 concentration, biomass production, and yield of plants will increase by one-third or more if other factors are not limiting. However, plant species differ in response to CO_2 concentration. Faster-growing species are more stimulated and produce more biomass than slow-growing species. Also, plants growing in better nutritional conditions respond more to increased CO_2 concentration than those that are exposed to nutritional stress [43]. Apart from the indirect effects of atmospheric elevated CO_2 concentration, CO_2 concentration directly affects C_3 plants if other factors are not limiting [44]. In research on potatoes in an open-growth chamber, it was found that the photosynthesis of plants grown under elevated CO_2 concentration (720 ppm) was 10 to 40% higher than those grown under ambient CO_2 concentration (400 ppm) [45]. In addition, leaf starch and sucrose content were higher in plants grown under CO_2 concentration conditions, especially in young leaves. This shows that the response of plants to the CO_2 concentration also depends on leaf age.

In general, increasing CO_2 concentration as a substrate for photosynthesis increases leaf area, biomass, and CO_2 fixation. The main reason for the increase in photosynthesis and subsequent increase in growth is the competitive effect of the Rubisco enzyme, which increases the carboxylation of this enzyme [46]. The results of the experiments showed that the rate of photosynthesis was significantly increased under elevated CO_2 concentration in two potato cultivars [10, 11, 47]. Chen and Setter [48] reported that cell division in physiological sinks is an important factor in increasing the photosynthesis of C_3 plants under CO_2 concentration. Increasing CO_2 concentration to 720 μmol mol^{-1} increased cotton canopy photosynthesis by 40% [49]. Also, the increased CO_2 concentration delayed the aging of sugarcane leaves [50]. Elevated CO_2 concentration also increased wheat production [51].

Potato plant leaves showed an 80–100% increase in photosynthetic rate when exposed to elevated CO_2 concentration [52]. However, long-term growth under elevated CO_2 concentration conditions led to plant acclimation to this environment and a relative decrease in photosynthesis [53]. Sicher and Bunce [54] reported that this acclimation is reversible by shifting plants to lower CO_2 concentration. Sicher and Bunce [55] stated that the acclimation response to higher CO_2 concentration is mainly due to a decrease in Rubisco activity than a decrease in the amount of this enzyme. In contrast, Schapendonk [56] found that photosynthetic acclimation, under elevated CO_2 concentration, was accompanied by a decrease in Rubisco and concluded that the acclimation is a complex mechanism resulting from the negative feedback of source-sink disequilibrium induced by high CO_2 concentration. In a study on two model tree species—coniferous Norway spruce and broadleaved sessile oak, A_n was increased in oak saplings under elevated CO_2 concentration (700 μmol CO_2 mol^{-1}), whereas in Norway spruce, Amax remained unchanged or slightly declined; indicating a down-regulation of photosynthesis. Such acclimation was associated with the acclimation of both J and V_C.

Transpiration rate and g_s were decreased with increasing CO_2 concentration, while WUE was increased [57]. Therefore, the beneficial effects of increased CO_2 concentration on yield may be due to changes in either A_n or WUE or both; on the other hand, the reduction of g_s can increase the temperature of the leaf, which further increases the speed of the developmental stages and shortens the grain filling period [58]. The increase in growth due to elevated CO_2 concentration has been attributed to the improvement of plant water relations or the increase of cell expansion [59]. An increase in C_i due to an increase in CO_2 concentration can trigger partial stomatal closure, although the process of how stomata respond to CO_2 signals remained uncertain [60].

An increase in CO_2 concentration accelerates aging in plants. One of the reasons for this is the effect of CO_2 on reducing g_s and increasing leaf temperature. Another reason is the increase in the demand for underground parts for nitrogen and the reduction of N supply to aerial organs [61]. Nitrogen redistribution from chlorophyll-binding proteins has been proposed as the main factor in chlorophyll degradation [62]. Chlorophyll is known as the first electron donor in the process of electron transfer and the photosynthesis apparatus and plays a fundamental role in absorbing light energy in the photosynthesis apparatus [63]. The results of various studies show that elevated CO_2 concentration causes a decrease [64, 65], an increase [66], or no change [52] in the chlorophyll content of potato leaves. Bindi [66] reported that the chlorophyll content of potato leaves under conditions of increased CO_2 concentration was on average 9.3% lower than that of plants under normal conditions.

Reducing g_s, oxidative stress, and decreasing the activity of Rubisco affect photosynthesis under environmental stresses [67]. In addition, PSI and PSII, ETR, and Chl biosynthesis are negatively influenced by abiotic stresses [68, 69]. The quantum efficiency of PSII is considered a quantitative indicator of electron transfer through PSII, which is related to the photochemical efficiency of PSII [69]. Non-photochemical quenching indicates how much excess energy is released as heat by the plant relative to linear electron transport. Under unfavorable conditions, *that is,* environmental stresses, more energy is required to be dissipated since qP is disrupted. Therefore, NPQ is strongly enhanced when physiological sinks are few and leaf physiology and biochemistry are adversely affected by environmental stresses [70]. Working on tomato and grape plants showed that elevated CO_2 concentration decreased NPQ of leaves, while *qP* was enhanced, indicating that higher CO_2 concentration probably stimulates the photosynthetic efficiency and improves the photochemistry of leaves [12, 16].

There are different reports on the effect of elevated CO_2 concentration on chlorophyll fluorescence. Hao [71] stated that the increase in CO_2 concentration increased the rate of photosynthesis and J_{max} with an increase in f_v/f_m, the efficiency of photoreceptors, and the transfer energy of PSII reaction centers (RC). Also, *qP* was reduced under those conditions. On the other hand, Pérez [72] and Ge [73] reported reduced leaf Chl content and factors related to chlorophyll fluorescence, including the photochemical efficiency of PSII and the ETR due to an increase in CO_2 concentration. Taub [74] also reported that in most of the species in their study, the efficiency of photosystem II (f_v/f_m) was significantly higher in plants grown under elevated CO_2 concentration. They stated that this higher efficiency was due to both higher F_m and lower F_o fluorescence. The results of a study showed that elevated CO_2 concentration (800 mmol mol^{-1}) improved leaf A_n, Vc_{max}, J_{max}, and f_v/f_m of tomato (*Solanum lycopersicum* L.) plants at a 24 h recovery [12]. Furthermore, the elevated CO_2 concentration also increased the absorption flux, trapped energy flux, ETR, energy dissipation per PSII cross-section, the concentration of $NADP^+$ and ratio of $NADP^+$/NADPH, and decreased photoinhibition, damage to PSs and ROS accumulation.

2.2 Extreme temperatures

Plants are exposed to frequent low and high-temperature stresses during their life [75]. Global warming induces temperature stress on plants and limits productivity and biomass production. Climate change is likely to increase extreme temperatures beyond the optimum temperatures for the growth of plants. Temperature above or below the optimal threshold disrupts plant cellular homeostasis, which further slows down plant growth, development, and metabolism [76]. The ideal temperature for plant growth and development is in the range of 10 to 35°C. Rising temperature to a specific point enhances plants to generate excess energy; however, heat stress adversely affects plant growth and diminishes the photosynthetic rate [77]. Elevated temperature increases respiration levels in plants. Raising the temperature from 15 to 40°C elevated the respiration rate and disturbed the morphological features of crop species [78].

Heat tolerance is directly related to the ability of plants to maintain the CO_2 assimilation rate. Stomatal conductance and transpiration rate are closely related to leaf temperature [79]. Stomatal conductance, substomatal CO_2 concentration, and leaf water status are affected by the temperature above the optimum levels for plant growth [80]. The concentration of substomatal CO_2 is altered at high temperatures upon stomatal closure and inhibits net photosynthesis [81]. Moreover, high temperatures directly affect the vapor pressure deficit that alters the plant's hydraulic conductance and water supply of the leaves [82]. Studies indicated that the net CO_2 assimilation rate in soybean decreased with an increase in temperature mainly due to the reduction in gs and C_i and lower biomass accumulation [83]. A reduction in photosynthetic ET diminished ATP production and A_n under high temperatures [84]. A significant decrease in the photosynthetic electron transport chain, ATP production, and NADPH under high temperatures led to a decrease in photosynthesis [85].

The negative effect of heat stress on photosynthesis might be due to the reduced Rubisco content and activity [86]. The reduced Rubisco thermal stability decreases its activation under higher temperatures [87]. Rubisco is activated by the RA at an optimum temperature. The catalytic activity of Rubisco is stimulated by an increase in temperature, but the RA fluctuates in response to high temperature [87]. While Rubisco is stable even at 50°C, the activity of RA is decreased at temperatures beyond the optimum [88]. The first step in photosynthetic and photorespiration pathways is catalyzed by Rubisco. The carboxylation efficiency of Rubisco is decreased at high temperatures because of the temperature sensitivity of the RA protein. An elevation in temperature leads to the deactivation of the Rubisco enzyme by the generation of inhibitory compounds such as xylulose-1,5-bisphosphate. Also, the RA breakdown at high temperatures causes the Rubisco disruption [89]. The RA is the main enzyme in the CO_2 fixation process in plants, but at higher temperatures, it is not sufficiently able to keep the balance of the inactivation [90].

Chlorophyll pigments are important for light harvesting; however, temperature stress negatively affects their biosynthesis in plastids [91]. High temperatures degrade the chlorophyll molecule due to different enzymatic impairments; the first enzyme in pyrrole biosynthesis (5-aminolevulinate dehydratase (ALAD)) is negatively affected by high temperature [92]. The decreased chlorophyll biosynthesis in celery leaves at high-temperature stress was likely due to the mRNA down-regulation of 15 genes involved in chlorophyll biosynthesis [93].

Plant productivity is restricted by temperature stress in different ways [94]. The photosynthetic apparatus is the first site of inhibition and is highly sensitive to heat stress. High-temperature alter the reduction-oxidation capacity of PSII acceptors and reduce the photosynthetic electron transport (ET) efficiency of both photosystems [76]. The important components of photosynthetic apparatus are the PSI and PSII, CO_2 reduction pathways, photosynthetic pigments, and ETR and any impairment inhibits overall photosynthesis [92].

High temperatures increase the permeability of membranes, damage PSII subunits, and the manganese complex, and limit ET. The increased permeability of thylakoid membranes leads to peroxidation of membranes, membrane protein changes, the opening of ionic channels, redistribution of specific lipids in thylakoid membranes, and the formation of single-layered membranes [76, 92]. The oxygen-evolving complex of plants grown at high temperatures is partially damaged. Kalaji [6] found that low and high temperatures decreased the reduced PSII electron acceptors pool (mainly Q_A) in barley seedlings. The Φ_{PSII} and the qP were decreased at high temperatures in oak leaves [95].

Kalaji [7] believed that the PI_{ABS} is the most sensitive indicator of various stressors including extreme temperatures. Damage to thylakoid membranes and a decrease in the PSII activity can be the reason for decreased fluorescence in response to high-temperature stress [89]. PSII thermostability is often calculated with the use of fluorescence methods by determining the relationships between F_o and leaf temperature. The fast fluorescence kinetics (JIP-test parameters) can also use to determine the effects of critical temperatures, which are often affected by a much lower temperature than the F_o [7, 96].

One of the crucial factors in predicting future global warming is the response of photosynthesis to temperature. Plant CO_2 assimilation is impaired under environmental stress conditions, such as temperature, while light absorption remains unaffected. Excessive light energy absorption leads to the production of ROS and the photosynthetic machinery, mainly PSII, which is highly sensitive to photodamage, is severely damaged. Although plants have various mechanisms to protect the PSII, photoinhibition occurs when the photodamage rate is exceeded the PSII repairment rate, leading to reduced photosynthetic efficiency [97].

High night temperature stress is increasing due to climate change, and it suppresses the net CO_2 assimilation rate in both C_3 and C_4 plants. The ratio of reduced plastoquinone (Q_B) to (Q_A) and the ratio of Q_A to RC is reduced under high night temperatures. Furthermore, f_v/f_m was decreased, and F_o was increased under high night temperatures [98]. High night temperature reduces qP, Φ_{PSII}, and ETR, increases NPQ, and inhibits the donation of electrons by the oxygen-evolving complex (OEC). Pan [12] observed that high temperature reduced tomato (*S. lycopersicum* L.) leaves photosynthesis by reducing the energy fluxes limitations, ET, and redox homeostasis. They observed that Vc_{max}, J_{max}, and f_v/f_m were diminished by high temperature (42°C for 24 h).

The saturation of fatty acids and membrane fluidity is induced by low temperatures, and it affects the efficiency of photosynthetic ET. Previous studies on various plant species elucidated that the leaf photosynthetic activity is affected by short-term or long-term high and low temperatures [7]. Plants by stimulating thermal energy dissipation and increasing the hydrophobic protein PsbS content, which participates in the thermal energy dissipation, try to reduce the generation of ROS and adapt to low temperatures [99]. Low temperatures inhibit sucrose synthesis, reduce photosynthetic ET, increase photoinhibition, and disturb the photophosphorylation process. Rapacz [100] found that mild frosts initially disturbed the energy transfer to the primary quinone electron acceptor of PSII, Q_A in wheat plants; however, lower temperatures, that is, freezing, may cease energy flow between the PSII RC, Chl, and Q_A, which these primary injuries could only be partially repaired. Consequently, further freezing hinders the ET between the PSII RCs and Q_A and the secondary damage may lead to PSII deactivation. They concluded that both primary and secondary freezing damages resulted in a decreased PI_{ABS}. Strauss [101] also observed that the PI_{ABS} was decreased at low temperatures in soybean plants. Working on faba bean (*Vicia faba* L.) landraces revealed that gas exchange variables are promising criteria for screening freezing-tolerant landraces at early growth stages [27]. The physiological, biochemical, and molecular modifications of chickpea (*Cicer arietinum* L.) seedlings were studied under freezing stress, and it was found that f_v'/f_m' and t Φ_{PSII} of the cold-tolerant genotype recovered faster compared to the cold-sensitive genotype [28, 29]. They found that f_v'/f_m' and Φ_{PSII} were significantly lower in freezing compared with higher temperatures. In a study on lentil (*Lens culinaris* Medik.) genotypes under freezing stress, Nabati [23] found that F_m', f_v'/f_m', and Φ_{PSII} were decreased at freezing temperatures. They concluded that the freezing-tolerant genotypes showed a high potential to restore PSII performance and survival rate.

2.3 Drought stress

Global climate change and lower availability of underground water induce a water crisis worldwide. The constant rise in the atmospheric global temperature induces frequent droughts around the world, which further impacts the biological systems [102]. Plants may experience different forms of abiotic stresses, such as drought during their life, which adversely affect plant growth, survival, and productivity [103]. Drought is a serious problem in arid and semiarid environments with precipitation deficiency [104].

Plant photosynthesis, growth, and yield are impaired by drought stress [105]. Photosynthesis is highly sensitive to drought stress and is the first-line process that is altered by drought stress. Lower photoassimilate production reduces leaf growth and crop yield [37]. Impaired photosynthesis under water deficit relates to either stomatal or non-stomatal limitations. Plants enhance their tolerance levels to survive under such a harsh environment by adopting different strategies, such as stomata closure and osmotic adjustment [106]. Closure of stomata as the primary response of leaves to drought conditions prevents water loss and decreases Tr and increases WUE of plants [92]. The primary response of plants to drought stress is closing the stomata. CO_2 and water exchange in plants are regulated by stomatal openings. Although stomatal closure limits water loss, CO_2 absorption and transportation of non-structural carbon (NSC) are also hindered by stomatal closure, leading to carbon starvation which further affects further processes [107].

Nonstomatal limitations of photosynthesis might be due to lower synthesis and supply of Rubisco and/or other metabolic responses [108]. The proteins D_1 and D_2 can also be damaged by drought stress [109]. Since the PSII is quite resistant to water stress, the photochemical reactions may only be influenced by severe water stress [110]. Lauriano [111] found that changes in the values of chlorophyll fluorescence parameters in peanut leaves were more pronounced under severe drought. Decreased leaf CO_2 transport rate under prolonged and severe water stress reduces CO_2 concentration in chloroplasts, thus weakening photosynthesis. The decrease in the cells CO_2 concentration reduces the activity of sucrose phosphate synthase, nitrate reductase, and capacity for ribulose bisphosphate (RuBP) regeneration, and deactivates Rubisco [49]. The chloroplast thylakoid membrane is degraded under water stress and adversely affects photosynthetic pigment and reduces the photosynthetic rate [112].

Water stress induces oxidative stress. Under water stress, a reduction in chloroplastic CO_2 concentration due to the stomatal closure leads to the impairment of the Calvin cycle and reduces the production of $NADP^+$, leading to excessive electron transport chain (ETC) reduction and directing the electrons to O_2 via Mehler reaction to form singlet O_2, and consequently, ROS [113]. Under drought conditions, triplet chlorophyll stages ($^3Chl^*$) may be overproduced if too much energy is delivered to antenna complexes. This promotes singleton oxygen ($^1O_2$) production, which is a highly reactive form of oxygen that can photo-oxidase chlorophyll (mainly P680) and cause peroxidation of membrane lipids [111]. Partial closure of the stomatal reduces CO_2 assimilation and might lead to an imbalance between PSII photochemical activity and NADPH demand, which in turn, the generation of ROS can be stimulated and lead to higher sensitivity to photodestruction. Under stressful conditions such as low water availability and high irradiance and temperature, photosynthetic efficiency decreases due to a probable high chronic photoinhibition [7].

Studies of the alterations in the chlorophyll fluorescence kinetics provide an in-depth understanding of the structure and functions of the photosynthetic apparatus, particularly PSII [114]. Drought can change the kinetics of chlorophyll fluorescence by affecting PSII. The photochemical efficiency of PSII is strongly influenced by the relative water content of the leaf. The reduction of photosynthesis and the accumulation of carbohydrates in the leaf decrease the quantum efficiency of PSII [7]. One of the consequences of drought is stomatal closure which reduces the heat exchange of leaves. High temperature affects PSII, photosynthetic ET, and ATP synthesis [7]. A decrease in f_v/f_m and yield are indicators of photoinhibition in plants under stressful conditions, indicating lower efficiency of photosynthetic conversion of PAR photon energy [108]. The f_v/f_m is decreased at advanced stages of stress. The f_v/f_m is directly related to chlorophyll activity in the PSs RC. Working on maize plants, Karvar [38] found that deficit irrigation decreased the f_v/f_m. A decrease in leaf Chl content was the likely reason for the diminished f_v/f_m. Carotenoids are non-enzymatic antioxidants that prevent Chl photooxidation under stressful conditions [103]. The stability of carotene and xanthophyll cycle pigments significantly contributed to the protection mechanism of PSII RCs. Furthermore, the cyclic electrons flow around PSI significantly contributed to the dissipation of excess energy in some plant species under water stress [111].

The PSII Φ_{PSII} and ETR_{PSII} are also important parameters to measure drought stress effects on leaves, which provide estimation for both stomatal and non-stomatal effects of drought stress. However, the relative fluorescence decreases ratio (Rfd) proposed by Lichtenthaler [115] as a more sensitive parameter correlated with photosynthetic assimilation than the PSII Φ_{PSII} or ETR_{PSII}. In sunflower plants, it was observed that water potential (Ψ), g_s, A_n, Φ_{PSII}, f_v/f_m, and daily accumulation of total non-structural carbohydrates (TNC) was decreased under drought, but NPQ, malondialdehyde concentration (MDA), and soluble carbohydrates content was increased [116]. The PI_{ABS} was also positively correlated with the water availability for plants. Van Heerden [104] found that a higher water supply increased PI_{ABS} in *Augea capensis* and *Zygophyllum prismatocarpum*.

3. Conclusions

Increasing greenhouse gases emission have led to global warming and climate change worldwide. The global climate change consequences, *that is,* elevated CO_2 concentration, water stress, and extreme temperatures, are serious problems affecting the photosynthetic efficiency and adaptation of plants and adversely affecting agricultural yields. Studies suggest that most plants will be more stressed and less productive in the future in response to climate change. Climate change reduces photosynthetic capacity directly by damaging photosynthetic structures and processes. The changes and modifications of the photosynthetic machinery under different stressful conditions can be evaluated by the chlorophyll fluorescence analysis. Analyses of chlorophyll fluorescence seem to be a promising tool for breeding crops with improved tolerance under stressful conditions. Therefore, the application of chlorophyll fluorescence can be useful to identify which part of the photosynthetic apparatus is affected by the stress and it might help identify good-performing genes by chlorophyll fluorescence to be used in breeding programs.

Abbreviations

A_n	net assimilation rate
Chl	chlorophyll
C_i	sub-stomatal CO_2 concentration
ETR	electron transport rate
ETR/A_n	photorespiration
F_m	maximal fluorescence
F_o	minimal fluorescence
fq'/F_m'	light-adapted operational efficiency of photosystem II
F_v'	light-adapted variable fluorescence
fv'/fm'	light-adapted maximum efficiency of photosystem II
g_m	mesophyll conductance
g_s	stomatal conductance

J	electron transport
J_{max}	maximum ribulose-1,5-bisphosphate (RuBP) regeneration rate
NPQ	non-photochemical quenching
OEC	oxygen-evolving complex
PI_{ABS}	performance index
PSI	photosystems I
PSII	photosystems II
qP	photochemical quenching
RA	Rubisco activase
R_D	dark respiration
R_L	photorespiration rate
ROS	reactive oxygen species
T_r	transpiration rate
V_C	Rubisco carboxylation rate
Vc_{max}	mximum carboxylation rate
WUE	water use efficiency
Φ_{CO2}	quantum yield of CO_2 assimilation
Φ_{PSII}	effective quantum yield

Author details

Mohammad Javad Ahmadi-Lahijani[1]* and Saeed Moori[2]

1 Ferdowsi University of Mashhad, Iran

2 Lorestan University, Iran

*Address all correspondence to: mjahmadi@um.ac.ir

References

[1] Hussain S, Liu T, Iqbal N, Brestic M, Pang T, Mumtaz M, et al. Effects of lignin, cellulose, hemicellulose, sucrose and monosaccharide carbohydrates on soybean physical stem strength and yield in intercropping. Photochemical & Photobiological Sciences. 2020;**19**(4):462-472

[2] Schimel DS. Terrestrial ecosystems and the carbon cycle. Global change biology. 1995;**1**(1):77-91

[3] Solomon S, Qin D, Manning M, Averyt K, Marquis M. Climate Change 2007-the Physical Science Basis: Working Group I Contribution to the Fourth Assessment Report of the IPCC. Switzerland: Cambridge University Press; 2007

[4] Rodziewicz P, Swarcewicz B, ChmielewskaK,WojakowskaA,StobieckiM. Influence of abiotic stresses on plant proteome and metabolome changes. Acta Physiologiae Plantarum. 2014;**36**(1):1-19

[5] Simkin AJ, López-Calcagno PE, Raines CA. Feeding the world: Improving photosynthetic efficiency for sustainable crop production. Journal of Experimental Botany. 2019;**70**(4):1119-1140

[6] Kalaji HM, Carpentier R, Allakhverdiev SI, Bosa K. Fluorescence parameters as early indicators of light stress in barley. Journal of Photochemistry and Photobiology B: Biology. 2012;**112**:1-6

[7] Kalaji MH, Goltsev VN, Żuk-Gołaszewska K, Zivcak M, Brestic M. Chlorophyll Fluorescence: Understanding Crop Performance—Basics and Applications. United States: CRC Press; 2017. p. 237

[8] Beddington JR, Asaduzzaman M, Clark ME, Fernández Bremauntz A, Guillou M, Howlett D, et al. What next for agriculture after Durban? Science. 2012;**335**(6066):289-290

[9] Lobell DB, Burke MB, Tebaldi C, Mastrandrea MD, Falcon WP, Naylor RL. Prioritizing climate change adaptation needs for food security in 2030. Science. 2008;**319**(5863):607-610

[10] Ahmadi-Lahijani MJ, Kafi M, Nezami A, Nabati J, Erwin JE. ABA and BAP improve the accumulation of carbohydrates and alter carbon allocation in potato plants at elevated CO_2. Physiology and Molecular Biology of Plants. 2021;**27**(2):313-325

[11] Ahmadi-Lahijani MJ, Kafi M, Nezami A, Nabati J, Zare Mehrjerdi M, Shahkoomahally S, et al. Variations in assimilation rate, photoassimilate translocation, and cellular fine structure of potato cultivars (*solanum Tuberosum* L.) exposed to elevated CO_2. Plant Physiology and Biochemistry: PPB. 2018;**130**:303-313

[12] Pan C, Ahammed GJ, Li X, Shi K. Elevated CO_2 improves photosynthesis under high temperature by attenuating the functional limitations to energy fluxes, electron transport and redox homeostasis in tomato leaves. Frontiers in plant science. 2018;**9**:1739

[13] Urban O, Klem K, Holišová P, Šigut L, Šprtová M, Teslová-Navrátilová P, et al. Impact of elevated CO_2 concentration on dynamics of leaf photosynthesis in *Fagus sylvatica* is modulated by sky conditions. Environmental Pollution. 2014;**185**:271-280

[14] Huxman TE, Hamerlynck EP, Loik ME, Smith S. Gas exchange and chlorophyll fluorescence responses of

three south-western yucca species to elevated CO_2 and high temperature. Plant, Cell & Environment. 1998;**21**(12):1275-1283

[15] Singh SK, Reddy VR. Combined effects of phosphorus nutrition and elevated carbon dioxide concentration on chlorophyll fluorescence, photosynthesis, and nutrient efficiency of cotton. Journal of Plant Nutrition and Soil Science. 2014;**177**(6):892-902

[16] Zhao X, Li W-F, Wang Y, Ma Z-H, Yang S-J, Zhou Q, et al. Elevated CO_2 concentration promotes photosynthesis of grape (*Vitis vinifera* L. cv.'pinot noir') plantlet in vitro by regulating RbcS and Rca revealed by proteomic and transcriptomic profiles. BMC plant biology. 2019;**19**(1):1-16

[17] Ofori-Amanfo KK, Klem K, Veselá B, Holub P, Agyei T, Marek MV, et al. Interactive effect of elevated CO_2 and reduced summer precipitation on photosynthesis is species-specific: The case study with soil-planted Norway spruce and sessile oak in a mountainous forest plot. Forests. 2020;**12**(1):42

[18] Abdulmajeed AM, Derby SR, Strickland SK, Qaderi MM. Interactive effects of temperature and UVB radiation on methane emissions from different organs of pea plants grown in hydroponic system. Journal of Photochemistry and Photobiology B: Biology. 2017;**166**:193-201

[19] Sattar A, Sher A, Ijaz M, Ul-Allah S, Rizwan MS, Hussain M, et al. Terminal drought and heat stress alter physiological and biochemical attributes in flag leaf of bread wheat. PLoS One. 2020;**15**(5):e0232974

[20] Jedmowski C, Ashoub A, Momtaz O, Brüggemann W. Impact of drought, heat, and their combination on chlorophyll fluorescence and yield of wild barley (*Hordeum spontaneum*). Journal of Botany. 2015;**2015**:1-9. DOI: 10.1155/2015/120868

[21] Camejo D, Rodríguez P, Morales MA, Dell'Amico JM, Torrecillas A, Alarcón JJ. High temperature effects on photosynthetic activity of two tomato cultivars with different heat susceptibility. Journal of Plant Physiology. 2005;**162**(3):281-289

[22] Wassie M, Zhang W, Zhang Q, Ji K, Chen L. Effect of heat stress on growth and physiological traits of alfalfa (*Medicago sativa* L.) and a comprehensive evaluation for heat tolerance. Agronomy. 2019;**9**(10):597

[23] Nabati J, Nezami A, Mirmiran SM, Hasanfard A, Ahmadi Lahijani MJ. The chlorophyll fluorescence parameters response of lentil (*Lens culinaris* Medik.) genotypes to freezing stress. Iranian journal of field. Crop Science. 2021;**53**(1):79-93

[24] Dashti M, Kafi M, Tavakoli H, Mirza M, Nezami A. Effects of freezing stress on Morpho-physiological indices and chlorophyll fluorescence of salvia leriifolia Benth. in seedling stage. Journal of Plant Research (Iranian Journal of Biology). 2016;**28**(5):962-973

[25] Nezami A, Khazaei H, Eshghizadeh H, Riahinia S. Evaluation of freezing temperature tolerance of lentil (*Lens culinaris* Medik.) genotypes with using chlorophyll fluorescence parameters. Agronomy Journal (Pajouhesh & Sazandegi). 2011;**99**:24-33

[26] Zhou R, Hyldgaard B, Yu X, Rosenqvist E, Ugarte RM, Yu S, et al. Phenotyping of faba beans (*Vicia faba* L.) under cold and heat stresses using chlorophyll fluorescence. Euphytica. 2018;**214**(4):1-13

[27] Nabati J, Nezami A, Hasanfard A, Haghighat SZ. The trend of changes in chlorophyll fluorescence parameters in two *Vicia faba* ecotype during freezing stresses. Iranian journal pulses. Research. 2018;**9**(2):139-150

[28] Karimzadeh Soureshjani H, Nezami A, Nabati J, Oskoueian E, Ahmadi-Lahijani MJ. The physiological, biochemical, and molecular modifications of chickpea (*Cicer arietinum* L.) seedlings under freezing stress. Journal of Plant Growth Regulation. 2022;**41**(3):1109-1124

[29] Karimzadeh Soureshjani H, Nezami A, Nabati J, Oskueian E, Ahmadi-Lahijani MJ. Genetic variations in antioxidant content and chlorophyll fluorescence of chickpea (*Cicer arietinum* L.) genotypes exposed to freezing temperatures. Acta Physiologiae Plantarum. 2022;**44**(12):1-13

[30] Dai F, Zhou M, Zhang G. The change of chlorophyll fluorescence parameters in winter barley during recovery after freezing shock and as affected by cold acclimation and irradiance. Plant Physiology and Biochemistry. 2007;**45**(12):915-921

[31] Rizza F, Pagani D, Stanca A, Cattivelli L. Use of chlorophyll fluorescence to evaluate the cold acclimation and freezing tolerance of winter and spring oats. Plant breeding. 2001;**120**(5):389-396

[32] Li R-h, Guo P-G, Michael B, Stefania G, Salvatore C. Evaluation of chlorophyll content and fluorescence parameters as indicators of drought tolerance in barley. Agricultural Sciences in China. 2006;**5**(10):751-757

[33] Zhang R, Zhang X, Camberato J, Xue J. Photosynthetic performance of maize hybrids to drought stress. Russian Journal of Plant Physiology. 2015;**62**(6):788-796

[34] Singh SK, Reddy KR. Regulation of photosynthesis, fluorescence, stomatal conductance and water-use efficiency of cowpea (*Vigna unguiculata* [L.] Walp.) under drought. Journal of Photochemistry and Photobiology B: Biology. 2011;**105**(1):40-50

[35] Oukarroum A, Schansker G, Strasser RJ. Drought stress effects on photosystem I content and photosystem II thermotolerance analyzed using Chl a fluorescence kinetics in barley varieties differing in their drought tolerance. Physiologia Plantarum. 2009;**137**(2):188-199

[36] Santos CM, Endres L, Ferreira VM, Silva JV, Rolim EV, Wanderley HC. Photosynthetic capacity and water use efficiency in *Ricinus communis* (L.) under drought stress in semi-humid and semi-arid areas. An. Acad. Brasil. Ciênc. 2017;**89**:3015-3029

[37] Ahmadi-Lahijani MJ, Emam Y. Post-anthesis drought stress effects on photosynthesis rate and chlorophyll content of wheat genotypes. Journal of Plant Physiology and Breeding. 2016;**6**(1):35-52

[38] Karvar M, Azari A, Rahimi A, Maddah-Hosseini S, Ahmadi-Lahijani MJ. Titanium dioxide nanoparticles (TiO_2-NPs) enhance drought tolerance and grain yield of sweet corn (*Zea mays* L.) under deficit irrigation regimes. Acta Physiologiae Plantarum. 2022;**44**(2):1-14

[39] Sage RF, Coleman JR. Effects of low atmospheric CO_2 on plants: More than a thing of the past. Trends in plant science. 2001;**6**(1):18-24

[40] Drake BG, González-Meler MA, Long SP. More efficient plants: A

consequence of rising atmospheric CO_2. Annual Review of Plant Biology. 1997;**48**:609-639

[41] Reddy AR, Rasineni GK, Raghavendra AS. The impact of global elevated CO_2 concentration on photosynthesis and plant productivity. Current Science. 2010;**99**(1):46-57

[42] Idso KE, Idso SB. Plant responses to atmospheric CO_2 enrichment in the face of environmental constraints: A review of the past 10 years' research. Agricultural and Forest Meteorology. 1994;**69**(3-4):153-203

[43] Poorter H. Do slow-growing species and nutrient-stressed plants respond relatively strongly to elevated CO_2? Global Change Biology. 1998;**4**(6):693-697

[44] Lawlor D, Mitchell R. The effects of increasing CO_2 on crop photosynthesis and productivity: A review of field studies. Plant, Cell & Environment. 1991;**14**(8):807-818

[45] Katny MAC, Hoffmann-Thoma G, Schrier AA, Fangmeier A, Jäger H-J, van Bel AJ. Increase of photosynthesis and starch in potato under elevated CO_2 is dependent on leaf age. Journal of Plant Physiology. 2005;**162**(4):429-438

[46] Lawlor DW, Mitchell RA. Crop ecosystem responses to climatic change: Wheat. Climate change and global crop productivity. 2000;**57**:80

[47] Ahmadi Lahijani MJ, Kafi M, Nezami A, Nabati J, Erwin J. Effect of CO_2 enrichment on gas exchanges, biochemical traits, and Minituber yield in potato (*Solanum tuberosum* L.) Cultivars. Journal of Agricultural Science and Technology. 2019;**21**(4):883-894

[48] Chen C-T, Setter TL. Response of potato dry matter assimilation and partitioning to elevated CO_2 at various stages of tuber initiation and growth. Environmental and experimental botany. 2012;**80**:27-34

[49] Reddy KR, Zhao D. Interactive effects of elevated CO_2 and potassium deficiency on photosynthesis, growth, and biomass partitioning of cotton. Field Crops Research. 2005;**94**(2-3):201, 213

[50] Vu JC, Allen LH Jr. Growth at elevated CO_2 delays the adverse effects of drought stress on leaf photosynthesis of the C_4 sugarcane. Journal of Plant Physiology. 2009;**166**(2):107-116

[51] Högy P, Fangmeier A. Effects of elevated atmospheric CO_2 on grain quality of wheat. Journal of Cereal Science. 2008;**48**(3):580-591

[52] Donnelly A, Craigon J, Black CR, Colls JJ, Landon G. Elevated CO_2 increases biomass and tuber yield in potato even at high ozone concentrations. New Phytologist. 2001;**149**(2):265-274

[53] Vandermeiren K, Black C, Lawson T, Casanova M, Ojanperä K. Photosynthetic and stomatal responses of potatoes grown under elevated CO_2 and/or O_3—Results from the European CHIP-programme. European Journal of Agronomy. 2002;**17**(4):337-352

[54] Sicher RC, Bunce JA. Adjustments of net photosynthesis in Solanum tuberosum in response to reciprocal changes in ambient and elevated growth CO_2 partial pressures. Physiologia Plantarum. 2001;**112**(1):55-61

[55] Sicher RC, Bunce JA. Photosynthetic enhancement and conductance to water vapor of field-grown *Solanum tuberosum* (L.) in response to CO_2 enrichment. Photosynthesis Research. 1999;**62**(2):155-163

[56] Schapendonk AH, van Oijen M, Dijkstra P, Pot CS, Jordi WJ, Stoopen GM. Effects of elevated CO_2 concentration on photosynthetic acclimation and productivity of two potato cultivars grown in open-top chambers. Functional Plant Biology. 2000;**27**(12):1119-1130

[57] Donnelly A, Jones MB, Burke JI, Schnieders B. Elevated CO_2 provides protection from O_3 induced photosynthetic damage and chlorophyll loss in flag leaves of spring wheat (*Triticum aestivum* L., cv. 'Minaret'). Agriculture, ecosystems & environment. 2000;**80**(1-2):159-168

[58] Amthor JS. Effects of atmospheric CO_2 concentration on wheat yield: Review of results from experiments using various approaches to control CO_2 concentration. Field Crops Research. 2001;**73**(1):1-34

[59] Mateos-Naranjo E, Redondo-Gómez S, Álvarez R, Cambrollé J, Gandullo J, Figueroa ME. Synergic effect of salinity and CO_2 enrichment on growth and photosynthetic responses of the invasive cordgrass *Spartina densiflora*. Journal of Experimental Botany. 2010;**61**(6):1643-1654

[60] Robredo A, Pérez-López U, de la Maza HS, González-Moro B, Lacuesta M, Mena-Petite A, et al. Elevated CO_2 alleviates the impact of drought on barley improving water status by lowering stomatal conductance and delaying its effects on photosynthesis. Environmental and experimental botany. 2007;**59**(3):252-263

[61] Miglietta F, Magliulo V, Bindi M, Cerio L, Vaccari F, Loduca V, et al. Free air CO_2 enrichment of potato (*Solanum tuberosum* L.): Development, growth and yield. Global change biology. 1998;**4**(2):163-172

[62] Matile P, Hortensteiner S, Thomas H, Krautler B. Chlorophyll breakdown in senescent leaves. Plant Physiology. 1996;**112**(4):1403

[63] Murray JW. Photosynthesis: Light and life. The Biochemist. 2013;**35**(5):4-7

[64] Bunce JA. Direct and acclimatory responses of stomatal conductance to elevated carbon dioxide in four herbaceous crop species in the field. Global change biology. 2001;**7**(3):323-331

[65] Lawson T, Craigon J, Tulloch A-M, Black CR, Colls JJ, Landon G. Photosynthetic responses to elevated CO_2 and O_3 in field-grown potato (*Solanum tuberosum*). Journal of Plant Physiology. 2001;**158**(3):309-323

[66] Bindi M, Hacour A, Vandermeiren K, Craigon J, Ojanperä K, Sellden G, et al. Chlorophyll concentration of potatoes grown under elevated carbon dioxide and/or ozone concentrations. European Journal of Agronomy. 2002;**17**(4):319-335

[67] Kohli SK, Handa N, Sharma A, Kumar V, Kaur P, Bhardwaj R. Synergistic effect of 24-epibrassinolide and salicylic acid on photosynthetic efficiency and gene expression in *Brassica juncea* L. under Pb stress. Turkish Journal of Biology. 2017;**41**(6):943-953

[68] Sharma A, Thakur S, Kumar V, Kanwar MK, Kesavan AK, Thukral AK, et al. Pre-sowing seed treatment with 24-epibrassinolide ameliorates pesticide stress in *Brassica juncea* L. through the modulation of stress markers. Frontiers in plant science. 2016;**7**:1569

[69] Kalaji HM, Jajoo A, Oukarroum A, Brestic M, Zivcak M, Samborska IA, et al. Chlorophyll a fluorescence as a tool to monitor physiological status of plants under abiotic stress conditions. Acta Physiologiae Plantarum. 2016;**38**(4):1-11

[70] Myers D, Thomas R, DeLucia E. Photosynthetic capacity of loblolly pine (*Pinus taeda* L.) trees during the first year of carbon dioxide enrichment in a forest ecosystem. Plant, Cell & Environment. 1999;**22**(5):473-481

[71] Hao X, Li P, Feng Y, Han X, Gao J, Lin E, et al. Effects of fully open-air CO2 concentration elevation on leaf photosynthesis and ultrastructure of Isatis indigotica fort. PLoS One. 2013;**8**(9):e74600

[72] Pérez P, Morcuende R, del Molino IMN, Martínez-Carrasco R. Diurnal changes of rubisco in response to elevated CO_2, temperature and nitrogen in wheat grown under temperature gradient tunnels. Environmental and experimental botany. 2005;**53**(1):13-27

[73] Ge Z-M, Zhou X, Kellomäki S, Wang K-Y, Peltola H, Martikainen P. Responses of leaf photosynthesis, pigments and chlorophyll fluorescence within canopy position in a boreal grass (*Phalaris arundinacea* L.) to elevated temperature and CO_2 under varying water regimes. Photosynthetica. 2011;**49**(2):172-184

[74] Taub DR, Seemann JR, Coleman JS. Growth in elevated CO_2 protects photosynthesis against high-temperature damage. Plant, Cell & Environment. 2000;**23**(6):649-656

[75] Mckersie BD, YaY L. Oxidative Stress. Stress and Stress Coping in Cultivated Plants. Switzerland: Springer Nature; 1994. pp. 15-54

[76] Mathur S, Agrawal D, Jajoo A. Photosynthesis: Response to high temperature stress. Journal of Photochemistry and Photobiology B: Biology. 2014;**137**:116-126

[77] Raza A, Razzaq A, Mehmood SS, Zou X, Zhang X, Lv Y, et al. Impact of climate change on crops adaptation and strategies to tackle its outcome: A review. Plants. 2019;**8**(2):34

[78] Jan SA, Bibi N, Shinwari ZK, Rabbani MA, Ullah S, Qadir A, et al. Impact of salt, drought, heat and frost stresses on morpho-biochemical and physiological properties of brassica species: An updated review. Journal of Rural Development and Agriculture. 2017;**2**(1):1-10

[79] Carmo-Silva AE, Gore MA, Andrade-Sanchez P, French AN, Hunsaker DJ, Salvucci ME. Decreased CO_2 availability and inactivation of rubisco limit photosynthesis in cotton plants under heat and drought stress in the field. Environmental and Experimental Botany. 2012;**83**:1-11

[80] Greer DH, Weedon MM. Modelling photosynthetic responses to temperature of grapevine (*Vitis vinifera* cv. Semillon) leaves on vines grown in a hot climate. Plant, Cell & Environment. 2012;**35**(6):1050-1064

[81] Hussain M, Khan TA, Yusuf M, Fariduddin Q. Silicon-mediated role of 24-epibrassinolide in wheat under high-temperature stress. Environmental Science and Pollution Research. 2019;**26**(17):17163-17172

[82] Yang Z, Sinclair TR, Zhu M, Messina CD, Cooper M, Hammer GL. Temperature effect on transpiration response of maize plants to vapour pressure deficit. Environmental and Experimental Botany. 2012;**78**:157-162

[83] Kumari VV, Roy A, Vijayan R, Banerjee P, Verma VC, Nalia A, et al. Drought and heat stress in cool-season food legumes in sub-tropical regions: Consequences, adaptation, and mitigation strategies. Plants. 2021;**10**(6):1038

[84] Cen Y-P, Sage RF. The regulation of rubisco activity in response to variation in temperature and atmospheric CO_2 partial pressure in sweet potato. Plant Physiology. 2005;**139**(2):979-990

[85] Wise R, Olson A, Schrader S, Sharkey T. Electron transport is the functional limitation of photosynthesis in field-grown Pima cotton plants at high temperature. Plant, Cell & Environment. 2004;**27**(6):717-724

[86] Jajoo A, Allakhverdiev SI. High-temperature stress in plants: Consequences and strategies for protecting photosynthetic machinery. Plant Stress Physiology. 2017;**2017**:138-154

[87] Salvucci ME, Crafts-Brandner SJ. Inhibition of photosynthesis by heat stress: The activation state of rubisco as a limiting factor in photosynthesis. Physiologia Plantarum. 2004;**120**(2):179-186

[88] Yamori W, von Caemmerer S. Effect of rubisco activase deficiency on the temperature response of CO_2 assimilation rate and rubisco activation state: Insights from transgenic tobacco with reduced amounts of rubisco activase. Plant Physiology. 2009;**151**(4):2073-2082

[89] Weng J-H, Lai M-F. Estimating heat tolerance among plant species by two chlorophyll fluorescence parameters. Photosynthetica. 2005;**43**(3):439-444

[90] Yamori W, Hikosaka K, Way DA. Temperature response of photosynthesis in C_3, C_4, and CAM plants: Temperature acclimation and temperature adaptation. Photosynthesis Research. 2014;**119**(1):101-117

[91] Efeoğlu B, Terzioğlu S. Photosynthetic responses of two wheat varieties to high temperature. EurAsian Journal of BioSciences (elektronik). 2009;**3**:97-106. DOI: 10.5053/ejobios. 2009.3.0.13

[92] Ashraf M, Harris PJ. Photosynthesis under stressful environments: An overview. Photosynthetica. 2013;**51**(2):163-190

[93] Huang W, Ma HY, Huang Y, Li Y, Wang GL, Jiang Q, et al. Comparative proteomic analysis provides novel insights into chlorophyll biosynthesis in celery under temperature stress. Physiologia Plantarum. 2017;**161**(4):468-485

[94] Allakhverdiev SI, Los DA, Mohanty P, Nishiyama Y, Murata N. Glycinebetaine alleviates the inhibitory effect of moderate heat stress on the repair of photosystem II during photoinhibition. Biochimica et Biophysica Acta (BBA)-Bioenergetics. 2007;**1767**(12):1363-1371

[95] Haldimann P, Feller U. Inhibition of photosynthesis by high temperature in oak (*Quercus pubescens* L.) leaves grown under natural conditions closely correlates with a reversible heat-dependent reduction of the activation state of ribulose-1, 5-bisphosphate carboxylase/oxygenase. Plant, Cell & Environment. 2004;**27**(9):1169-1183

[96] Srivastava A, Strasser RJ. How do land plants respond to stress temperature and stress light. Archives des Sciences. 1995;**48**:135-146

[97] Muñoz P, Munné-Bosch S. Photo-oxidative stress during leaf, flower and fruit development. Plant Physiology. 2018;**176**(2):1004-1014

[98] Brestic M, Zivcak M. PSII fluorescence techniques for measurement of drought and high temperature stress signal in crop plants: Protocols and applications. In: Molecular Stress Physiology of Plants. Switzerland: Springer Nature; 2013. pp. 87-131

[99] Demmig-Adams B, Adams WW III, Barker DH, Logan BA,

Bowling DR, Verhoeven AS. Using chlorophyll fluorescence to assess the fraction of absorbed light allocated to thermal dissipation of excess excitation. Physiologia Plantarum. 1996;**98**(2):253-264

[100] Rapacz M. Chlorophyll a fluorescence transient during freezing and recovery in winter wheat. Photosynthetica. 2007;**45**(3):409-418

[101] Strauss A, Krüger G, Strasser R, Van Heerden P. Ranking of dark chilling tolerance in soybean genotypes probed by the chlorophyll a fluorescence transient OJIP. Environmental and Experimental Botany. 2006;**56**(2): 147-157

[102] Parry ML, Canziani O, Palutikof J, Van der Linden P, Hanson C. Climate Change 2007-Impacts, Adaptation and Vulnerability: Working Group II Contribution to the Fourth Assessment Report of the IPCC. Switzerland: Cambridge University Press; 2007

[103] Farooq M, Wahid A, Kobayashi N, Fujita D, Basra S. Plant drought stress: Effects, mechanisms and management. In: Sustainable Agriculture. Springer; 2009. pp. 153-188

[104] Van Heerden P, Swanepoel J, Krüger G. Modulation of photosynthesis by drought in two desert scrub species exhibiting C_3-mode CO_2 assimilation. Environmental and Experimental Botany. Netherlands. 2007;**61**(2):124-136

[105] Meng L-L, Song J-F, Wen J, Zhang J, Wei J-H. Effects of drought stress on fluorescence characteristics of photosystem II in leaves of *Plectranthus scutellarioides*. Photosynthetica. 2016;**54**(3):414-421

[106] Sharma A, Shahzad B, Kumar V, Kohli SK, Sidhu GPS, Bali AS, et al. Phytohormones regulate accumulation of osmolytes under abiotic stress. Biomolecules. 2019;**9**(7):285

[107] Sevanto S. Phloem transport and drought. Journal of Experimental Botany. 2014;**65**(7):1751-1759

[108] Yin C, Berninger F, Li C. Photosynthetic responses of *Populus przewalski* subjected to drought stress. Photosynthetica. 2006;**44**(1):62-68

[109] Oukarroum A, El Madidi S, Schansker G, Strasser RJ. Probing the responses of barley cultivars (*Hordeum vulgare* L.) by chlorophyll a fluorescence OLKJIP under drought stress and re-watering. Environmental and Experimental Botany. 2007;**60**(3): 438-446

[110] Souza R, Machado E, Silva J, Lagôa A, Silveira J. Photosynthetic gas exchange, chlorophyll fluorescence and some associated metabolic changes in cowpea (*Vigna unguiculata*) during water stress and recovery. Environmental and Experimental Botany. 2004;**51**(1):45-56

[111] Lauriano J, Ramalho J, Lidon F. Mechanisms of energy dissipation in peanut under water stress. Photosynthetica. 2006;**44**(3):404-410

[112] Bertioli DJ, Cannon SB, Froenicke L, Huang G, Farmer AD, Cannon EK, et al. The genome sequences of *Arachis duranensis* and *Arachis ipaensis,* the diploid ancestors of cultivated peanut. Nature genetics. 2016;**48**(4):438-446

[113] Noctor G, Mhamdi A, Foyer CH. The roles of reactive oxygen metabolism in drought: Not so cut and dried. Plant Physiology. 2014;**164**(4):1636-1648

[114] Longenberger PS, Smith C, Duke S, McMichael B. Evaluation of chlorophyll fluorescence as a tool for

the identification of drought tolerance in upland cotton. Euphytica. 2009;**166**(1):25-33

[115] Lichtenthaler H, Buschmann C, Knapp M. How to correctly determine the different chlorophyll fluorescence parameters and the chlorophyll fluorescence decrease ratio RFd of leaves with the PAM fluorometer. Photosynthetica. 2005;**43**(3):379-393

[116] Correia MJ, Osório ML, Osório J, Barrote I, Martins M, David MM. Influence of transient shade periods on the effects of drought on photosynthesis, carbohydrate accumulation and lipid peroxidation in sunflower leaves. Environmental and Experimental Botany. 2006;**58**(1-3):75-84

Chapter 6

Molecular Mechanisms and Strategies Contributing toward Abiotic Stress Tolerance in Plants

Aimen Nasir, Irum Shahzadi and Ismat Nawaz

Abstract

Plants respond to climate change *via* sensing the extreme environmental conditions at cell level, which initiated significant changes in their physiology, metabolism, and gene expression. At the cell membrane, plants activate certain genes (like GRP, PRP, AGP) to provide strengthening to cell wall. Drought and salinity stress tolerance attained by osmotic adjustments, activation of transcriptional factors (like AREB, ABF, DREB2), and regulation of Na^+ homeostasis *via* transporters (like NSCC, NHX1, SOS1, HKT1, LTC1). For adaptations to chilling and frost stress, plants use hydrophobic barriers (waxes/cuticles), antinucleator (cryoprotective glycoprotein), and antifreeze proteins. Higher expression of HSPs (heatshock proteins such as HSP70, HSP100, HSP90, HSP60) is important for thermal tolerance. Tolerance to heavy metal (HM) stress can be achieved *via* vacuolar sequestration and production of phytochelatin, organic acids and metallothionein. ROS generated due to abiotic stresses can be alleviated through enzymatic (APX, CAT, POD, SOD, GR, GST) and nonenzymatic (ascorbate, glutathione, carotenoids, flavonoids) antioxidants. Genetic manipulation of these genes in transgenic plants resulted in better tolerance to various abiotic stresses. Genetic engineering of plants through various genome editing tools, such as CRISPR/Cas9, improve the abiotic stress tolerance as well as enhance the crops' quality, texture, and shelf life.

Keywords: climate change, molecular mechanisms, abiotic stress, adaptations, genome editing

1. Introduction

Climatic conditions are important for the well-functioning of any ecosystem. Altered climatic conditions have direct effects on plant health, productivity, and yield. Earth's climate is changing very rapidly mainly due to human activities, having negative effects on human and the ecosystem (particularly on the agriculture sector) [1]. Altered climatic conditions contribute negatively due to increased levels of greenhouse gases, global warming, higher emission of CO_2, deforestation, and excessive use of fossil fuels. Plants are the primary producers and a very important component of the ecosystem. Temperature fluctuations, particularly due to altered climate conditions, may trigger other factors, such as drought, flood, soil erosion, waterlogging, and salinity, which ultimately led to lower crop productivity and yield.

As the world population is increasing rapidly, demand for the food is also increasing. We need good quality and quantity of crops to fulfill the feeding requirements of the world population. Natural stress factors (light intensity, temperature, water stress, and nutrient availability) and anthropogenic stress factors (mainly HM pollution, excessive use of herbicides, acid rain, and enhanced UV-B radiations) contribute strongly to deteriorate crop health and productivity. Continuously changing climatic conditions induce higher stress on crops due to irregular patterns of moisture contents, more pest and disease infection, more waterlogging conditions, increased soil erosion, and global warming [2]. Climate change and food shortage are the most challenging factors of this century, which need our serious efforts and attention. It is very important to develop crops that are better able to tolerate abrupt climate changes and associated abiotic stresses to keep a balance between environment and agricultural crop production [3]. Among other techniques, the plant genetic engineering approach could also be used for abiotic stress management in crops. Transgenic plants have better tolerance level to various kinds of abiotic stress. They also have improved fruit quality, shelf life, and plant architecture. Genetic engineering of plants also results in reduced postharvest losses, which improve productivity and yield [4]. Induction of the expression of stress-related TFs (MYC, bzip, DREB1A, DREB1B, DREB1C, CBF1, CBF2), stress-responsive genes, signaling pathway kinases (MAPK, CDPK, S6K, PIP5K) hormonal biosynthesis (ABA, ethylene), antioxidant and ROS scavenging mechanism (APX, GSH, GR, GST, SOD, flavonoids, carotenoids), regulatory proteins (HSPs, LEA, dehydrins, aquaporins, metallothioneins, phytochelatins) osmolytes, and compatible solutes (proline, sorbitol, mannitol, polyamines, amino acids, glycine betaine), transporters (NHX, HKT, HMAs) improve the crop performance under altered environmental conditions. Through genetic engineering and genome editing tools, transgenic plants have developed, which are better able to adapt to climate changes without affecting their productivity and yield. Genotyping, sequencing, transcriptomics, proteomics, metabolomics, and functional genomics can be integrated collectively for the identification of stress-responsive genes/gene products and their expression in targeted plants to develop abiotic stress-tolerant cultivars [3]. Different genome editing tools are being used like CRIPR/Cas9, which is of prime importance due to its rapid and effective outcomes. It is an environment-friendly technique to produce transgenic plants, which are better adapted to stress conditions that emerges due to climate change. The CRISPR system is based on candidate gene knockout/insertion or gene replacement, which results in either loss of function, downregulated, or over-expression of gene for abiotic stress tolerance [5].

2. Drought stress

Decreased water availability in soil or excessive loss of water from plants due to high transpiration rate causes drought stress. At the vegetative stage, it affects growth, development, turgidity, and stomatal conductance. At the cellular level, drought stress causes damage to cell division, expansion, nutrient uptake, chlorophyll content, and CO_2 assimilation [6].

2.1 Molecular responses of genetically engineered plants

Genetic engineering basically focuses on the identification of key genes involved in drought stress resistance mechanisms and their potential transfer to different crops through recombinant DNA technology. These candidate genes are involved in osmotic adjustments, induction of dehydrins, synthesis of abscisic acid (ABA), indole-3-acetic acid (IAA), polyamines for maintaining turgidity, tissue water potential, net photosynthesis, stomatal conductance, growth of plants, and development of deeper and prolific root system [7].

2.1.1 Induction of trehalose biosynthesis

Trehalose is a compatible osmoprotectant that plays its role in improving root architecture, maintaining the integrity of thylakoid membrane and sugar signaling pathway, and increasing photosynthesis, stomatal conductance, and drought resistance [8]. The bacterial OtsB gene is introduced in tobacco for better growth and development under drought stress by promoting a deeper and more prolific root system. Such transgenic tobacco plants show more photosynthetic and water-retaining capacity [9]. The yeast TPS1 gene is involved in developing drought tolerance in transgenic plants by regulating carbohydrate levels through SnRK1and ABA signaling pathways to regulate stomatal conductance [10].

2.1.2 Induction of LEA protein

LEA (late embryogenic abundant protein) plays a role in the sequestration and compartmentalization of ions. They also protect other proteins from degradation during cellular dehydration. HVA1 gene isolated from barley is introduced in rice for higher growth rates, hydraulic conductivity, and water permeability under drought stress [11]. Transgenic wheat containing barley HVA1 gene showed overexpression of aquaporins (PIP1, PIP2,NIP, TIP3, XIP) for cell proliferation, ions transport, germination, and morphogenesis under drought stress [9].

2.1.3 Induction of proline and polyamines biosynthetic pathways

Proline and polyamines are compatible solutes, whose expression is regulated under drought stress. They are of low molecular weight and are highly soluble in the cytosol. P5CS gene isolated from moth bean is inserted in tobacco plants for detoxification of ROS and stabilizing structures of membranes, enzymes, and proteins [10]. Transgenic soybean plants modified with P5CR gene showed increased proline accumulation for plant growth and development under drought stress. It also protects the lipid bilayer from damage during cellular dehydration [11]. Genetic engineering of the induction of polyamine biosynthetic pathway mainly focuses on two species rice and tobacco. Transgenic rice modified with oats ADC (arginine decarboxylase) shows increased biomass due to regulated plant growth, antioxidant defense, and metabolism [12].

2.1.4 Induction of transcriptional factors (TFs)

Transcription factors are proteins that bind to the promoter of the respective gene to regulate its expression. Under drought stress, different transcription factors are activated, such as DREB (DREB1A, DREB1B, DREB1C) and CBF (CBF1, CBF2, CBF3). Transgenic wheat modified with DREB1A gene from *Arabidopsis thaliana* showed increased drought resistance by overexpressing drought-tolerant genes [13].

Gene	Source	Transgenic plant	Adaptations	References
P5CS1	*Vigna aconitifolia*	*Nicotiana tabacum*	Delayed wilting of transgenic plants under drought stress.	[15]
Delta OAT1	*Arabidopsis thaliana*	*Oryza sativa*	Transgenic plants showed higher growth rate under drought stress.	[16]
DREB1A CBF3	*A. thaliana*	*Festuca arundinacea*	ABA dependent signaling pathway that activates drought-responsive genes.	[17]
betA	*E.coli*	*Gossypium hirsutum*	Higher RWC and photosynthesis rate, reduced ion leakage, and lipid membrane peroxidation.	[18]
TPS1	*Pichia angusta*	*Solanum tuberosum*	Increased yield and metabolite production, such as sugar, proteins, osmolytes, and hormones.	[10]
OtsA OtsB	*E.coli*	*O. sativa*	Less photo-oxidative damage and improved mineral balance under drought stress.	[19]
codA	*E.coli*	*S. tuberosum*	Drought stress-inducible glycine-betaine (GB) production for protecting proteins, enzymes, and biological membranes.	[18]

Table 1.
Molecular responses of transgenic plants to drought stress.

2.1.5 Induction of MAPK pathway

MAPK is involved in phosphorylating other protein molecules that initiate a downstream oxidative signaling cascade. NPK1 gene isolated from tobacco is inserted in maize to increase drought tolerance of transgenic maize by protecting photosynthetic machinery under drought stress [14].

2.1.6 Induction of ABA signaling pathway

Under drought stress, the level of ABA increases to cause the closure of stomata and to prevent water loss through transpiration. Genetically engineered plants modified with ERA1gene isolated from *A. thaliana* showed better growth, development, and net photosynthesis under drought stress as compared to control. Identification and transference of candidate genes from donor to transgenic plant and their molecular responses have been summarized in **Table 1**.

3. Temperature stress (chilling/freezing and heat stress)

Temperature variations across the world have a direct effect on plant productivity. There is a prominent change in the growth and survival patterns of plants under temperature stress. Temperature stress is divided into two categories, that is, low-temperature stress (chilling injury) and high-temperature stress (heat stress).

3.1 Chilling injury

Morphologically, chilling injured leaves become purple or red in color, and wilting of leaves is also observed. Growth is retarded and foliage of leaves appears soggy. At the cellular level, changes in membrane structure and composition due to decreased fluidity and permeability of plasma membrane were observed. Decreased photosynthesis activity under low temperature is mainly due to distorted and swollen thylakoids, reduction in size and number of starch granules, unstacking of grana, and disappearance of the chloroplast envelope. Condensation of chromatin, alternation in the appearance of nucleolus, Golgi apparatus, and endoplasmic reticulum has also been reported under low temperatures.

3.2 Molecular responses of genetically engineered plants

3.2.1 Induction of dehydration-responsive element (DRE)

DREB1A is a transcription factor that interacts with DRE to induce the expression of cold-responsive genes (COR). At low-temperature, ICE transcription factor turn on the expression of CBF/DREB genes, which induces freezing tolerance [20]. Transgenic *A. thaliana* over-expressing ICE-1 showed increased tolerance to chilling stress by regulating CBF and other cold-responsive regulons [21]. In wheat, ICE-1 homologs TaICE141 and TaICE187 are overexpressed to activate the wheat CBF family. Transgenic *A. thaliana* modified with these homologs showed increased freezing tolerance by transcriptional and post-transcriptional changes [22].

3.2.2 Induction of cold-responsive LEA proteins

LEA proteins act as antifreeze proteins that prevent ice nucleation and the formation of ice crystals. They slowed the growth and recrystallization of ice. PmLEAS is a cold-responsive gene, which is expressed in *Prunus mume* under chilling stress. Transgenic tobacco modified with PmLEAS showed increased freezing tolerance by modifying the composition of the lipid bilayer to increase the proportion of unsaturated fatty acid. It also increases the activity of the desaturase enzyme [23].

3.2.3 Expression of LcFIN1 gene in transgenic A. thaliana

LcFIN1 gene is overexpressed in sheep grass to provide adaptation to cold stress. Transgenic *A. thaliana* modified with LcFIN1 gene showed high germination rates and long survival time period due to the accumulation of compatible solutes, membrane stabilization, reduced ROS generation, and expression of COR genes [24].

3.2.4 Induction of dehydrins

Dehydrins are thermostable, hydrophilic, and cryoprotective protein molecules. They are molecular chaperone. Chilling stress results in the formation of secondary structures of RNA. Dehydrins prevent the formation of secondary structures by acting as molecular chaperons. They also protect other proteins and enzymes from denaturation. ABA treatment induces the expression of dehydrins. PmLEA

Gene	Source	Transgenic plant	Adaptations	References
CodA	*Arthrobacter globiformis*	*Brassica campestris*	Transgenic plants showed an increased net photosynthetic rate.	[29]
CodA	*A. globiformis*	*Lycopersicon esculentum*	Chloroplastic synthesis of glycine betaine showed increased tolerance against chilling stress.	[30]
GmTCF1a	*Glycine max*	*Arabidopsis thaliana*	Improved survival rate and decreased electrolyte leakage.	[31]
BoCRP1	*Brassica oleracea*	*L. esculentum*	Increased accumulation of osmoprotectants and increased activity of ROS-scavenging enzymes.	[32]
DaCB4/ DaCBF7	*Deschampsia antactica*	*Oryza sativa*	Increased expression of cold-responsive genes and Ca signaling pathway.	[33]

Table 2.
Molecular responses of transgenic plants to chilling stress.

is a cold-responsive gene of *P. mume,* which is overexpressed under chilling stress. Transgenic tobacco modified with PmLEA showed increased chilling tolerance due to reduced lipid peroxidation and electrolyte leakage [25]. Similarly, the maize ZmDHN2B gene inserted in tobacco provides cold adaptations by preventing the destabilization of membranes. It also increases the unsaturated to saturated fatty acid ratio to prevent ice crystals formation [26].

3.2.5 Induction of compatible solutes

Compatible solutes, such as amino acids, proline, polyamines, and sugars, provide molecular adaptations under chilling stress. Glycine betaine is very important for osmotic adjustments and subcellular functions. CodA (choline oxidase) isolated from *Arthrobacter globiformis* is inserted in transgenic *A. thaliana* provides cold acclimation [27]. Zinc finger protein gene OSISAP1 from rice is inserted into tobacco plants that showed increased growth and survival rates under chilling stress. OSISAP1 encodes for proline biosynthesis enzyme [28]. A short review of molecular responses of some transgenic plants has been summarized in **Table 2**.

4. Heat stress

Heat stress means temperature above a threshold level, which causes irreversible damage to plant growth and development. Scorching of leaves, leaf senescence, abscission, fruit discoloration, reflective leaf hair, leaf curling, and vertical leaf orientation are the main morphological effects of higher temperature. Heat stress induces the production of NH_3 within plant tissues. It leads to ammonia toxicity. CAM pathway is responsible for the high production of organic acids, such as pyruvate, citrate, malate, PEP, and oxaloacetic acid. These organic acids prevent ammonia toxicity within the cytoplasm under heat stress.

4.1 Molecular responses of genetically engineered plants

4.1.1 Induction of heat-shock proteins (HSPs)

Expression of genes for the synthesis of various hormones, such as ABA, ethylene, salicylic acid, and brassinosteroids, are very important for thermotolerance. These hormones stabilize the heat-shock transcription factors and help them to bind with heat shock-related genes. DcHSP17.7 is a heat shock-related gene of carrots. It was inserted into the potato under the control of 35S promoter. Transgenic tomatoes showed increased tolerance to heat stress by stabilization of the tertiary structure of proteins and enzymes [34]. Fad8 is a cytosolic protein of *Brassica napus,* which is over-expressed in the tobacco plant. Transgenic tobacco showed much more heat sensitivity, which shows that silencing of fad8 is important for heat stress tolerance because *fad8* encodes for desaturase enzyme [35]. OsHSFA2e was isolated from *oryza sativa* and was introduced in *A. thaliana.* Resultant transgenic *A. thaliana* showed increased thermotolerance by upregulation of HSF-related genes [35]. Slhsp gene isolated from *Solanum lycopersicon* was introduced in *N. tabacum*. hsp101 gene was isolated from *A. thaliana* and was introduced in *O. sativa.* Both of the transgenic plants showed increased thermotolerance due to prevented protein aggregation [36]. *A. thaliana* AtPLC9 gene is responsible for heat tolerance as it induces the expression of HSPs and HSFAs. AtPLC9 gene was inserted into *O. sativa.* Transformed rice plants showed increased heat stress tolerance due to the over-expression of OsHSFAs, calcium ions, and calmodulin-related genes [37].

4.1.2 Induction of membrane associated lipid metabolism

Heat stress increases membrane fluidity, which causes disruption of cellular functions and membrane permeability. Plants achieve adaptation to heat stress by

Gene	Source	Transgenic plant	Adaptations	Reference
hsp21	*A. thaliana*	*S. lycopersicon*	Protection of PSII from heat-induced oxidative stress and detoxification of photo-induced H_2O_2.	[39]
OsHsfA2e	*O. sativa*	*A. thaliana*	Transformed *A. thaliana* showed increased thermotolerance.	[40]
TaHsfA6f	*Triticum aestivum*	*A. thaliana*	Increased thermotolerance by inducing the expression of ABA metabolism and other heat stress-responsive genes (*APX, LEA3, LTP3*).	[40]
Badh	*Spinacia oleracea*	*N. tabacum*	Overexpression of chloroplastic glycine betaine providing thermotolerance during vegetative stage.	[40]
TaHSP23.9	*T. aestivum*	*A. thaliana*	High thermotolerance by preventing irreversible protein unfolding and aggregation.	[41]

Table 3.
Molecular responses of transgenic plants to heat stress.

increasing saturated fatty acids in membrane composition. Fad7 gene isolated from *A. thaliana* was introduced in *N. tabacum*. This gene encodes for the desaturase enzyme. Transformants that showed silencing of fad7 gene were able to adapt to heat stress more effectively. Similarly, fad8 isolated from *B. napus* was introduced in *N. tabacum.* Transformants with silenced fad8 gene showed better growth, chlorophyll content, and photochemical efficiency [38]. A short review of molecular responses of transgenic plants to heat stress has been summarized (**Table 3**).

5. Heavy metal stress

Plants required a small amount of HM for the proper functioning of their physiological processes. When the concentration exceeds the threshold value, these HM become toxic for plants. Excess of arsenic (As) causes photosynthesis inhibition, and decreases biomass and yield; cadmium (Cd) toxicity causes chlorosis, reduced water, and nutrient uptake, browning of root tips, and ultimate death; chromium (Cr) and lead (Pb) stress cause reduced nutrient uptake and disturbance in metabolic pathways, respectively. Mercury (Hg) and zinc (Zn) toxicity cause reduced photosynthesis due to the inhibition of photosystems I & II. Excess of nickel (Ni) causes retarded seed germination, reduced plant height, reduced root length, and reduced chlorophyll content [42].

5.1 Molecular responses of genetically engineered plants

5.1.1 Induction of the organic acid biosynthetic pathway

TaALMT1 gene isolated from *Triticum aestivum* was inserted into tobacco and barley crops. Transgenic tobacco and barley showed increased tolerance to HM stress because TaALMT1 induces the expression of the malate biosynthetic pathway. Malate acts as a metal chelator and causes metal efflux. SbMATE gene isolated from sorghum was inserted into *A. thaliana.* Transgenic *A. thaliana* showed increased HM-stress tolerance. SbMATE induces the expression of citrate transporter for metal efflux [43].

5.1.2 Genetic engineering for Cd toxicity tolerance

Cd is a highly toxic metal due to its fast mobility and persistency. A very small concentration of Cd is lethal to plants. Different genetic engineering approaches have been implied to develop transgenic plants that can withstand Cd toxicity. *gsh1* isolated from *E. coli* was inserted into *Brassica juncia,* which showed increased Cd tolerance. gsh1 gene encodes for γ-glutamylcysteine synthetase for the synthesis of glutathione (GSH) and phytochelatins (PCs). GSH plays role in HM-induced ROS scavenging by initiating the ascorbate-glutathione cycle. PCs form a complex with HM (HM-PC), which is transported to the vacuole for detoxification. *N. tabacum* modified with RCS1 gene of *O. sativa* showed higher cysteine synthase activity [44]. CDna-LTC1 a nonspecific transporter of Cd was introduced in tobacco that showed increased Cd tolerance due to less storage of Cd in roots [45].

5.1.3 Induction of the expression of metallothioneins (MT)

MTs act as chelator that binds with free metals and releases them slowly. MT1 gene isolated from chickpea was inserted into *A.thaliana, which* showed increased HM-stress

Gene	Source	Transgenic plant	Adaptations	References
NAS1	*Arabidopsis thaliana*	*Nicotiana tabacum*	Increased Ni tolerance and accumulation.	[51]
MTA1	*Pisum sativum*	*Populus alba*	Increased Cu and Zn tolerance.	[52]
HMA4/ MT2	*A. thaliana*	*N. tabacum*	Increased Cd tolerance and Cd/Zn translocation efficiency.	[51]
MRP7	*A. thaliana*	*N. tabacum*	Increased tolerance to Cd and increased metal accumulation in roots.	[51]
merA	*E.coli*	*N. tabacum*	More efficient Hg volatilization and tolerance.	[53]
ECS/ arsC	*E.coli*	*A. thaliana*	Enhanced As tolerance and hyperaccumulation.	[53]

Table 4.
Molecular responses of transgenic plants to HM stress.

tolerance by upregulation of antioxidative enzymes (APX, GPX, GSH, GR) and reduced electrolyte leakage [46]. OsMT1e-P, a MT gene of *Oryza sativa* was inserted into tobacco that showed improved HM-stress tolerance (Cu and Zn) by metal ions compartmentalization and vacuolar sequestration [47]. Human MT2 was inserted into tobacco and oil seed crops. Transference of various MT genes (human MTIA, human MTII, yeast CUPI, pea PsMTA, and TaMT) into *A. thaliana, Brassica compestris,* and *N. tabacum* showed increased HM-stress tolerance due to overexpression of GSH-S- transferase activity. BcMT1 and BcMT2 genes from *Brassica compestris* were inserted into *A. thaliana and* showed improved tolerance to HM stress by upregulation of the activity of anti-oxidative enzymes [48].

5.1.4 Induction of the expression of metal transporter genes

Metal transporters are important for the transportation and compartmentalization of free metal ions. Genetic engineering mainly focuses on the expression of metal transporter genes in plants. Induction of AtPHT1/AtPHT7 genes isolated from *A. thaliana* along with YCF1 gene of *Saccharomyces cerevisiae* in tobacco showed much more As tolerance and accumulation. TgMTP1 gene from *N. Goesingense* was genetically engineered into *A. thaliana* that showed improved Zn tolerance [49]. *Znta* gene isolated from *E.coli* was inserted into *A. thaliana* that showed increased resistance to Pb and Cd. Znta gene encodes for V-type ATPase metal transporter that transports free metal ions from cytoplasm to vacuole for sequestration [50]. PvACR3 transporter gene isolated from *Pteris vittata* was introduced in *A. thaliana,* which showed increased tolerance to As [49]. A brief review of the molecular responses of transgenic plants to HM stress has been summarized in **Table 4**.

6. Salinity stress

Increased concentration of soluble salts in the soil causes salinity stress. Salinity stress causes high-level accumulation of Na + and Cl- ions within the cytoplasm, which disturb enzyme activities and photosynthetic processes. It also causes ROS-induced oxidative damage to lipids, proteins, and nucleic acids. Other adverse effects include decreased nutritional value of plants, salinity-induced osmotic stress, decreased rate of seed germination, and decreased plant growth and productivity.

6.1 Molecular responses of genetically engineered plants

6.1.1 Induction of the expression of Na+/H+ antiporter

Genetic engineering approaches focus on the identification of various genes that encode ion transporters, antiporters, cationic channels, compatible solutes, osmoprotectants, etc. Ion transporters play important role in the selective transport of ions and maintain the optimal level of these ions. Vacuolar Na+/H+ antiporter catalyzes the exchange of Na + from the cytoplasm to vacuole for sequestration. It helps in maintaining cellular homeostasis, pH, and cell turgidity. *B. napus* modified with AtNHX gene from *A. thaliana.* AtNHX gene encodes Na+/H+ antiporter. The transgenic plant showed increased salt tolerance, growth, and photosynthetic rate. Similar results were observed when *Brassica juncia* was transformed with pgNHX1 gene [54]. *T. aestivum* was modified by vacuolar Na+/H+ antiport gene AtNHX1 from *A. thaliana*. The transgenic wheat plant showed a lower accumulated level of Na + in leaves. Transformed *A. thaliana* with a high expression level of AtNHX1 gene showed high salt tolerance. AtNHX1 gene is responsible for the compartmentalization and sequestration of Na + into the vacuole [55]. *O. sativa* was modified by Na+/H+ antiport gene nhaA from *E.coli*. Transgenic rice showed better salt tolerance, seed germination rates, growth, and productivity [56].

6.1.2 Induction of the expression of SOS gene

The high salt level is detected by receptors, which increases the cytosolic level of calcium. SOS3 binds to free Ca and activates the expression of SOS3 protein kinase. SOS3-SOS2 complex induces the expression of SOS1 gene, which encodes Na+/H+ antiporter. *ThSOS1-ThSOS5* genes were isolated from *T. hispidia* and inserted into *A. thaliana.* The transformed plant showed increased salt tolerance due to increased ROS scavenging activity, and lower MDA and H202 levels [57]. *SOS1* and *AHA* genes were isolated from *Sesuvium portulacastrum* and coexpressed in *A. thaliana*. Transgenic *A. thaliana* showed increased salt tolerance due to rapid Na + extrusion and regulated cellular homeostasis [58].

6.1.3 Induction of the expression of HKT1-type transporters

HKT1 transporters are responsible for regulating Na homeostasis by keeping a balance between Na and K in the cytoplasm. PpHKT1 gene isolated from almond rootstock was inserted into *A. thaliana*. The transgenic plant showed reduced electrolyte leakage, longer lateral roots, and increased salt tolerance [59]. McHKT2 gene isolated from *Mesembryanthemum crystallinum* was inserted into *A. thaliana*. Transgenic *A. thaliana* showed increased salt tolerance due to lower root Na uptake and lower Na concentration in xylem sap. A short review of molecular responses of transgenic plants to salt stress has been summarized in **Table 5**.

Gene	Source	Transgenic plant	Adaptations	References
Vacuolar Na^+/ H^+ antiporter MsNHX1	Alfalfa (*Medicago sativa*)	*A. thaliana*	Increased osmotic balance. Reduced aggregation of Na + and increased accumulation of K+ in leaves.	[60]
Vacuolar Na^+/ H^+ antiporter AlNHXI	*Aeluropus littoralis*	*Nicotiana tabacum*	Compartmentalization of Na in roots. Maintenance of K^+/Na^+ ratio in the leaf.	[60]
GhNHX1	*Gossypium hirsutum*	*N. tabacum*	Na + extrusion and vacuolar sequestration.	[54]
AgNHX1	*Atriplex gmelini*	*Oryza sativa*	Na + vacuolar sequestration.	[54]
ProDH	*A. thaliana*	*A. thaliana*	Proline biosynthesis for osmotic homeostasis.	[54]
Avp1	*A. thaliana*	*O. sativa*	Vacuolar H+ pyrophosphatase synthesis. Vacuolar acidification.	[54]

Table 5.
Molecular responses of transgenic plants to salt stress.

7. Genome editing strategy (CRISPR/Cas9) for transgenic plants

Different genome editing strategies are being used to develop transgenic-resistant crops to deal with food insecurity issues. CRISPR/Cas9 is the most powerful system to develop a genetically engineered crop system that is able to adapt to different climate change-induced stresses more effectively. It also helps in producing high-yielding and stress-resistant crops. It is environment friendly, fast, rapid, accurate and economical. SgRNA/Cas9 construct is established to target the specific genome sequence of a plant. This construct is introduced in the targeted crop by means of various transformation methods, such as Agrobacterium-mediated transformation and protoplast transfection, to produce abiotic stress-resistant variety [61]. *B. napus* was modified with CRISPR technology to downregulate the expression of CLVTA3. Transgenic *B. napus* showed more seed production. TaGW2 gene's expression was knocked down using the CRISPR system in wheat. This modification helped in increasing the seed size of wheat. Overexpression of SlMAPK3 gene by CRISPR/Cas9 in *S. lycopersicon* showed improved drought tolerance. Knocking down the expression of Osann3 by CRISPR/Cas9 in rice produced more cold stress tolerance [62]. CRISPR/Cas9-mediated genome editing technology is very efficient to develop transgenic crops, which are environment-friendly and able to adapt to climate changes efficiently. A short review of the application of CRISPR/Cas9 in abiotic stress tolerance has been summarized in **Table 6**.

8. Conclusion

In this chapter, molecular responses of transgenic plants to different types of abiotic stresses have been discussed. This review has thrown light on the effects of

Specie	Targeted gene/s	Genome editing strategy	Improved trait	References
Oryza sativa	OsbHLH024	Knockdown expression by CRISPR/Cas9.	Salt tolerance	[5]
O. sativa	OsHKT1;3, SOS1, OsHAK7	Enhanced expression of ion transporter genes by CRISPR/Cas9.	Salt tolerance	[5]
Zea mays	ARGOS8	Replacing *ARGOS8* with *GOS2* by CRISPR/Cas9 to improve yield.	Drought tolerance	[5]
S. lycopersicon	SlLBD40	CRISPR/Cas9-mediated gene mutation.	Drought tolerance	[5]
Z. mays	ZmWRKY106	Enhanced expression by CRISPR/Cas9.	Heat stress tolerance	[63]
O. sativa	OsMYB30	CRISPR/Cas9-mediated gene knockout.	Cold tolerance	[63]
O. sativa	OsNramp, OsCd1, and OsNramp5	Overexpression of metal transporter genes by CRISPR/Cas9.	HM-stress tolerance.	[63]
A. thaliana	AtPDF2.6	Overexpression by CRISPR/Cas9.	HM-stress tolerance by chelation of cytoplasmic Cd.	[64]

Table 6.
Genetic engineering of plants through CRISPR/Cas9 to adapt abiotic stress.

climate change-induced stress factors for plants and how genetic engineering can help to develop transgenic plants that are able to respond to these stresses at the molecular level. Climate change is a major threat factor for the agricultural sector as it causes lower yield and productivity. It causes increased food demand leading to hunger and starvation. Drought stress, salinity stress, temperature stress, and heavy metal stress are major types of abiotic stresses that cause retarded and stunted growth, lower yield, and productivity. Molecular adaptations through genetic engineering can be achieved by expressing stress-related genes, accumulation of compatible solutes, activation of signaling pathways, activation of transcriptional factors, action of various transporters, synthesis of heat-shock proteins and secondary metabolites, and enzymatic and nonenzymatic defense mechanisms, etc. through various genome editing tools, such as CRISPR/Cas9. This review has covered a detailed analysis of each type of abiotic stress and responses of transgenic plants at the molecular level.

Most of the studies of transgenic plants are based on Arabidopsis, rice, and tobacco. Further studies are required by using model plants on different staple crops, such as wheat, cereals, and legumes. to meet food requirements. Climate is changing very rapidly and it has adverse effects on plant growth, productivity, and yield. Climate change-induced stress factors can cause famine and starvation. Different technologies and genome editing tools are being used worldwide to produce resistant transgenic crops, such as CRISPR/Cas9, OMICS, TALEN, QTL, nanobiotechnology, and miRNAs/siRNAs. These technologies focus on genome editing of transcription factors and stress-responsive genes to introduce novel modifications in the plant genome. The development of transgenic crops through genetic engineering is a need of time and demands serious efforts.

Author details

Aimen Nasir[1], Irum Shahzadi[2] and Ismat Nawaz[1]*

1 Department of Biosciences, COMSATS University Islamabad, Islamabad, Pakistan

2 Deptarment of Biotechnology, COMSATS University Islamabad, Abbottabad Campus, Pakistan

*Address all correspondence to: saraneva1109@gmail.com

References

[1] Chaudhry S, Sidhu GP. Climate change regulated abiotic stress mechanisms in plants: A comprehensive review. Plant Cell Reports. 2021;**41**(1):1-31

[2] Miura K, Shiba H, Ohta M, Kang SW, Sato A, Yuasa T, et al. SlICE1 encoding a MYC-type transcription factor controls cold tolerance in tomato, *Solanum lycopersicum*. Plant Biotechnology. 2012;**29**(3):253-260

[3] Raza A, Razzaq A, Mehmood S, Zou X, Zhang X, Lv Y, et al. Impact of climate change on crops adaptation and strategies to tackle its outcome: A review. Plants. 2019;**8**(2):34

[4] Sánchez-Bermúdez M, Del Pozo JC, Pernas M. Effects of combined abiotic stresses related to climate change on root growth in crops. Frontiers in Plant Science. 2022;**13**:918537

[5] Li Y, Wu X, Zhang Y, Zhang Q. CRISPR/Cas genome editing improves abiotic and biotic stress tolerance of crops. Frontiers in Genome Editing. 2022;**4**:987817

[6] Hussain S, Hussain S, Qadir T, Khaliq A, Ashraf U, Parveen A, et al. Drought stress in plants: An overview on implications, tolerance mechanisms and agronomic mitigation strategies. Plant Science Today. 2019;**6**(4):389-402

[7] Trujillo LE, Sotolongo M, Menéndez C, Ochogavía ME, Coll Y, Hernández I, et al. SodERF3, a novel sugarcane ethylene responsive factor (ERF), enhances Salt and drought tolerance when overexpressed in tobacco plants. Plant and Cell Physiology. 2008;**49**(4):512-525

[8] Shao J, Wu W, Rasul F, Munir H, Huang K, Awan MI, et al. Trehalose induced drought tolerance in plants: Physiological and molecular responses. Notulae Botanicae Horti Agrobotanici. 2022;**50**(1):12584

[9] Pilon-Smits E, Ebskamp M, Paul MJ, Jeuken M, Weisbeek PJ, Smeekens S. Improved performance of transgenic Fructan-accumulating tobacco under drought stress. Plant Physiology. 1995;**107**(1):125-130

[10] Romero C, Bellés JM, Vayá JL, Serrano R, Culiáñez-Macià FA. Expression of the yeast trehalose-6-phosphate synthase gene in transgenic tobacco plants: Pleiotropic phenotypes include drought tolerance. Planta. 1997;**201**(3):293-297

[11] Xu D, Duan X, Wang B, Hong B, Ho T, Wu R. Expression of a late embryogenesis abundant protein gene, HVA1, from barley confers tolerance to water deficit and Salt stress in transgenic Rice. Plant Physiology. 1996;**110**(1):249-257

[12] Zhang Y, Wang C, Wang C, Yun L, Song L, Idrees M, et al. OsHsfB4b confers enhanced drought tolerance in transgenic Arabidopsis and Rice. IJMS. 2022;**23**(18):10830

[13] Pellegrineschi A, Reynolds M, Pacheco M, Brito RM, Almeraya R, Yamaguchi-Shinozaki K, et al. Stress-induced expression in wheat of the *Arabidopsis thaliana* DREB1A gene delays water stress symptoms under greenhouse conditions. Genome. 2004;**47**(3):493-500

[14] Shou H. Expression of the Nicotiana protein kinase (NPK1) enhanced drought tolerance in transgenic maize. Journal of Experimental Botany. 2004;**55**(399):1013-1019

[15] Kishor P, Hong Z, Miao GH, Hu C, Verma D. Overexpression of [delta]-Pyrroline-5-carboxylate Synthetase increases proline production and confers Osmotolerance in transgenic plants. Plant Physiology. 1995;**108**(4):1387-1394

[16] Wu L. Over-expression of an Arabidopsis d -OAT gene enhances salt and drought tolerance in transgenic rice. Chinese Science Bulletin. 2003;**48**(23):2594

[17] Yang Z, Eticha D, Albacete A, Rao IM, Roitsch T, Horst WJ. Physiological and molecular analysis of the interaction between aluminium toxicity and drought stress in common bean (Phaseolus vulgaris). Journal of Experimental Botany. 2012;**63**(8):3109-3125

[18] Lv W, Lin B, Zhang M, Hua X. Proline accumulation is inhibitory to Arabidopsis seedlings during heat stress. Plant Physiology. 2011;**156**(4):1921-1933

[19] Garg AK, Kim J, Owens TG, Ranwala AP, Choi YD, Kochian LV, et al. Trehalose accumulation in rice plants confers high tolerance levels to different abiotic stresses. Proceedings of the National Academy Sciences of the USA. 2002;**99**(25):15898-15903

[20] Gilmour SJ, Zarka DG, Stockinger EJ, Salazar MP, Houghton JM, Thomashow MF. Low temperature regulation of the ArabidopsisCBF family of AP2 transcriptional activators as an early step in cold-induced CORgene expression. The Plant Journal. 1998;**16**(4):433-442

[21] Chinnusamy V, Ohta M, Kanrar S, Lee B, Hong X, Agarwal M, et al. ICE1: A regulator of cold-induced transcriptome and freezing tolerance in *Arabidopsis*. Genes & Development. 2003;**17**(8):1043-1054

[22] Badawi M, Reddy YV, Agharbaoui Z, Tominaga Y, Danyluk J, Sarhan F, et al. Structure and functional analysis of wheat ICE (inducer of CBF expression) genes. Plant and Cell Physiology. 2008;**49**(8):1237-1249

[23] Smith MA, Graether SP. The disordered Dehydrin and its role in plant protection: A biochemical perspective. Biomolecules. 2022;**12**(2):294

[24] Gao Q, Li X, Jia J, Zhao P, Liu P, Liu Z, et al. Overexpression of a novel cold-responsive transcript factor*LcFIN1*from sheep grass enhances tolerance to low temperature stress in transgenic plants. Plant Biotechnology Journal. 2016;**14**(3):861-874

[25] Du D, Zhang Q, Cheng T, Pan H, Yang W, Sun L. Genome-wide identification and analysis of late embryogenesis abundant (LEA) genes in Prunus mume. Molecular Biology Reports. 2013;**40**(2):1937-1946

[26] Ju H, Li D, Li D, Yang X, Liu Y. Overexpression of ZmDHN11 could enhance transgenic yeast and tobacco tolerance to osmotic stress. Plant Cell Reports. 2021;**40**(9):1723-1733

[27] Hayashi H, Alia ML, Deshnium P, Ida M, Murata N. Transformation of Arabidopsis thaliana with the codA gene for choline oxidase; accumulation of glycinebetaine and enhanced tolerance to salt and cold stress. The Plant Journal. 1997;**12**(1):133-142

[28] Mukhopadhyay A, Vij S, Tyagi AK. Overexpression of a zinc-finger protein gene from rice confers tolerance to cold, dehydration, and salt stress in transgenic tobacco. Proceedings of the National Academy Sciences of the USA. 2004;**101**(16):6309-6314

[29] Wang Q, Xu W, Xue Q, Su W. Transgenic Brassica chinensis plants

expressing a bacterial codA gene exhibit enhanced tolerance to extreme temperature and high salinity. Journal of Zhejiang University. Science. B. 2010;**11**(11):851-861

[30] Park E, Jeknić Z, Pino M, Murata N, Chen TH. Glycinebetaine accumulation is more effective in chloroplasts than in the cytosol for protecting transgenic tomato plants against abiotic stress. Plant, Cell & Environment. 2007;**30**(8):994-1005

[31] Dong Z, Wang H, Li X, Ji H. Enhancement of plant cold tolerance by soybean RCC1 family gene GmTCF1a. BMC Plant Biology. 2021;**21**(1):369

[32] Wani UM, Majeed ST, Raja V, Wani ZA, Jan N, Andrabi KI, et al. Ectopic expression of a novel cold-resistance protein 1 from Brassica oleracea promotes tolerance to chilling stress in transgenic tomato. Scientific Reports. 2021;**11**(1):16574

[33] Byun MY, Cui LH, Lee J, Park H, Lee A, Kim WT, et al. Identification of Rice genes associated with enhanced cold tolerance by comparative transcriptome analysis with two transgenic Rice plants overexpressing DaCBF4 or DaCBF7, isolated from Antarctic flowering plant Deschampsia antarctica. Frontiers in Plant Science. 2018;**9**:601

[34] Ahn Y, Zimmerman JL. Introduction of the carrot HSP17.7 into potato (Solanum tuberosum L.) enhances cellular membrane stability and tuberization in vitro. Plant, Cell & Environment. 2006;**29**(1):95-104

[35] Wang C, Ru J, Liu Y, Li M, Zhao D, Yang J, et al. Maize WRKY transcription factor ZmWRKY106 confers drought and heat tolerance in transgenic plants. IJMS. 2018;**19**(10):3046

[36] Katiyar-Agarwal S, Agarwal M, Grover A. None. Plant Molecular Biology. 2003;**51**(5):677-686

[37] Liu Y, Liu X, Wang X, Gao K, Qi W, Ren H, et al. Heterologous expression of heat stress-responsive AtPLC9 confers heat tolerance in transgenic rice. BMC Plant Biology. 2020;**20**(1):514

[38] Sohn SO, Back K. Transgenic rice tolerant to high temperature with elevated contents of dienoic fatty acids. Biologia Plantarum. 2007;**51**(2):340-342

[39] Neta-Sharir I, Isaacson T, Lurie S, Weiss D. Dual role for tomato heat shock protein 21: Protecting photosystem II from oxidative stress and promoting color changes during fruit maturation. The Plant Cell. 2005;**17**(6):1829-1838

[40] Yokotani N, Ichikawa T, Kondou Y, Matsui M, Hirochika H, Iwabuchi M, et al. Expression of rice heat stress transcription factor OsHsfA2e enhances tolerance to environmental stresses in transgenic Arabidopsis. Planta. 2008;**227**(5):957-967

[41] Wang J, Gao X, Dong J, Tian X, Wang J, Palta JA, et al. Over-expression of the heat-responsive wheat gene TaHSP23.9 in transgenic Arabidopsis conferred tolerance to heat and Salt stress. Front. Plant Science. 2020;**11**:243

[42] Ghori N, Ghori T, Hayat MQ, Imadi SR, Gul A, Altay V, et al. Heavy metal stress and responses in plants. International journal of Environmental Science and Technology. 2019;**16**(3):1807-1828

[43] Magalhaes JV, Liu J, Guimarães CT, Lana UGP, Alves VMC, Wang Y, et al. A gene in the multidrug and toxic compound extrusion (MATE) family confers aluminum tolerance in sorghum. Nature Genetics. 2007;**39**(9):1156-1161

[44] Harada E, Choi Y, Tsuchisaka A, Obata H, Sano H. Transgenic tobacco plants expressing a rice cysteine synthase gene are tolerant to toxic levels of cadmium. Journal of Plant Physiology. 2001;**158**(5):655-661

[45] Roy B, Noren S, Mandal AB, Basu AK. Genetic engineering for abiotic stress tolerance in agricultural crops. Biotechnology. 2010;**10**(1):1-22

[46] Dubey AK, Kumar A, Kumar N, Kumar S, Meenakshi A, Gautam MA, et al. Over-expression of chickpea metallothionein 1 gene confers tolerance against major toxic heavy metal stress in Arabidopsis. Physiology and Molecular Biology of Plants. 2021;**27**(12):2665-2678

[47] Kumar G, Kushwaha HR, Panjabi-Sabharwal V, Kumari S, Joshi R, Karan R, et al. Clustered metallothionein genes are co-regulated in rice and ectopic expression of OsMT1e-P confers multiple abiotic stress tolerance in tobacco via ROS scavenging. BMC Plant Biology. 2012;**12**:107

[48] Stomp AM, Han KH, Wilbert S, Gordon MP, Cunningham SD. Genetic strategies for enhancing phytoremediation. Annals of the New York Academy of Sciences. 1994;**721**:481-491

[49] Gustin JL, Loureiro ME, Kim D, Na G, Tikhonova M, Salt DE. MTP1-dependent Zn sequestration into shoot vacuoles suggests dual roles in Zn tolerance and accumulation in Zn-hyperaccumulating plants. The Plant Journal. 2009;**57**(6):1116-1127

[50] Lee J, Bae H, Jeong J, Lee JY, Yang YY, Hwang I, et al. Functional expression of a bacterial heavy metal transporter in Arabidopsis enhances resistance to and decreases uptake of heavy metals. Plant Physiology. 2003;**133**(2):589-596

[51] Mizuno D, Higuchi K, Sakamoto T, Nakanishi H, Mori S, Nishizawa NK. Three nicotianamine synthase genes isolated from maize are differentially regulated by iron nutritional status. Plant Physiology. 2003;**132**(4):1989-1997

[52] Balestrazzi A, Botti S, Zelasco S, Biondi S, Franchin C, Calligari P, et al. Expression of the PsMT A1 gene in white poplar engineered with the MAT system is associated with heavy metal tolerance and protection against 8-hydroxy-2′-deoxyguanosine mediated-DNA damage. Plant Cell Reports. 2009;**28**(8):1179-1192

[53] He YK, Sun JG, Feng XZ, Czakó M, Márton L. Differential mercury volatilization by tobacco organs expressing a modified bacterial merA gene. Cell Research. 2001;**11**(3):231-236

[54] Rajagopal D, Agarwal P, Tyagi W, Singla-Pareek SL, Reddy MK, Sopory SK. Pennisetum glaucum Na+/H+ antiporter confers high level of salinity tolerance in transgenic Brassica juncea. Molecular Breeding. 2007;**19**(2):137-151

[55] Xue Z, Zhi D, Xue G, Zhang H, Zhao Y, Xia G. Enhanced salt tolerance of transgenic wheat (Tritivum aestivum L.) expressing a vacuolar Na+/H+ antiporter gene with improved grain yields in saline soils in the field and a reduced level of leaf Na+. Plant Science. 2004;**167**(4):849-859

[56] Wu L, Fan Z, Guo L, Li Y, Chen Z, Qu L. Over-expression of the bacterial nhaA gene in rice enhances salt and drought tolerance. Plant Science. 2005;**168**(2):297-302

[57] Liu Z, Xie Q, Tang F, Wu J, Dong W, Wang C, et al. The ThSOS3 gene improves the Salt tolerance of transgenic Tamarix hispida and Arabidopsis thaliana. Frontiers in Plant Science. 2020;**11**:597480

[58] Fan Y, Yin X, Xie Q, Xia Y, Wang Z, Song J, et al. Co-expression of SpSOS1 and SpAHA1 in transgenic Arabidopsis plants improves salinity tolerance. BMC Plant Biology. 2019;**19**(1):74

[59] Kaundal A, Sandhu D, Duenas M, Ferreira JFS. Expression of the high-affinity K+ transporter 1 (PpHKT1) gene from almond rootstock 'Nemaguard' improved salt tolerance of transgenic Arabidopsis. PLoS One. 2019;**14**(3):e0214473

[60] Gupta B, Huang B. Mechanism of salinity tolerance in plants: Physiological, biochemical, and molecular characterization. International Journal of Genomics. 2014;**2014**:1-18

[61] Khurshid H, Jan SA, Shinwari ZK, Jamal M, Shah SH. An era of CRISPR/Cas9 mediated plant genome editing. Current Issues in Molecular Biology. 2018;**26**:47-54

[62] Shen C, Que Z, Xia Y, Tang N, Li D, He R, et al. Knock out of the annexin gene OsAnn3 via CRISPR/Cas9-mediated genome editing decreased cold tolerance in rice. Journal of Plant Biology. 2017;**60**(6):539-547

[63] Rahman M, Zulfiqar S, Raza MA, Ahmad N, Zhang B. Engineering abiotic stress tolerance in crop plants through CRISPR genome editing. Cell. 2022;**11**(22):3590

[64] Rai KK, Pandey N, Meena RP, Rai SP. Biotechnological strategies for enhancing heavy metal tolerance in neglected and underutilized legume crops: A comprehensive review. Ecotoxicology and Environmental Safety. 2021;**208**:111750

Chapter 7

Transgenic Plants in Heat Stress Adaptation: Present Achievements and Prospects

Alp Ayan, Sinan Meriç, Tamer Gümüş and Çimen Atak

Abstract

Global warming, which was rhetorical in the previous century, is a preeminent issue in multiple scientific areas today. Global warming has increased the frequency of extreme high temperature events all around the globe and expanded heat zones from tropic areas through both poles and even changed frigid poles to temperate zones. In the terrestrial earth, plants are the major CO_2 consumers. The emergence and evolution of plants on earth decreased the global temperatures dramatically from mid-Devonian to mid-Carboniferous Era; however, the human factors as industrialization were not in equation. Today, plants are still main actors of the nature-based solutions to global warming through afforestation and reforestation solutions. However, high temperature is a major deleterious abiotic stress for plant growth and productivity. Plant heat stress adaptation has been a focus of research for both environmental and agricultural purposes. Plant heat stress adaptation requires utilization of complex physiological traits and molecular networks combined. The present chapter summarizes recent progress in transgenic approach through five main targets as heat shock proteins, osmoprotectants, antioxidants, transcription factors, and miRNAs. Additionally, miscellaneous novel transgenic attempts from photosynthetic machinery to signal transduction cascades are included to cover different physiological, transcriptional, and post-transcriptional regulation of the plant heat responses.

Keywords: global warming, heat shock proteins, heat shock factors, antioxidants, osmolytes

1. Introduction

Plants are subjected to various biotic (insects, parasites, nematodes, weeds, bacteria, fungi, viruses, etc.) and abiotic stress factors in their natural environment due to the stationary lifestyle. A major part of the abiotic stresses is caused by factors related to the physical and chemical composition of the soil, while the rest may be related to climate properties such as cold and heat, UV exposure, and light intensity. Among these, heat stress is particularly important since all the anabolism and catabolism reactions require particular cardinal temperatures for enzyme activities. Average surface temperature of the earth increased roughly 1°C since the beginning of pre-industrial era 120–140 years ago. In local terms, it may seem insignificant; however, globally accumulated heat has vast effects. Various independent research groups measure and calculate global average surface temperatures through absolute temperature observations and temperature anomalies from different locations [1]. According to the Annual Global Climate Report 2021 statistics of The National Oceanic and Atmospheric Administration (NOAA), 2021 was the sixth warmest recorded year of the earth since 1880 by the rate of 0.84°C higher than average of twentieth century. It was also the 45th consecutive year in which the average global temperature surpassed the average of twentieth century, which means it was never colder than average since 1977. Each year in the last decade takes place among the top ten warmest years. The average temperature increase per decade was in range of 0.08°C since 1880; however, the rate was increased 2.25-fold to rate of 0.18°C since 1981. Worse than this rate, the earth is expected to warm roughly 1.5°C within the next two decades [2]. Moreover, major crops in tropical and subtropical regions present 2.5–16% yield losses for every 1°C increase in seasonal temperatures. Global temperature rises also lead to reduction in land and sea, more frequent heavy rains, increase on habitat ranges of some plants and animals and decrease on some others, regionally [3].

Heat stress changes diverse molecular pathways and causes physiological and morphological alterations. Various stages of plant development such as germination, seedling emergence, tillering, floral initiation, pollination, fertilization, and consequently yield and grain quality are in range of heat effects. There are multiple factors which may divert the heat effects to a more dramatic or mild direction. Length, abruptness, and magnitude of heat are the major factors along with relatively minor factors as soil moisture and atmospheric CO_2 concentrations [4]. Anther and pollen development stages are considered as the most heat vulnerable stages; however, exposure during earlier stages may also lead to inadequate germination through reduced root and shoot growth. Heat exposure after the germination stage reduces green leaf area and the number of tillers per plant to ease the effects through reducing exposure surface. Prolonged exposure after anthesis may lead to flower abortion. Heat stress after flowering stage is referred as terminal heat stress which effects early meiosis to tetrad stages of pollen production and utterly reduce grain number, filling, and maturity. Terminal heat stress does not only reduce quantitative traits but also reduce qualitative traits such as dry matter accumulation and grain quality. Developmental stage-specific treatments and breeding strategies against various heat regimes are still under investigation [5, 6].

Physiological functions are mediated through enzymatic processes in all living organisms. Even though, there are thermophile organisms in lower evolutionary branches as archaea, bacteria, and fungus which all have resilient enzyme systems. Any deviations over the optimum temperature hamper enzymatic processes of plants. Photosynthesis is one of the most vital but fragile metabolic processes which is severely affected by heat stress. Some heat acclimation adaptations including reduced spiky leaf shape, altered leaf orientation, rolled leaves or small surface hairs, thick waxy cuticle, and stomatal crypts are present in high heat climate plants to reduce drastic effects; however, crop plants do not possess most of these structures. Symptoms as lower stomatal conductance, reduced CO_2 assimilation, and water loss utilize non-photorespiratory processes. Heat stress also directly alters enzyme and protein structure and cell membrane permeability leading to photochemical modifications in chloroplasts, damage on thylakoid membrane, and reduction of soluble proteins as Rubisco and Rubisco binding proteins. Damaged chloroplasts cripple the photosynthetic capacity of plants and lead to leaf senescence, while disturbed thylakoid membranes elevate cellular reactive oxygen species levels. Respiration is crucial for leaf surface cooling in trade-off water loss. Leaf water potential and transpiration pull is a driving force for nutritional uptake and transport of photosynthesis assimilates from leaves to the grains. Heat stress also disturbs nitrate and ammonium assimilation. Factors as decreased root mass, surface area, and/or a decrease in nutrient uptake per unit root or direct heat damage to roots are plausible for nutrient acquisition decrease. Uptake of most of the nutritional elements is mediated by specific influx or efflux protein activity. Therefore, reduced proteins per unit root rate directly affect the mineral content [7, 8].

Heat stress causes all the above-mentioned damage through direct (primary) and indirect (secondary) effects at different levels. Weakening and damaging bio membrane integrity, altering fluidity, leading to electrolyte leakage, denaturing and misfolding proteins are among the most deleterious direct damage effects. Indirect effects can be listed as oxidative stress, methylglyoxal (MG) stress, and osmotic stress. Therefore, plant heat stress tolerance and adaptation mechanisms include heat shock proteins (HSPs), antioxidant systems, osmolytes, fortification of membrane lipids, and MG detoxification, in general. The present chapter will summarize the current knowledge on heat stress tolerance/adaptation approaches and will discuss transgenic approach contribution to these mechanisms with the emphasis on prospects.

2. Plant heat stress adaptation and tolerance targets

2.1 Heat shock proteins (HSPs)

HSPs are a highly conserved group of proteins which are expressed abundantly following the sudden increase of temperature in wide variety of evolutionary branches as bacteria, fungi, plants, and animals. Even though plants are more responsive to temperature changes and react to fluctuation as small as 1°C, HSPs are expressed in response to sudden 8–10°C temperature increases. HSPs expression may increase within a few seconds following the temperature increase and reach the maximum level of transcripts within one to two hours of exposure. In high temperatures, protein synthesis is reduced to prevent misfolded protein production and protein denaturation which may present toxic properties for cells. Likewise, HSP expression is reduced following the cooldown of environment to optimum temperatures. Expressed HSPs are detectable for approximately 20 hours and generate thermotolerance for further temperature increases. Plant HSPs can be categorized under five conserved families based on their molecular weights as HSP100, HSP90, HSP70, HSP60, and small HSPs (sHSPs) [9].

HSP100 family members which are found in prokaryotes as well as eukaryotes are 75–100 kDa proteins. They can be further divided into two classes based on their ATP-binding sites as class I contains two while class II contains one site. HSP100 family protein takes part in acquisition of thermotolerance through preventing and unfolding of protein aggregations in association with chaperons by ATP-dependent manner. Their expression increases in different developmental stages as well as in response to heat shock. High salt, desiccation, abscisic acid (ABA), and cold stress-induced expressions are also reported. HSP100 protein accumulation initiates as soon as heat stress begins and is retained for prolonged durations during recovery. Hence, the crucial role of Hsp100 family is generally speculated for recovery instead of prevention. However, early accumulation of these proteins as stress initiates suggests that HSP100 members may play important during stress as well [10].

HSP90s are evolutionarily conserved essential molecular chaperones in eukaryotic cells, undertaking key functions in signal transduction networks, cell-cycle control, folding of newly synthesized proteins as well as re-folding and stabilizing tertiary structures of already folded proteins, and protein trafficking. HSP90s, which are constitutively expressed and abundant as 1–2% of total proteins in cell, are induced during stress conditions particularly in response to heat. They involve root, hypocotyl, shoot apical meristem, and stomatal development as well as fertilization and embryo formation. HSP90 is an ATP-dependent chaperone, which constitutes HSP90 chaperon complex in cooperation with other chaperons and co-chaperons to maintain its function. For an instance, proteins which require HSP90 chaperon activity to re-gain their functional conformation called client proteins as newly synthesized or misfolded proteins, initially bind to general protein folding chaperones such as HSP40 and HSP70 which can recognize unfolded proteins. Then, HSP90/HSP70-organizing protein (HOP) mediates binding of the client protein to HSP90. Role of the HSP40/HSP70 chaperone machinery during abiotic stress response is well documented. Acute heat shock temporarily reduces the cytoplasmic HSP90 activity, as it is recruited to stress-labile proteins hence releasing inhibition on stress response induction [11–13].

HSP70s are the most structurally and functionally conserved members of the whole protein family. Hsp70s are the most ubiquitous class of ATP-dependent chaperone proteins which are present in the cytosol of all eubacteria and eukaryotes, and some archaea, as well as within mitochondria, ER, and plastids of eukaryotic cells. In plants and other higher eukaryotes, they have constitutive expression for undertaking the cellular protein quality control and degradation system roles. In other organisms, they are stress-inducible for cyto-protective functions under several different conditions. As the most abundant HSPs, Hsp70 holds hydrophobic regions of misfolded proteins and prevents protein aggregation that can present toxicity to cells. They utilize ubiquitin-mediated proteasomal degradation pathway. Under heat shock and other abiotic stress conditions, heat shock transcription factors are triggered by the signal transduction from misfolded or unfolded outer membrane proteins to inner targets. One of the most notorious trans-acting elements are heat shock transcription factors (HSFs) which are associated with cis-acting heat shock elements (HSEs) in promoter regions of heat stress responsive genes [3, 14, 15].

HSP60s are ATP-dependent mitochondrial chaperones which are involved in importing mitochondrial proteins and macromolecule assembly. They can be categorized into structurally similar two groups which differ in amino acid sequences. Group I HSP60s are found in mitochondria and chloroplasts as well as prokaryotes. This group includes chaperonin 60 and its co-chaperon chaperonin 10. Chloroplast chaperonins have effects on growth, embryo development, flowering, and chlorosis of plants. In unstressed conditions, HSP60s utilize appropriate folding of the key proteins, while under heat stress they take part in prevention of protein misfolding and promote re-assembling and refolding of mitochondrial matrix proteins. Group II chaperonins are found in archaea and eukaryote cytosols, in general [16, 17].

Small HSPs are in the range of 15–42 kDa. They have highly conserved sequences in C-terminal; hence they are found in all domains of life. They interact with higher HSPs as co-chaperons in response to heat stress. Individually, they constitute the first line of maintenance for misfolding of proteins. Contrary to the higher HSPs, sHSPs are not ATP-dependent and have high specificity and capacity to bind disordered proteins during primarily as heat, oxidative, and salinity stress conditions. They do not possess the ability to fold unfolded proteins; however, they can prevent irreversible unfolding and protein aggregations by re-folding denaturated or already folded proteins to some extent. This large protein family consists of six classes based on their cellular localizations, immunological properties, and sequence alignments. Cytoplasmic and nuclear groups are clustered in classes I, II, and III, while classes IV, V, and VI are the groups found in chloroplast, ER, and mitochondria, respectively [18, 19].

In past decade, substantial knowledge has been accumulated on mechanism of HSPs and chaperones as they are regulatory molecules that participation in stress sensing, signal transduction, and transcription activation of stress responsive genes in heat stress management. Therefore, transgenic plant approach is widespread among the studies which aim to improve crop productivity during consistently increasing heat stress worldwide [20]. **Table 1** summarizes the recent progress of transgenic approach regarding HSPs to improve heat stress tolerance in crop and model plants.

2.2 Antioxidants

Different plants present variations in temperature response depending on species, organs, and developmental stages. Disturbance in equilibrium between ROS scavenging capacity and ROS production during heat stress leads to major indirect effects in plants [30]. Perception of heat is a crucial step for induction of stress responsive gene expression. Beside its deleterious cellular effects, ROS has significant intra- and inter-cellular signaling properties for local and holistic control in plants. Through signal transduction, they contribute to the acquisition of thermotolerance along with HSPs, molecular chaperones, and phytohormones. Hyper-activation of the ROS scavenging components is also a viable strategy since it prevents cellular damage caused by ROS to membranes, organelles, and critical biomolecules as DNA, proteins, lipids, and more.

Enzymatic antioxidant defense in plants is composed of super oxide dismutase (SOD), catalase (CAT), ascorbate peroxidase (APX), monohydro ascorbate reductase (MDHAR), dehydroascorbate reductase (DHAR), glutathione reductase (GR), glutathione peroxidase (GPx), glutathione-S-transferase (GST), guaiacol peroxidase (GPX), peroxiredoxine (Prx), and thioredoxine (Trx). On the other hand, non-enzymatic antioxidant defense is divided into two categories as water solubles as ascorbic acid (AsA), glutathione (GSH), polyphenol, and lipid solubles as α-tocopherols, carotenoids, flavonoids, and retinoids [31].

In cellular processes, superoxide ions (O_2^-) are converted to H_2O_2 by SOD in chlorophyll, cytosol, apoplast, mitochondria, and peroxisomes. H_2O_2 is detoxified into H_2O by CAT in peroxisomes, chlorophyll, and mitochondria, APX in chlorophyll, cytosol, apoplast, mitochondria, peroxisomes, and GPX in mitochondria and cytosol. GST also contributes to the process in chlorophyll, cytosol, and mitochondria. The oxidized form of GSH is produced through the DHAR activity in chlorophyll,

Gene	Gene action	Gene source	Transformed plant	Transfer method	Agrobacterium strain	Heat tolerance (°C)	Reference
Hsp90A2	Heat shock protein 90s	*Glycine max*	*Glycine max*	Cotyledonary node co-cultivation infection	*Agrobacterium tumefaciens* EHA101	50	[21]
Hsp90	Heat shock protein 90s	*Dianthus caryophyllus*	*Arabidopsis thaliana*	Floral dip	*Agrobacterium tumefaciens* GV3101	42	[22]
Hsp90.5	Heat shock protein 90s	*Chrysanthemum morifolium*	*Arabidopsis thaliana*	Floral dip	*Agrobacterium tumefaciens*	40	[23]
Hsp70	Heat shock protein 70s	*Agave sisalana*	*Gossypium hirsutum*	Embryos co-cultivation infection	*Agrobacterium tumefaciens* LBA4404	45	[24]
Hsp70	Heat shock protein 70s	*Zostera japonica*	*Arabidopsis thaliana*	Floral dip	*Agrobacterium tumefaciens* GV3101	40	[25]
Hsp40, J3	Heat shock protein 40s	*Arabidopsis thaliana*	*Arabidopsis thaliana*	Floral dip	*Agrobacterium tumefaciens* GV3101	37	[26]
Hsp17.2	Cytosolic class II small heat shock protein	*Primula forrestii*	*Arabidopsis thaliana*	Particle bombardment	—	42	[27]
Hsp23	Small heat shock protein	*Medicago sativa*	*Medicago sativa*	Hypocotyls segments co-cultivation infection	*Agrobacterium tumefaciens* EHA105	42	[28]
Hsp25.9	Small heat shock protein	*Capsicum annuum* (R9)	*Arabidopsis thaliana* (Col-0)	Infiltration	*Agrobacterium tumefaciens* CV3101	38	[29]

Table 1.
Recent HSP gene transfers to improve heat stress tolerance in plants.

cytosol, mitochondria, and GPx activity in mitochondria and cytosol. GR reduced this by-product back into the reduced form of GSH. Monodehydroascorbate (MDHA) and dehydroascorbate (DHA) are produced as a result of APX activity, and both are reduced to AsA by MDHAR and DHAR, respectively, in chlorophyll, cytosol, and mitochondria [32]. Beside the notorious antioxidant enzymes as CAT and GPX, kinetic studies point out that Prxs reduce more than 90% of cellular peroxides [33]. Trxs function as cysteine reductases in plants. Eukaryotic cells utilize sensor proteins with redox-sensitive cysteine residues that function as signaling switches. Cysteines also provide signaling complexity through allowing reversible redox-based modifications such as S-nitrosylation, S-sulfenation, S-thiolation, and S-glutathionylation [34]. Therefore, redox sensors as Prx and redox transmitters as Trx take part crucial roles in posttranscriptional/translational regulation and initiation of signaling cascades during stress conditions [35].

Basal heat tolerance is significantly stronger in enhanced ROS scavenger species since their ROS scavenging gene expression is rapidly induced during heat stress. Therefore, fortification of antioxidant machinery is preferable option for reverse genetic applications as transgenic approaches. **Table 2** summarizes the recent progress of transgenic approach regarding antioxidants to improve heat stress tolerance in crops and model plants.

2.3 Osmolytes

Osmolytes, also known as cytoprotectants, osmoprotectants, or compatible solutes, are low molecular weight (LMW) compounds or metabolites that play important roles in balancing cellular redox, maintaining membrane integrity and protein stability, scavenging ROS, defending antioxidant compounds, and easing toxicity, and protecting cellular components in total. There are numerous samples which can be categorized as sugars, polyamines, secondary metabolites, amino acids, and polyols as proline, glycine betaine, trehalose, sorbitol, gamma-aminobutyric acid (GABA) which are widely used in bioengineering applications named as osmolyte induced stress tolerance [46]. Polyols as mannitol, D-ononitol, trehalose, sucrose, and fructane have been proven to accumulate in distinct evolutionary groups in response to various osmotic stress factors. They interact with the glutathione-ascorbate cycle enzymes which were mentioned earlier in this chapter to protect cellular membranes and enzyme complexes [47]. Proline, as one of the most studied amino acid type compatible solutes, has high water solubility and stable structure. Besides its essential structural roles, it plays well-known osmotic adjustment roles in plant cells. By these fundamental properties, its accumulation is observed in different kingdoms from bacteria to marine invertebrates. Most of the osmoprotectants are localized in cytoplasm during osmotic stress as it initiates. Osmoprotectants are suggested to ease osmotic imbalance through regulating osmotic potential within the cell. Reduced osmotic pressure maintains turgor pressure under heat stress conditions in which water potential is low as well as the conditions of high ionic strength. They also stabilize protein complexes and cellular membranes by protecting the hydration shell of proteins [48]. Genetic transformation technologies allow deliberate transfer of genes precisely in predictable manner. Therefore, transgenic approach is a viable option to manipulate the osmoprotectant biosynthesis pathways for enhanced accumulation of such molecules [49]. **Table 3** summarizes the recent improvements in osmolytes overexpressing transgenic plant approaches to provide protection by the osmotic action alone.

Gene	Gene source	Transformed plant	Transfer method	Agrobacterium strain	Heat tolerance (°C)	Reference
Apx1 (cytosolic ascorbate peroxidase)	*Hordeum vulgare* (Haruna-nijyo)	*Arabidopsis thaliana* (Col-0)	Floral dip	—	35	[36]
Apx	*Cephalotaxus fortunei* [37]	*Arabidopsis thaliana* (Col-0)	Floral dip	—	40	[38]
Grxs17 (glutaredoxin S17)	*Arabidopsis thaliana*	*Zea mays* (B104)	Zygotic embryos co-cultivation infection	*Agrobacterium tumefaciens* EHA101	37	[39]
Grxs17	*Arabidopsis thaliana* (Col-0)	*Chrysanthemum morifolium* (Ramat)	Leaf disc infection	*Agrobacterium tumefaciens* GV3101	45	[40]
Grx5	*Pteris vittata*	*Arabidopsis thaliana* (Col-0)	Floral dip	*Agrobacterium tumefaciens* C58	36	[41]
Mdar (Monodehydro ascorbate reductase)	*Lycopersicon esculentum* (Zhongshu 4)	*Lycopersicon esculentum*	Leaf disc infection	*Agrobacterium tumefaciens* LBA4404	40	[42]
Noa1 (Nitric Oxide-Associated Protein 1)	*Arabidopsis thaliana* (Col-0)	*Arabidopsis thaliana* (bohB/D Mutant)	Floral dip	—	45	[43]
Prx 2-Cys (2-Cys peroxiredoxin)	*Arabidopsis thaliana*	*Festuca arundinacea*	Callus co-cultivation infection	*Agrobacterium tumefaciens* EHA105	42	[44]
Sod (Superoxide dismutase)	*Allium sativum* (Daeseo)	*Nicotiana benthamiana*	Infiltration	*Agrobacterium tumefaciens* GV3101	45	[45]

Table 2.
Recent antioxidant gene transfers to improve heat stress tolerance in plants.

Gene	Gene action	Gene source	Transformed plant	Transfer method	Agrobacterium strain	Heat tolerance (°C)	Reference
codA	Choline oxidase, Glycine betaine accumulation	*Arthrobacter globiformis*	*Solanum lycopersicum*	Cotyledon and hypocotyl co-cultivation infection	*Agrobacterium tumefaciens* EHA101	42	[50]
codA and BADH	Choline oxidase- Glycine betaine accumulation, Betaine aldehyde dehydrogenase	*Arthrobacter globiformis* for codA, *Spinacia oleracea* for BADH	*Solanum lycopersicum*	Cotyledon and hypocotyl co-cultivation infection for codA, Leaf disc infection for BADH	*Agrobacterium tumefaciens* EHA101 for codA, *Agrobacterium tumefaciens* LBA4404 for BADH	42	[51]
MIPS2	Osmoprotectants, myo-inositol-1-phosphate synthase 2	*Triticum aestivum*	*Arabidopsis*	Floral dip	*Agrobacterium tumefaciens*	45	[52]
PRP6	Proline-rich proteins (PRPs), cell wall proteins	*Malus domestica*	*Nicotiana nudicaulia*	Leaf disc infection	*Agrobacterium tumefaciens* EHA105	48	[53]
SlSPS	Sucrose phosphate synthase, sucrose synthesis pathway	*Solanum lycopersicum*	*Solanum lycopersicum*	Cotyledon explants co-cultivation infection	*Agrobacterium tumefaciens* GC3101	42	[54]
SSI	Soluble starch synthase gene, starch biosynthesis	*Oryza sativa*	*Triticum aestivum,*	Biolistic transformation	—	34	[55]
TPSP	Trehalose-6-phosphate synthase-phosphatase fusion gene, enhancing the level of trehalose	*Escherichia coli*	*Solanum lycopersicum*	Cotyledon explants co-cultivation infection	*Agrobacterium tumefaciens* LBA4404	55	[56]

Table 3.
Recent osmoprotectant gene transfers to improve heat stress tolerance in plants.

2.4 Transcription factors (TFs)

Heat stress adversely affects the vegetative and reproductive stages of crop plants and leads to vast yield losses. Response to heat stress requires alterations on regular metabolic pathways through changes in gene expression profiles which are mainly regulated by various types of transcription factors (TFs). TFs are trans-acting elements which interact with cis-acting elements in promoter region of target stress responsive genes through genome. They are important signal transducers which convert perceived stress signals to stress specific responses. Many TFs including WRKY, MYB, NAC, bZIP, zinc finger protein, AP2/ERF, DREB, ERF, bHLH, and brassinosteroids transcription factors are associated with families of heat stress transcription factors or heat shock factors (HSFs). Approximately 7% of protein coding sequences of plant genomes consist of TFs. HSFs are among the largest gene families in plants compared to the other eukaryotes. The multiplicity of HSFs in plants is suggested to be related to the gene duplications and whole-genome duplications at different stages of evolution [57]. Plant HSFs have highly conserved modular structure. Their N-terminal domain has DNA binding properties. Promoter sequences of heat stress responsive genes include heat stress elements (HSEs) and are specific targets for central helix-turn-helix motif of HSPs. The C-terminal is an activation domain for plant HSFs. It contains short peptide motifs which play important roles in transcription activation of stress-inducible genes. Depending on hydrophobic amino acid residues linked to the DNA binding domain, plant HSFs are classified into three classes as HSFA, B, and C. HSFBs present common properties to the HSFs of other domains of life. On the other hand, HSFAs have 21 additional amino acid residues, while HSFCs have seven amino acid residue extensions [37]. HSFs are responsive to various abiotic stresses as drought, heat, and salinity. In nature, plants are constantly subjected to combination of different biotic and abiotic stresses. Therefore, it is considerably challenging to extrapolate the tolerance contribution of individual HSFs directly. Nevertheless, each TFs regulates many genes and thus are good candidates for engineering crop plants with enhanced heat stress tolerance due to their regulatory role. **Table 4** summarizes the recent progress of transgenic approach regarding TFs to improve heat stress tolerance in crop and model plants.

2.5 MicroRNAs (miRNAs)

In recent years, as an inevitable result of global climate change, there has been a significant increase in the number and severity of abiotic stress factors that plants are exposed to. Plants are vulnerable to the effects of heat, drought, salinity, cold, heavy metals, diseases, and pests due to their sessile nature. Therefore, the importance of developing plants tolerant to stress is increasing day by day. One of the most powerful methods for producing tolerant plants stands out as transgenic plants. MicroRNA (miRNA) transfer to plants is used as an important tool for thermotolerance.

miRNAs are RNA molecules with a length of 19–24 nucleotides (nt), not encoded by genes and involved in the regulation of gene expression. miRNAs are synthesized in the nucleus by RNA polymerase II as pri-miRNAs called precursor miRNAs. These pri-miRNAs are in hairpin structure and contain the mature miRNA sequence. The pri-miRNA structure is cleaved by the RNAase III enzyme to form the pre-miRNA

Gene	Gene source	Transformed plant	Transfer method	Agrobacterium strain	Heat Tolerance (°C)	Reference
AF1 and ANAC055	*Arabidopsis thaliana* (Col-0)	*A. thaliana* (knock-out mutants ataf1 and anac055)	Floral dip	*Agrobacterium tumefaciens* GV3101	44	[58]
bZIP60	*Triticum aestivum* (TAM107)	*A. thaliana* (Col-0)	Floral dip	*A. tumefaciens* GV3101	50	[59]
BZR1 (Brassinazole Resistant)	*Solanum lycopersicum mutant* (Condine Red)	*Solanum lycopersicum* (Condine Red)	CRISPR/Cas9	*A. tumefaciens* EHA105	42	[60]
CBF1 (C-Repeat Binding Factor)	*A. thaliana* (Col-0)	*A. thaliana* (cbf1 Mutant)	CRISPR/Cas9	*A. tumefaciens* GV3101	42	[61]
DREB20 (DRE-binding transcription factor)	*Musa acuminata* (Grand Nain and Hill Banana)	*A. thaliana* (Col-0)	Floral dlp	*A. tumefaciens* GV3101	44	[62]
DREB2A	*Pennisetum glaucum*	*A. thaliana* (Col-0)	Floral dip	*A. tumefaciens* GV3101	37	[63]
ERF110 (APETALA2/ Ethylene-responsive factor)	*Lilium longiflorum* (White heaven)	*A. thaliana* (Col-0) and *Nicotiana benthamiana*	Floral dip	*A. tumefaciens* GV3101	45	[64]
HB4 (HomeoBox)	*Helianthus annuus*	*Glycine max* (Williams 82)	Agrobacterium-mediated	*A. tumefaciens* EHA101	25–30	[65]
Hsf05 (Heat shock factor)	*Zea mays* (H21)	*A. thaliana* (athsfa2)	Floral dip	*A. tumefaciens* GV3101	42–45	[66]
HsfA1	*Amorphophallus albus*	*A. thaliana*	Infiltration	*A. tumefaciens* GV3101	41	[67]
HsfA1d	*A. thaliana*	*Pisum sativum*	Cotyledons co-cultivation infection	*A. tumefaciens* GV3101	37	[68]
HsfA2	*Populus euphratica*	*Populus tomentosa* (YiXianCiZhu B385)	Leaf disc infection	*A. tumefaciens* EHA105	45	[69]

Gene	Gene source	Transformed plant	Transfer method	Agrobacterium strain	Heat Tolerance (°C)	Reference
HsfA2c	*Festuca arundinacea* (Barlexas)	*A. thaliana* (Col-0)	Floral dip/ PEG-mediated transformation	*A. tumefaciens* EHA105	45	[70]
HsfA4	*Lilium Longiflorum* (White heaven)	*A. thaliana* (Col-0)	Infiltration	*A. tumefaciens* GV3101	45	[71]
HsfA6b	*Triticum aestivum* (PBW343)	*A. thaliana* (Col-0)	Floral dip	*A. tumefaciens* GV3101	42	[72]
HsfA6b	*Triticum aestivum*	*Hordeum vulgare*	Particle bombardment	—	35	[73]
HsfB1	*Solanum peruvianum,*	*Solanum lycopersicum*	Leaf disc infection	*A. tumefaciens* GV3101	39	[74]
HsfC1b	*Lolium perenne*	*A. thaliana*	Floral dip	*A. tumefaciens*	42	[75]
JA2 (Jasmonic acid)	*Solanum lycopersicum*	*Nicotiana tabacum* (NC 89) and *A. thaliana* (Col-0)	Leaf disc infection	—	42	[76]
MADS114 and MADS115	*Dactylis glomerata* (Donata)	*A. thaliana* (Col-0)	Floral dip	*A. tumefaciens* EHA105	40	[77]
NAC56	*Prunus persica* (chunjie)	*Solanum lycopersicum* (Micro Tom)	Infiltration	*A. tumefaciens* GV3101	42	[78]
NAC074	*Zea mays* (B73)	*A. thaliana* (Col-0)	Floral dip	*A. tumefaciens* GV3101	42	[79]
NTL3 (Plasma Membrane-Associated NAC Transcription Factor)	*Oryza Sativa* (Nipponbare)	*Oryza sativa* (Nipponbare)	CRISPR/Cas9	—	—	[80]
SHN1	*H. vulgare*	*N. tabacum*	Leaf disc infection	*A. tumefaciens* GV3101	42	[81]

DOI: http://dx.doi.org/10.5772/intechopen.111791

Gene	Gene source	Transformed plant	Transfer method	Agrobacterium strain	Heat Tolerance (°C)	Reference
WRKY30	*A. thaliana* (Col-0)	*Triticum aestivum* (Sakha-61)	Agrobacterium-mediated	*A. tumefaciens* EHA105	40	[82]
WRKY40	*Capsicum annuum*	*N. tabacum* (K326)	Leaf disc infection	*A. tumefaciens* LBA105	42	[83]
WRKY149	*Brassica napus* (ZS11)	*A. thaliana* (Col-0)	Agrobacterium-mediated	*A. tumefaciens* GV3101	40	[84]
ZnF	*Triticum aestivum*	*A. thaliana* (Col-0)	Floral dip	—	42	[85]

Table 4.
Recent TF gene transfers to improve heat stress tolerance in plants.

molecule. The resulting pre-miRNA is transported to the cytoplasm via Exportin-5 (XPO5). 19–24 nt long duplex miRNA is formed by being cut again by Dicer, a ribonuclease enzyme in the cytoplasm. Argonaute in the RNA-induced silencing complex (RISC) complex, which will form the mature miRNA sequence, is loaded. The miRNA-loaded RISC complex regulates transcriptional repression or degradation of mRNA. miRNAs play a role in the regulation of many biological processes in the cell, such as plant growth, development, stress responses, and control of the correct folding of proteins [86, 87].

High temperatures reduce the efficiency of photosynthetic activity in plants, cause negative effects on growth, damage to cell membranes, cell death due to senescence, protein misfolding, decrease in germination percentage, and release of weak pollen by preventing decomposition of anthers. Transfer of miRNAs to plants for tolerance to abiotic stresses is an important tool for plant tolerance. miRNAs increase the tolerance to stress factors by acting on the expressed genes at the transcriptional and post-transcriptional levels, inhibiting or regulating them. Temperature-sensitive miRNAs provide refolding of proteins, regulation of flowering, protection of reproductive tissues, repair of photosynthetic damage and regulating the antioxidant defense mechanism to alleviate the effects of stress. **Table 5** summarizes the recent progress of transgenic approach regarding miRNAs to improve heat stress tolerance in crop and model plants.

2.6 Other approaches

For producing temperature-tolerant genetically modified plants, it is a prerequisite to figure out how plants respond and adapt to heat stress and to characterize and identify novel heat stress-related genes. Heat stress (HS) can affect almost all aspects of plant processes such as germination, growth, development, reproduction, and yield, particularly by disturbing metabolic homeostasis, protein folding and processing capacity. In response to this challenge, plants utilize pathways/molecular mechanisms in complex and diverse systems, including photosynthetic metabolism, chaperones, signal transduction, epigenetic regulation, hormone signaling, lipid biosynthesis, plant growth regulation and additional intracellular actions. This radius of influence has allowed the development of a wide variety of strategies for the improvement of thermotolerance enhanced crop plants using genetic engineering approaches. Previous parts of the chapter presented that heat stress proteins (HSPs), heat stress factors (HSFs), transcription factors, osmoprotectants, ROS scavenging enzymes, and miRNAs are vital players in the plant's response to heat stress. In addition to all these responses, numerous studies have been reported to increase thermotolerance of plants by transferring genes that play a key role in plant metabolism to heat sensitive plants. Among these genes involved in stress management, genes encoding energy-dependent proteases, intramembrane proteases, calcium-dependent protein kinases, methyltransferases responsible for histone methylation, rubisco-related enzymes involved in carbon assimilation, enzymes involved in RNA metabolism, proteins acting as transcriptional regulators, molecular chaperones such as disulfide isomerases, 14-3-3 and DnaJ-like proteins, phytohormones, proteins participated in metal hemostasis, the ubiquitin-proteasome system, carotenoid and flavonoid accumulation, and late embryogenesis abundant proteins come to the forefront as an effective targets. **Table 6** summarizes the recent progress of transgenic approach regarding miscellaneous targets to improve heat stress tolerance in crop and model plants.

Gene	Gene source	Transformed plant	Transfer method	Agrobacterium strain	Heat Tolerance (°C)	Reference
miR156	*Arabidopsis thaliana* (Col-0)	*A. thaliana* (AGO1)	Floral dip	—	44	[88]
miR160	*A. thaliana* (Col-0)	*A. thaliana* (arf10, arf16, and arf17 Mutant)	Floral dip	*Agrobacterium tumefaciens* LBA4404	44–50	[89]
miR164	*A. thaliana* (Col-0)	*A. thaliana* (Overexpressors 164OE)	Floral dip	*A. tumefaciens* LBA4404	35	[90]
miR167	*Vitis vinifera* (Thompson Seedless)	*A. thaliana* (Col-0)	Floral dip	*A. tumefaciens* GV3101	45	[91]
miR172b-3p	*Solanum tuberosum* (Unica)	*S. tuberosum* (Russet Burbank)	Leaf and internodal co-cultivation infection	*A. tumefaciens* LBA4404	39	[92]
miR319d	*Solanum habrochaites* *(LA1777)*	*Solanum lycopersicum* (Micro-Tom)	Leaf disc infection	*A. tumefaciens* GV3101	40	[93]
miR398	*A. thaliana* (Col-0)	*A. thaliana* (csd2 Knockdown Mutant)	Floral dip	*A. tumefaciens* GV3101	37	[94]
miR447A	*A. thaliana* (Col-0)	*A. thaliana* Flower (Col-0)	Agrobacterium-mediated	*A. tumefaciens* GV3101	27	[95]
miR398a, miR398b, miR398c	*A. thaliana* (Col-0)	*A. thaliana* (Col-0)	—	*A. tumefaciens* GV3101	38	[96]
Novel_105 miRNA	*S. tuberosum* (Unica)	*S. tuberosum* (Russet Burbank)	Internode co-cultivation infection	*A. tumefaciens* LBA4404	39	[97]

Table 5.
Recent miRNA gene transfers to improve heat stress tolerance in plants.

Gene	Gene action	Gene source	Transformed plant	Transfer method	Agrobacterium strain	Heat Tolerance (°C)	Reference
ALA6	Aminophospholipid ATPase6 membrane systems	*Arabidopsis thaliana*	*A. thaliana*	Floral dip	*Agrobacterium tumefaciens* GV3101	37.5–43	[98]
ChiVI2	Chitin-binding proteins (CBP)	*Capsicum annuum*	*A. thaliana*	Floral dip	*A. tumefaciens* GV3101	45	[99]
CLPB1	Clp ATPases, energy-dependent proteases and molecular chaperones.	*Ziziphus nummularia*	*Nicotiana tabacum*	Leaf disc infection	*A. tumefaciens* EHA105	42	[100]
CDPK7	Calcium-dependent protein kinase, Ca2 + −mediated signaling pathways	*Zea mays*	*Z. mays*	Embryo co-cultivation infection	*A. tumefaciens* GV3101	28–42	[101]
CPL1	RNA polymerase II CTD phosphatase-like 1 enzyme, transcriptional regulator of the plant response to various abiotic stresses	*Chrysanthemum morifolium*	*C. morifolium*	Leaf disc infection	*Agrobacterium tumefaciens* EHA105	45	[102]
DJLP	DnaJ-like protein, molecular chaperone	*Medicago sativa*	*N. tabacum*	Leaf disc infection	*A. tumefaciens* EHA105	42	[103]
DOG1L-T	Delay of germination 1-like,	*N. tabacum*	*N. tabacum*	Infiltration	*A. tumefaciens*	45	[104]
F3H	flavanol 3-hydroxylase flavonoid biosynthesis	*Oryza sativa*	*O. sativa*	Calli co-cultivation infection	*A. tumefaciens* LBA4404	40	[105]
FBA1	F-Box protein gene, core component of the Skp1-Cullin-F-box (SCF) E3 ligase complex	*Triticum aestivum*	*N. tabacum*	Leaf disc infection	*A. tumefaciens* LBA440	45	[106]
FER-5B	Ferritin gene, iron storage protein, sequestering or releasing iron	*T. aestivum*	*A. thaliana T. aestivum*	Floral dip for Arabidopsis, Immature embryos particle bombardment for *T. aestivum*	*A. tumefaciens* GV3101	40	[107]

Gene	Gene action	Gene source	Transformed plant	Transfer method	Agrobacterium strain	Heat Tolerance (°C)	Reference
FT-1	14–3-3 protein, molecular chaperone	*Haloxylon ammodendron*	*A. thaliana*	—	—	37–47	[108]
GLP	hydrogen peroxide-producing germin-like protein	*Solanum tuberasum*	*Solanum tuberasum*	Infiltration	*A. tumefaciens* GV3101	25–45	[109]
Golden SNP-Carrying Orange Gene	Carotenoid accumulation	*Ipomoea batatas*	*I. batatas*	Embryogenic calli co-cultivation infection	*A. tumefaciens* GV3101	47	[110]
HIRP1	E3 ligase, heat-induced RING finger protein 1	*O. sativa*	*A. thaliana*	Floral dip	*A. tumefaciens* GV3101	45	[111]
HVA1	Late embryogenesis abundant (Lea) protein, leading to the accumulation of small Hsps	*T. aestivum*	*T. aestivum*	Anther culture-based approach	*A. tumefaciens* LBA4404	42	[112]
OEP16–2-5B	Wheat plastid outer envelope protein gene	*T. aestivum*	*A. thaliana*	Floral dip	*A. tumefaciens* GV3101	40	[113]
PhyCYS	Phytocystatins, proteinaceous inhibitors of the papain-like (C1A) and legumain (C13) families of plant cysteine proteases (CPs)	*A. thaliana*	*A. thaliana*	Floral dip	*A. tumefaciens* GV3101	37–50	[114]
PRMT1	Protein arginine methyltransferases (PRMTs)	*Z. mays*	*A. thaliana*	Floral dip	*Agrobacterium-mediated*	42	[115]
Rca1β	Rubisco activase B,	*T. aestivum,*	*O. sativa*	Scutellum tissue co-cultivation infection	*A. tumefaciens*	42°C	[116]

Gene	Gene action	Gene source	Transformed plant	Transfer method	Agrobacterium strain	Heat Tolerance (°C)	Reference
Rubisco and Rubisco activase	Catalyzing the binding of CO_2 to ribulose-1,-5-bisphosphate (RuBP) and rubisco activation	*Z. mays*	*O. sativa*		*Agrobacterium-mediated*	36–40	[117]
SAMS	S- adenosyl methionine synthetase, Spd biosynthesis	*Trifolium repens*	*A. thaliana*	Floral dip	*A. tumefaciens* EHA105	33–38	[118]
SlEGY2	Metalloprotease	*Solanum lycopersicum*	*S. lycopersicum (antisense transgenic tomato plant)*	Leaf disc, anti-sens	*A. tumefaciens* LBA4404	42	[119]
PDI	Protein disulfide isomerase gene, chaperone function and disulfide isomerase activity	*Methanothermobacter thermautotrophicus*	*O. sativa*	Tissues co-cultivationinfection	*A. tumefaciens* EHA105	42	[120]
TOGR1	DEAD-Box RNA helicase, RNA metabolism	*O. sativa*	*Brassica rapa*	Vacuum infiltration	*A. tumefaciens* EHA105	38–46	[121]

Table 6.
Miscellaneous recent gene transfers to improve heat stress tolerance in plants.

3. Conclusions

Proteins undertake important structural and functional properties in cells. Among all abiotic stress factors, heat affects biological activity of proteins more directly by leading to aggregation and/or misfolding. HSPs constitute the frontal zone of defense against heat stress-induced accumulation of aggregated/misfolded proteins which may induce heat shock responses (HSR) in plant cells. Hsps are main targets for gene transfer approaches due to their chaperone roles to co-operate functional networks as well as re-solubilization roles for the recovery phase of aggregated/misfolded proteins. Along with the definite evidence to succession of HSP gene transfer-related thermotolerance, osmolytes as members of non-enzymatic antioxidative system contribute to the process through habilitating cellular environment to more reductive state due to higher energy status. Hence, by binding to the cellular proteins, they protect them from denaturation/aggregation. Likewise, enzymatic antioxidant systems as cell detoxification components undertake the major role in regulation of reductive cellular environment and minimizing the loss of active proteins. Besides, classification and association of different HSFs and HSPs as functional candidates in heat stress tolerance and other developmental pathways are extremely crucial. Even though structural and functional association of Hsps/Hsfs have been widely established, they are still not mainstream targets in crop plant applications against heat stress. However, applicability is improving impetuously. On the other hand, transgenic approaches in heat stress tolerance through miRNAs in plants mainly involve model plants such as *Arabidopsis* or rice at present. Moreover, stress and species-specific miRNAs still require further discovery. A large number of miRNAs and their target genes related to heat stress have not been discovered yet. Since individual miRNAs may also play multiple roles in other various development regulatory pathways and biotic and/or abiotic stresses as well as heat, it is necessary to explore new miRNAs, reveal their target genes, and further evaluate the miRNA-mediated regulatory networks before announcing them as designated targets. Other than protecting protein stability, it is also a viable approach to sustain cellular membranes as their fluidity is vital to maintain cell volume. The physical state of the cellular membranes influences gene expression by initiating signal transduction. Altering membrane structures can also affect interactions of membrane lipids with proteins. Hereby, we can conclude transgenic approaches may still offer vast number of opportunities to heat stress tolerance area as in recent years we can follow miscellaneous novel targets from photosynthetic machinery to signal transduction cascades, despite the fact that there is still no biotech/GM crop events that have been approved for commercialization/planting and importation (food and feed) in International Service for the Acquisition of Agribiotech Applications (ISAAA) GM Approval Database for heat stress tolerance.

Abbreviations

ABA	abscisic acid
ALA	aminophospholipid ATPase
AP2/ERF	APETALA2 (AP2)/ethylene responsive element binding factor (EREB)
APX	ascorbate peroxidase
AsA	ascorbic acid
ATP	adenosine triphosphate

BADH	betaine aldehyde dehydrogenase
bHLH	basic helix-loop-helix
bZIP	Basic Leucine Zipper
CAT	catalase
CBF	C-repeat binding factor
CBP	chitin-binding proteins
CDPK	calcium-dependent protein kinase
Cod	choline oxidase
Col-0	Columbia-0
CRISPR	clustered regularly interspaced short palindromic repeats
DHA	dehydroascorbate
DHAR	dehydroascorbate reductase
DJLP	DnaJ-like protein
DNA	deoxyribonucleic acid
DOG1L	delay of germination 1-like
DREB	dehydration-responsive element-binding protein
ER	endoplasmic reticulum
F3H	flavanol 3-hydroxylase
GABA	gamma-aminobutyric acid
GLP	Germin-like protein
GM	genetically modified
GPx	glutathione peroxidase
GPX	guaiacol peroxidase
GR	glutathione reductase
GSH	glutathione
GST	glutathione-S-transferase
HB	HomeoBox
HIRP	heat-induced RING finger protein
HOP	HSP90/HSP70-organizing protein
HS	heat stress
HSEs	heat shock elements
HSFs	heat shock transcription factors
HSPs	heat shock proteins
ISAAA	International Service for the Acquisition of Agri-biotech Applications
JA	jasmonic acid
kDa	kilodaltons
MDHA	monodehydroascorbate
MDHAR	monohydro ascorbate reductase
MG	methylglyoxal
MIPS2	Myo-inositol-1-phosphate synthase 2
miRNAs	micro RNAs
MYB	Myb-related protein B
NAC	NAM (No Apical Meristem), ATAF1/2 (*Arabidopsis thaliana* Transcription Activator Factor 1/2) and CUC2 (Cup-shaped Cotyledon 2)
Noa1	nitric oxide-associated protein 1
NOAA	The National Oceanic and Atmospheric Administration
nt	nucleotide
OEP	outer envelope protein

PDI	protein disulfide isomerase
PRMT	protein arginine methyltransferases
PRPs	proline-rich proteins
Prx	peroxiredoxine
Rca	rubisco activase
RISC	RNA-induced silencing complex
ROS	reactive oxygene species
Rubisco	ribulose-1,5-bisphosphate carboxylase/oxygenase
SAMS	S-adenosyl methionine synthetase
SISPS	sucrose phosphate synthase
SNP	single nucleotid polymorphism
SOD	super oxide dismutase
SSI	starch synthase I
TFs	transcription factors
TOGR	thermotolerant growth required
TPSP	trehalose-6-phosphate synthase
Trx	thioredoxine
UV	ultraviolet
ZnF	Zinc Finger

Author details

Alp Ayan*, Sinan Meriç, Tamer Gümüş and Çimen Atak
Department of Molecular Biology and Genetic, Faculty of Science and Letters, Istanbul Kultur University, Istanbul, Turkey

*Address all correspondence to: a.ayan@iku.edu.tr

References

[1] Lindsey R, Dahlman L. Climate change: Global temperature. climate. gov. 2020; 16

[2] Lounasheimo J, Cederlöf M, Mäntylä, I. Annual Climate Report 2021. Publications of the Ministry of the Environment 2021; 13

[3] Ray D, Ghosh A, Mustafi SB, Raha S. Plant stress response: Hsp70 in the spotlight. Heat Shock Proteins and Plants. 2016;**2016**:123-147. DOI: 10.1007/978-3-319-46340-7_7

[4] Gümüş T, Meriç S, Ayan A, Atak Ç. Plant Abiotic Stress Factors: Current Challenges of Last Decades and Future Threats. London, UK: InTechOpen; 2023. DOI: 10.5772/intechopen.110367

[5] Xu J, Lowe C, Hernandez-Leon SG, Dreisigacker S, Reynolds MP, Valenzuela-Soto EM, et al. The effects of brief heat during early booting on reproductive, developmental, and chlorophyll physiological performance in common wheat (*Triticum aestivum* L.). Frontiers in Plant Science. 2022;**2022**:13. DOI: 10.3389/fpls.2022.886541

[6] Taratima W, Chuanchumkan C, Maneerattanarungroj P, Trunjaruen A, Theerakulpisut P, Dongsansuk A. Effect of heat stress on some physiological and anatomical characteristics of Rice (*Oryza sativa* L.) cv. KDML105 callus and seedling. Biology. 2022;**11**(11):1587. DOI: 10.3390/biology11111587

[7] Giri A, Heckathorn S, Mishra S, Krause C. Heat stress decreases levels of nutrient-uptake and-assimilation proteins in tomato roots. Plants. 2017;**6**(1):6. DOI: 10.3390/plants6010006

[8] Driesen E, Van den Ende W, De Proft M, Saeys W. Influence of environmental factors light, CO_2, temperature, and relative humidity on stomatal opening and development: A review. Agronomy. 2020;**10**(12):1975. DOI: 10.3390/agronomy10121975

[9] Chen J, Gao T, Wan S, Zhang Y, Yang J, Yu Y, et al. Genome-wide identification, classification and expression analysis of the HSP gene superfamily in tea plant (*Camellia sinensis*). International Journal of Molecular Sciences. 2018;**19**(9):2633. DOI: 10.3390/ijms19092633

[10] Agarwal M, Katiyar-Agarwal S, Grover A. Plant Hsp100 proteins: Structure, function and regulation. Plant Science. 2002;**163**(3):397-405. DOI: 10.1016/S0168-9452(02)00209-1

[11] Tichá T, Samakovli D, Kuchařová A, Vavrdová T, Šamaj J. Multifaceted roles of HEAT SHOCK PROTEIN 90 molecular chaperones in plant development. Journal of Experimental Botany. 2020;**71**(14):3966-3985. DOI: 10.1093/jxb/eraa177

[12] di Donato M, Geisler M. HSP 90 and co-chaperones: A multitaskers' view on plant hormone biology. FEBS Letters. 2019;**593**(13):1415-1430. DOI: 10.1002/1873-3468.13499

[13] Xu ZS, Li ZY, Chen Y, Chen M, Li LC, Ma YZ. Heat shock protein 90 in plants: Molecular mechanisms and roles in stress responses. International Journal of Molecular Sciences. 2012;**13**(12):15706-15723. DOI: 10.3390/ijms131215706

[14] Usman MG, Rafii MY, Martini MY, Yusuff OA, Ismail MR, Miah G. Molecular analysis of Hsp70 mechanisms in plants and their function in response to stress. Biotechnology and Genetic

Engineering Reviews. 2017;**33**(1):26-39. DOI: 10.1080/02648725.2017.1340546

[15] Batool F, Agossa BA, Sandhu ZY, Sarwar MB, Hassan S, Rashid B. Heat shock proteins (HSP70) gene: Plant transcriptomic oven in the hot desert. In: Advances in Plant Defense Mechanisms. London, UK: IntechOpen; 2022. DOI: 10.5772/intechopen.105391

[16] Caruso Bavisotto C, Alberti G, Vitale AM, Paladino L, Campanella C, Rappa F, et al. Hsp60 post-translational modifications: Functional and pathological consequences. Frontiers in Molecular Biosciences. 2020;7:95. DOI: 10.3389/fmolb.2020.00095

[17] Haq S, Khan A, Ali M, Khattak AM, Gai WX, Zhang HX, et al. Heat shock proteins: Dynamic biomolecules to counter plant biotic and abiotic stresses. International Journal of Molecular Sciences. 2019;**20**(21):5321. DOI: 10.3390/ijms20215321

[18] Waters ER, Vierling E. Plant small heat shock proteins–evolutionary and functional diversity. New Phytologist. 2020;**227**(1):24-37. DOI: 10.1111/nph.16536

[19] Santhanagopalan I, Basha E, Ballard KN, Bopp NE, Vierling E. Model chaperones: Small heat shock proteins from plants. The Big Book on Small Heat Shock Proteins. 2015;**2015**:119-153. DOI: 10.1007/978-3-319-16077-1_5

[20] Ayan A, Meriç S, Gümüş T, Atak Ç. Current strategies and future of mutation breeding in soybean improvement. In: Soybean-Recent Advances in Research and Applications. London, UK: IntechOpen; 2022. DOI: 10.5772/intechopen.104796

[21] Huang Y, Xuan H, Yang C, Guo N, Wang H, Zhao J, et al. GmHsp90A2 is involved in soybean heat stress as a positive regulator. Plant Science. 2019;**285**:26-33. DOI: 10.1016/j.plantsci.2019.04.016

[22] Xue P, Sun Y, Hu D, Zhang J, Wan X. Genome-wide characterization of DcHsp90 gene family in carnation (*Dianthus caryophyllus* L.) and functional analysis of DcHsp90-6 in heat tolerance. Protoplasma. 2022;**2022**:1-13. DOI: 10.1007/s00709-022-01815-5

[23] Wang X, Wu J, Wang Y, Jiang Y, Li F, Chen Y, et al. Chrysanthemum CmHSP90.5 as a tool to regulate heat and salt stress tolerance. Horticulturae. 2022;**8**(6):532. DOI: 10.3390/horticulturae8060532

[24] Batcho AA, Sarwar MB, Rashid B, Hassan S, Husnain T. Heat shock protein gene identified from Agave sisalana (As HSP70) confers heat stress tolerance in transgenic cotton (*Gossypium hirsutum*). Theoretical and Experimental Plant Physiology. 2021;**33**:141-156

[25] Chen S, Qiu G. Overexpression of Zostera japonica heat shock protein gene ZjHsp70 enhances the thermotolerance of transgenic Arabidopsis. Molecular Biology Reports. 2022;**49**(7):6189-6197. DOI: 10.1007/s11033-022-07411-3

[26] Wu JR, Wang TY, Weng CP, Duong NKT, Wu SJ. AtJ3, a specific HSP40 protein, mediates protein farnesylation-dependent response to heat stress in Arabidopsis. Planta. 2019;**250**:1449-1460. DOI: 10.1007/s00425-019-03239-7

[27] Zhang L, Hu W, Gao Y, Pan H, Zhang Q. A cytosolic class II small heat shock protein, PfHSP17.2, confers resistance to heat, cold, and salt stresses in transgenic Arabidopsis. Genetics and Molecular

Biology. 2018;**41**:649-660. DOI: 10.1590/1678-4685-GMB-2017-0206

[28] Lee KW, Rahman M, Choi GJ, Kim KY, Ji HC, Hwang TY, et al. Expression of small heat shock protein23 enhanced heat stress tolerance in transgenic alfalfaplants. JAPS: Journal of Animal & Plant Sciences. 2017;**27**(4):1238-1244

[29] Feng XH, Zhang HX, Ali M, Gai WX, Cheng GX, Yu QH, et al. A small heat shock protein CaHsp25. 9 positively regulates heat, salt, and drought stress tolerance in pepper (*Capsicum annuum* L.). Plant Physiology and Biochemistry. 2019;**142**:151-162. DOI: 10.1016/j.plaphy.2019.07.001

[30] Çelik Ö, Ayan A, Atak Ç. Enzymatic and non-enzymatic comparison of two different industrial tomato (*Solanum lycopersicum*) varieties against drought stress. Botanical Studies. 2017;**58**:1-13. DOI: 10.1186/s40529-017-0186-6

[31] Hong-Bo S, Xiao-Yan C, Li-Ye C, Xi-Ning Z, Gang W, Yong-Bing Y, et al. Investigation on the relationship of proline with wheat anti-drought under soil water deficits. Colloids and Surfaces B: Biointerfaces. 2006;**53**(1):113-119. DOI: 10.1016/j.colsurfb.2006.08.008

[32] Rajput VD, Singh RK, Verma KK, Sharma L, Quiroz-Figueroa FR, Meena M, et al. Recent developments in enzymatic antioxidant defence mechanism in plants with special reference to abiotic stress. Biology. 2021;**10**(4):267. DOI: 10.3390/biology10040267

[33] Perkins A, Nelson KJ, Parsonage D, Poole LB, Karplus PA. Peroxiredoxins: Guardians against oxidative stress and modulators of peroxide signaling. Trends in Biochemical Sciences. 2015;**40**(8):435-445

[34] Mata-Pérez C, Spoel SH. Thioredoxin-mediated redox signalling in plant immunity. Plant Science. 2019;**279**:27-33

[35] Martí MC, Jiménez A, Sevilla F. Thioredoxin network in plant mitochondria: Cysteine S-posttranslational modifications and stress conditions. Frontiers in Plant Science. 2020;**11**:571288. DOI: 10.3389/fpls.2020.571288

[36] Shi WM, Muramoto Y, Ueda A, Takabe T. Cloning of peroxisomal ascorbate peroxidase gene from barley and enhanced thermotolerance by overexpressing in *Arabidopsis thaliana*. Gene. 2001;**273**(1):23-27. DOI: 10.1016/S0378-1119(01)00566-2

[37] Baniwal SK, Bharti K, Chan KY, Fauth M, Ganguli A, Kotak S, et al. Heat stress response in plants: A complex game with chaperones and more than twenty heat stress transcription factors. Journal of Biosciences. 2004;**29**:471-487. DOI: 10.1007/BF02712120

[38] Zhang Y, Yang L, Zhang M, Yang J, Cui J, Hu H, et al. CfAPX, a cytosolic ascorbate peroxidase gene from *Cryptomeria fortunei*, confers tolerance to abiotic stress in transgenic Arabidopsis. Plant Physiology and Biochemistry. 2022;**172**:167-179. DOI: 10.1016/j.plaphy.2022.01.011

[39] Sprague SA, Tamang TM, Steiner T, Wu Q, Hu Y, Kakeshpour T, et al. Redox-engineering enhances maize thermotolerance and grain yield in the field. Plant Biotechnology Journal. 2022;**20**(9):1819-1832. DOI: 10.1111/pbi.13866

[40] Kang BC, Wu Q, Sprague S, Park S, White FF, Bae SJ, et al. Ectopic overexpression of an Arabidopsis monothiol glutaredoxin AtGRXS17

affects floral development and improves response to heat stress in chrysanthemum (Chrysanthemum morifolium Ramat.). Environmental and Experimental Botany. 2019;**167**:103864. DOI: 10.1016/j.envexpbot.2019.103864

[41] Sundaram S, Rathinasabapathi B. Transgenic expression of fern Pteris vittata glutaredoxin PvGrx5 in Arabidopsis thaliana increases plant tolerance to high temperature stress and reduces oxidative damage to proteins. Planta. 2010;**231**:361-369. DOI: 10.1007/s00425-009-1055-7

[42] Li F, Wu QY, Sun YL, Wang LY, Yang XH, Meng QW. Overexpression of chloroplastic monodehydroascorbate reductase enhanced tolerance to temperature and methyl viologen-mediated oxidative stresses. Physiologia Plantarum. 2010;**139**(4):421-434. DOI: 10.1111/j.1399-3054.2010.01369.x

[43] Wang L, Guo Y, Jia L, Chu H, Zhou S, Chen K, et al. Hydrogen peroxide acts upstream of nitric oxide in the heat shock pathway in Arabidopsis seedlings. Plant Physiology. 2014;**164**(4):2184-2196. DOI: 10.1104/pp.113.229369

[44] Kim KH, Alam I, Lee KW, Sharmin SA, Kwak SS, Lee SY, et al. Enhanced tolerance of transgenic tall fescue plants overexpressing 2-Cys peroxiredoxin against methyl viologen and heat stresses. Biotechnology Letters. 2010;**32**:571-576. DOI: 10.1007/s10529-009-0185-0

[45] Ji HS, Bang SG, Ahn MA, Kim G, Kim E, Eom SH, et al. Molecular cloning and functional characterization of heat stress-responsive superoxide dismutases in Garlic (*Allium sativum* L.). Antioxidants. 2021;**10**(5):815. DOI: 10.3390/antiox10050815

[46] Ghosh UK, Islam MN, Siddiqui MN, Khan MAR. Understanding the roles of osmolytes for acclimatizing plants to changing environment: A review of potential mechanism. Plant Signaling & Behavior. 2021;**16**(8):1913306. DOI: 10.1080/15592324.2021.1913306

[47] Meriç S, Ayan A, Atak Ç. Molecular abiotic stress tolerans strategies: From genetic engineering to genome editing era. In: Abiotic Stress in Plants. London, UK: IntechOpen; 2020. DOI: 10.5772/intechopen.94505

[48] Vajda T, Perczel A. Role of water in protein folding, oligomerization, amyloidosis and miniprotein. Journal of Peptide Science. 2014;**20**(10):747-759. DOI: 10.1002/psc.2671

[49] Khan MS, Ahmad D, Khan MA. Utilization of genes encoding osmoprotectants in transgenic plants for enhanced abiotic stress tolerance. Electronic Journal of Biotechnology. 2015;**18**(4):257-266. DOI: 10.1016/j.ejbt.2015.04.002

[50] Li D, Wang M, Zhang T, Chen X, Li C, Liu Y, et al. Glycinebetaine mitigated the photoinhibition of photosystem II at high temperature in transgenic tomato plants. Photosynthesis Research. 2021;**147**:301-315. DOI: 10.1007/s11120-020-00810-2

[51] Zhang T, Li Z, Li D, Li C, Wei D, Li S, et al. Comparative effects of glycinebetaine on the thermotolerance in codA-and BADH-transgenic tomato plants under high temperature stress. Plant Cell Reports. 2020;**39**:1525-1538. DOI: 10.1007/s00299-020-02581-5

[52] Khurana N, Sharma N, Khurana P. Overexpression of a heat stress inducible, wheat myo-inositol-1-phosphate synthase 2 (TaMIPS2) confers tolerance to various abiotic stresses in Arabidopsis thaliana. Agri Gene. 2017;**6**:24-30. DOI: 10.1016/j.aggene.2017.09.001

[53] Zhang X, Gong X, Li D, Yue H, Qin Y, Liu Z, et al. Genome-wide identification of PRP genes in apple genome and the role of MdPRP6 in response to heat stress. International Journal of Molecular Sciences. 2021;**22**(11):5942. DOI: 10.3390/ijms22115942

[54] Zhang Y, Zeng D, Liu Y, Zhu W. SlSPS, a sucrose phosphate synthase gene, mediates plant growth and thermotolerance in tomato. Horticulturae. 2022;**8**(6):491. DOI: 10.3390/horticulturae8060491

[55] Tian B, Talukder SK, Fu J, Fritz AK, Trick HN. Expression of a rice soluble starch synthase gene in transgenic wheat improves the grain yield under heat stress conditions. In Vitro Cellular & Developmental Biology-Plant. 2018;**54**:216-227. DOI: 10.1007/s11627-018-9893-2

[56] Lyu JI, Park JH, Kim JK, Bae CH, Jeong WJ, Min SR, et al. Enhanced tolerance to heat stress in transgenic tomato seeds and seedlings overexpressing a trehalose-6-phosphate synthase/phosphatase fusion gene. Plant Biotechnology Reports. 2018;**12**:399-408. DOI: 10.1007/s11816-018-0505-8

[57] Scharf KD, Berberich T, Ebersberger I, Nover L. The plant heat stress transcription factor (Hsf) family: structure, function and evolution. Biochimica et Biophysica Acta (BBA)-Gene Regulatory Mechanisms. 2012;**1819**(2):104-119. DOI: 10.1016/j.bbagrm.2011.10.002

[58] Alshareef NO, Otterbach SL, Allu AD, Woo YH, de Werk T, Kamranfar I, et al. NAC transcription factors ATAF1 and ANAC055 affect the heat stress response in Arabidopsis. Scientific Reports. 2022;**12**(1):11264. DOI: 10.1038/s41598-022-14429-x

[59] Geng X, Zang X, Li H, Liu Z, Zhao A, Liu J, et al. Unconventional splicing of wheat TabZIP60 confers heat tolerance in transgenic Arabidopsis. Plant Science. 2018;**274**:252-260. DOI: 10.1016/j.plantsci.2018.05.029

[60] Yin Y, Qin K, Song X, Zhang Q, Zhou Y, Xia X, et al. BZR1 transcription factor regulates heat stress tolerance through FERONIA receptor-like kinase-mediated reactive oxygen species signaling in tomato. Plant and Cell Physiology. 2018;**59**(11):2239-2254. DOI: 10.1093/pcp/pcy146

[61] Yun SD, Kim MH, Oh SA, Soh MS, Park SK. Overexpression of C-repeat binding factor1 (CBF1) gene enhances heat stress tolerance in Arabidopsis. Journal of Plant Biology. 2022;**65**(3):253-260. DOI: 10.1007/s12374-022-09350-9

[62] Chaudhari RS, Jangale BL, Krishna B, Sane PV. Improved abiotic stress tolerance in Arabidopsis by constitutive active form of a banana DREB2 type transcription factor, MaDREB20. CA, than its native form, MaDREB20. Protoplasma. 2022;**2022**:1-20. DOI: 10.1007/s00709-022-01805-7

[63] Meena RP, Ghosh G, Vishwakarma H, Padaria JC. Expression of a Pennisetum glaucum gene DREB2A confers enhanced heat, drought and salinity tolerance in transgenic Arabidopsis. Molecular Biology Reports. 2022;**49**(8):7347-7358. DOI: 10.1007/s11033-022-07527-6

[64] Li T, Wu Z, Xiang J, Zhang D, Teng N. Overexpression of a novel heat-inducible ethylene-responsive factor gene LlERF110 from Lilium longiflorum decreases thermotolerance. Plant Science. 2022;**319**:111246. DOI: 10.1016/j.plantsci.2022.111246

[65] Ribichich KF, Chiozza M, Ávalos-Britez S, Cabello JV,

Arce AL, Watson G, et al. Successful field performance in warm and dry environments of soybean expressing the sunflower transcription factor HB4. Journal of Experimental Botany. 2020;**71**(10):3142-3156. DOI: 10.1093/jxb/eraa064

[66] Li GL, Zhang HN, Shao H, Wang GY, Zhang YY, Zhang YJ, et al. ZmHsf05, a new heat shock transcription factor from Zea mays L. improves thermotolerance in Arabidopsis thaliana and rescues thermotolerance defects of the athsfa2 mutant. Plant Science. 2019;**283**:375-384. DOI: 10.1016/j.plantsci.2019.03.002

[67] Yue Z, Wang Y, Zhang N, Zhang B, Niu Y. Expression of the Amorphophallus albus heat stress transcription factor AaHsfA1 enhances tolerance to environmental stresses in Arabidopsis. Industrial Crops and Products. 2021;**174**:114231. DOI: 10.1016/j.indcrop.2021.114231

[68] Shah Z, Shah SH, Ali GS, Munir I, Khan RS, Iqbal A, et al. Introduction of Arabidopsis's heat shock factor HsfA1d mitigates adverse effects of heat stress on potato (*Solanum tuberosum* L.) plant. Cell Stress & Chaperones. 2020;**25**(1):57-63. DOI: 10.1007/s12192-019-01043-6

[69] Li HG, Yang Y, Liu M, Zhu Y, Wang HL, Feng CH, et al. The in vivo performance of a heat shock transcription factor from Populus euphratica, PeHSFA2, promises a prospective strategy to alleviate heat stress damage in poplar. Environmental and Experimental Botany. 2022;**201**:104940. DOI: 10.1016/j.envexpbot.2022.104940

[70] Wang X, Huang W, Liu J, Yang Z, Huang B. Molecular regulation and physiological functions of a novel FaHsfA2c cloned from tall fescue conferring plant tolerance to heat stress. Plant Biotechnology Journal. 2017;**15**(2):237-248. DOI: 10.1111/pbi.12609

[71] Wang C, Zhou Y, Yang X, Zhang B, Xu F, Wang Y, et al. The heat stress transcription factor LlHsfA4 enhanced basic Thermotolerance through regulating ROS metabolism in lilies (Lilium Longiflorum). International Journal of Molecular Sciences. 2022;**23**(1):572. DOI: 10.3390/ijms23010572

[72] Meena S, Samtani H, Khurana P. Elucidating the functional role of heat stress transcription factor A6b (Ta HsfA6b) in linking heat stress response and the unfolded protein response in wheat. Plant Molecular Biology. 2022;**108**(6):621-634. DOI: 10.1007/s11103-022-01252-1

[73] Poonia AK, Mishra SK, Sirohi P, Chaudhary R, Kanwar M, Germain H, et al. Overexpression of wheat transcription factor (TaHsfA6b) provides thermotolerance in barley. Planta. 2020;**252**:1-14. DOI: 10.1007/s00425-020-03457-4

[74] Fragkostefanakis S, Simm S, El-Shershaby A, Hu Y, Bublak D, Mesihovic A, et al. The repressor and co-activator HsfB1 regulates the major heat stress transcription factors in tomato. Plant, Cell & Environment. 2019;**42**(3):874-890. DOI: 10.1111/pce.13434

[75] Sun T, Shao K, Huang Y, Lei Y, Tan L, Chan Z. Natural variation analysis of perennial ryegrass in response to abiotic stress highlights LpHSFC1b as a positive regulator of heat stress. Environmental and Experimental Botany. 2020;**179**:104192. DOI: 10.1016/j.envexpbot.2020.104192

[76] Liu ZM, Yue MM, Yang DY, Zhu SB, Ma NN, Meng QW. Over-expression

of SlJA2 decreased heat tolerance of transgenic tobacco plants via salicylic acid pathway. Plant Cell Reports. 2017;**36**:529-542. DOI: 10.1007/s00299-017-2100-9

[77] Yang Z, Nie G, Feng G, Xu X, Li D, Wang X, et al. Genome-wide identification of MADS-box gene family in orchardgrass and the positive role of DgMADS114 and DgMADS115 under different abiotic stress. International Journal of Biological Macromolecules. 2022;**223**:129-142

[78] Meng X, Wang N, He H, Tan Q, Wen B, Zhang R, et al. Prunus persica transcription factor PpNAC56 enhances heat resistance in transgenic tomatoes. Plant Physiology and Biochemistry. 2022;**182**:194-201. DOI: 10.1016/j.plaphy.2022.04.026

[79] Xi Y, Ling Q, Zhou Y, Liu X, Qian Y. ZmNAC074, a maize stress-responsive NAC transcription factor, confers heat stress tolerance in transgenic *Arabidopsis*. Frontiers in Plant Science. 2022;**2022**:13. DOI: 10.3389/fpls.2022.986628

[80] Liu XH, Lyu YS, Yang W, Yang ZT, Lu SJ, Liu JX. A membrane-associated NAC transcription factor OsNTL3 is involved in thermotolerance in rice. Plant Biotechnology Journal. 2020;**18**(5):1317-1329. DOI: 10.1111/pbi.13297

[81] Djemal R, Khoudi H. The barley SHN1-type transcription factor HvSHN1 imparts heat, drought and salt tolerances in transgenic tobacco. Plant Physiology and Biochemistry. 2021;**164**:44-53. DOI: 10.1016/j.plaphy.2021.04.018

[82] El-Esawi MA, Al-Ghamdi AA, Ali HM, Ahmad M. Overexpression of AtWRKY30 transcription factor enhances heat and drought stress tolerance in wheat (*Triticum aestivum* L.). Genes. 2019;**10**(2):163. DOI: 10.3390/genes10020163

[83] Dang FF, Wang YN, Yu L, Eulgem T, Lai Y, Liu ZQ, et al. CaWRKY40, a WRKY protein of pepper, plays an important role in the regulation of tolerance to heat stress and resistance to Ralstonia solanacearum infection. Plant, Cell & Environment. 2012;**36**(4):757-774. DOI: 10.1111/pce.12011

[84] Chen H, Wang Y, Liu J, Zhao T, Yang C, Ding Q, et al. Identification of WRKY transcription factors responding to abiotic stresses in *Brassica napus* L. Planta. 2022;**255**:1-17. DOI: 10.1007/s00425-021-03733-x

[85] Agarwal P, Khurana P. Characterization of a novel zinc finger transcription factor (TaZnF) from wheat conferring heat stress tolerance in Arabidopsis. Cell Stress and Chaperones. 2018;**23**:253-267. DOI: 10.1007/s12192-017-0838-1

[86] Çelik Ö, Ayan A, Meriç S, Atak Ç. Heavy metal stress-responsive Phyto-miRNAs. In: Faisal M, Saquip Q, Alatar A, Khedhairy A, editors. Cellular and Molecular Phytotoxicity of Heavy Metals. 1st ed. Cham: Springer; 2020. pp. 137-155. DOI: 10.1007/978-3-030-45975-8_9

[87] Celik Ö, Meriç S, Ayan A, Atak Ç. Bioticstress-tolerant plants through small RNA technology. In: Guleria P, Kumar V, editors. Plant Small RNA. 1st ed. Academic Press; 2020. pp. 435-468. DOI: 10.1016/B978-0-12-817112-7.00020-1

[88] Stief A, Altmann S, Hoffmann K, Pant BD, Scheible WR, Bäurle I. Arabidopsis miR156 regulates tolerance to recurring environmental stress through SPL transcription factors.

The Plant Cell. 2014;**26**(4):1792-1807. DOI: 10.1105/tpc.114.123851

[89] Lin JS, Kuo CC, Yang IC, Tsai WA, Shen YH, Lin CC, et al. MicroRNA160 modulates plant development and heat shock protein gene expression to mediate heat tolerance in Arabidopsis. Frontiers in Plant Science. 2018;**9**:68. DOI: 10.3389/fpls.2018.00068

[90] Tsai WA, Sung PH, Kuo YW, Chen MC, Jeng ST, Lin JS. Involvement of microRNA164 in responses to heat stress in Arabidopsis. Plant Science. 2023;**329**(1):111598. DOI: 10.1016/j.plantsci.2023.111598

[91] Zhang L, Fan D, Li H, Chen Q, Zhang Z, Liu M, et al. Characterization and identification of grapevine heat stress-responsive microRNAs revealed the positive regulated function of vvi-miR167 in thermostability. Plant Science. 2023;**329**:111623. DOI: 10.1016/j.plantsci.2023.111623

[92] Asim A, Gökçe ZNÖ, Bakhsh A, Çaylı İT, Aksoy E, et al. Individual and combined effect of drought and heat stresses in contrasting potatocultivars overexpressing miR172b-3p. Turkish Journal of Agriculture and Forestry. 2021;**45**(5):651-668. DOI: 10.3906/tar-2103-60

[93] Shi X, Jiang F, Wen J, Wu Z. Overexpression of Solanum habrochaites microRNA319d (sha-miR319d) confers chilling and heat stress tolerance in tomato (*S. lycopersicum*). BMC Plant Biology. 2019;**19**:1-17. DOI: 10.1186/s12870-019-1823-x

[94] Guan Q, Lu X, Zeng H, Zhang Y, Zhu J. Heat stress induction of miR398 triggers a regulatory loop that is critical for thermotolerance in Arabidopsis. The Plant Journal. 2013;**74**(5):840-851. DOI: 10.1111/tpj.12169

[95] Lin Y, Zhang L, Zhao Y, Wang Z, Liu H, Zhang L, et al. Comparative analysis and functional identification of temperature-sensitive miRNA in Arabidopsis anthers. Biochemical and Biophysical Research Communications. 2020;**532**(1):1-10. DOI: 10.1016/j.bbrc.2020.05.033

[96] Li Y, Li X, Yang J, He Y. Natural antisense transcripts of MIR398 genes suppress microR398 processing and attenuate plant thermotolerance. Nature Communications. 2020;**11**(1):5351. DOI: 10.1038/s41467-020-19186-x

[97] Yalçin M, Öztürk Gökçe ZN. Investigation of the effects of overexpression of Novel_105 miRNA in contrasting potatocultivars during separate and combined drought and heat stresses. Turkish Journal of Botany. 2021;**45**(5):397-411. DOI: 10.3906/bot-2103-39

[98] Niu Y, Qian D, Liu B, Ma J, Wan D, Wang X, et al. ALA6, a P4-type ATPase, is involved in heat stress responses in arabidopsis thaliana. Frontiers in Plant Science. 2017;**8**:1732. DOI: 10.3389/fpls.2017.01732

[99] Ali M, Muhammad I, et al. The CaChiVI2 gene of *Capsicum annuum* L. confers resistance against heat stress and infection of Phytophthora capsici. Frontiers in Plant Science. 2020;**11**:219. DOI: 10.3389/fpls.2020.00219

[100] Panzade KP, Vishwakarma H, Padaria JC. Heat stress inducible cytoplasmic isoform of ClpB 1 from Z. nummularia exhibits enhanced thermotolerance in transgenic tobacco. Molecular Biology Reports. 2020;**47**:3821-3831. DOI: 10.1007/s11033-020-05472-w

[101] Zhao Y, Du H, Wang Y, Wang H, Yang S, Li C, et al. The

calcium-dependent protein kinase ZmCDPK7 functions in heat-stress tolerance in maize. Journal of Integrative Plant Biology. 2021;**63**(3):510-527. DOI: 10.1111/jipb.13056

[102] Qi Y, Liu Y, Zhang Z, Gao J, Guan Z, Fang W, et al. The over-expression of a chrysanthemum gene encoding an RNA polymerase II CTD phosphatase-like 1 enzyme enhances tolerance to heat stress. Horticultural Research. 2018;**5**(1):37-47. DOI: 10.1038/s41438-018-0037-y

[103] Lee K, Rahman MA, Kim KY, Choi GJ, Cha JY, Cheong MS, et al. Overexpression of the alfalfa DnaJ-like protein (MsDJLP) gene enhancestolerance to chilling and heat stresses in transgenic tobacco plants. Turkish Journal of Biology. 2018;**42**(1):12-22. DOI: 10.3906/biy-1705-30

[104] Dai X, Wang Y, Chen Y, Li H, Xu S, Yang T, et al. Overexpression of NtDOG1L-T improves heat stress tolerance by modulation of antioxidant capability and defense-, heat-, and ABA-related gene expression in tobacco. Frontiers in Plant Science. 2020;**11**:568489. DOI: 10.3389/fpls.2020.568489

[105] Jan R, Kim N, Lee SH, Khan MA, Asaf S, et al. Enhanced flavonoid accumulation reduces combined salt and heat stress through regulation of transcriptional and hormonal mechanisms. Frontiers in Plant Science. 2021;**12**:2999. DOI: 10.3389/fpls.2021.796956

[106] Li Q, Wang W, Wang W, Zhang G, Liu Y, Wang Y, et al. Wheat F-box protein gene TaFBA1 is involved in plant tolerance to heat stress. Frontiers in Plant Science. 2018;**9**(521):15. DOI: 10.3389/fpls.2018.00521

[107] Zang X, Geng X, Wang F, Liu Z, Zhang L, Zhao Y, et al. Overexpression of wheat ferritin gene TaFER-5B enhances tolerance to heat stress and other abiotic stresses associated with the ROS scavenging. BMC Plant Biology. 2017;**17**(1):1-13. DOI: 10.1186/s12870-016-0958-2

[108] Pan R, Ren W, Zhang H, Deng X, Wang B. Ectopic over-expression of HaFT-1, a 14-3-3 Protein from Haloxylon ammodendron, leads to enhanced acquired thermotolerance of transgenic Arabidopsis. Plant Molecular Biology. 2022;**29**(1):17. DOI: 10.21203/rs.3.rs-1857782/v1

[109] Gangadhar BH, Mishra RK, Kappachery S, Baskar V, Venkatesh J, et al. Enhanced thermo-tolerance in transgenic potato (*Solanum tuberosum* L.) overexpressing hydrogen peroxide-producing germin-like protein (GLP). Genomics. 2021;**113**(5):3224-3234. DOI: 10.1016/j.ygeno.2021.07.013

[110] Kim SE, Lee CJ, Par SU, Lim YH, et al. Overexpression of the golden SNP-carrying Orange gene enhances carotenoid accumulation and heat stress tolerance in sweetpotato plants. Antioxidants. 2021;**10**(1):51. DOI: 10.3390/antiox10010051

[111] Kim JH, Lim SD, Jang CS. Oryza sativa heat-induced RING finger protein 1 (OsHIRP1) positively regulates plant response to heat stress. Plant Molecular Biology. 2019;**99**(6):545-559. DOI: 10.1007/s11103-019-00835-9

[112] Samtani H, Sharma A, Khurana P. Overexpression of HVA1 enhances drought and heat stress tolerance in Triticum aestivum doubled haploid plants. Cell. 2022;**11**(5):912. DOI: 10.3390/cells11050912

[113] Zang X, Geng X, Liu K, Wang F, Liu Z, Zhang L, et al. Ectopic expression

of TaOEP16-2-5B, a wheat plastid outer envelope protein gene, enhances heat and drought stress tolerance in transgenic Arabidopsis. Plant Science. 2017;**258**:1-11. DOI: 10.1016/j.plantsci.2017.01.011

[114] Song C, Kim T, Chung WS, Lim CO. The Arabidopsis phytocystatin AtCYS5 enhances seed germination and seedling growth under heat stress conditions. Molecules and Cells. 2017;**40**(8):577. DOI: 10.14348/molcells.2017.0075

[115] Ling Q, Liao J, Liu X, Zhou Y, Qian Y. Genome-wide identification of maize protein arginine Methyltransferase genes and functional analysis of ZmPRMT1 reveal essential roles in Arabidopsis flowering regulation and abiotic stress tolerance. International Journal of Molecular Sciences. 2022;**23**(21):12793. DOI: 10.3390/ijms232112793

[116] Chaudhary C, Sharma N, Khurana P. Decoding the wheat awn transcriptome and overexpressing TaRca1β in rice for heat stress tolerance. Plant Molecular Biology. 2021;**105**:133-146. DOI: 10.1007/s11103-020-01073-0

[117] Qu Y et al. Overexpression of both Rubisco and Rubisco activase rescues rice photosynthesis and biomass under heat stress. Plant, Cell & Environment. 2021;**44**(7):2308-2320. DOI: 10.1111/pce.14051

[118] Li Z, Cheng B, Zhao Y, Luo L, Zhang Y, Feng G, et al. Metabolic regulation and lipidomic remodeling in relation to spermidine-induced stress tolerance to high temperature in plants. International Journal of Molecular Sciences. 2021;**23**(20):12247. DOI: 10.3390/ijms232012247

[119] Zhang S, Chen C, Dai S, Yang M, Meng Q, Lv W, et al. A tomato putative metalloprotease SlEGY2 plays a positive role in Thermotolerance. Agriculture. 2022;**12**(7):940. DOI: 10.3390/agriculture12070940

[120] Wang X, Chen J, Liu C, Luo J, Yan X, Ai A, et al. Over-expression of a protein disulfide isomerase gene from *Methanothermobacter thermautotrophicus*, enhances heat stress tolerance in rice. Gene. 2022;**684**:124-130. DOI: 10.1016/j.gene.2018.10.064

[121] Yarra R, Xue Y. Ectopic expression of nucleolar DEAD-Box RNA helicase OsTOGR1 confers improved heat stress tolerance in transgenic Chinese cabbage. Plant Cell Reports. 2020;**39**:1803-1814. DOI: 10.1007/s00299-020-02608-x

Chapter 8

Molecular Basis of Plant Adaptation against Aridity

Kinjal Mondal, Shani Raj, Kalpna Thakur, Anjali Verma, Neerja Kharwal, Animesh Chowdhury, Supratim Sadhu, Mala Ram, Pooja Bishnoi, Sukanya Dutta, Ayush G Jain and Saroj Choudhary

Abstract

Environment fluctuations have become the greatest threat to global food security. Of various abiotic stress factors, aridity hampers the most yield contributing attributes. In the context of agriculture, term "aridity" refers to a protracted period of insufficient precipitation, having detrimental influence on crop development and overall biological output. A sustained drought has considerable negative effects on crops and livestock, including the reduced production, destruction of property, and livestock sell-offs. Consequently, plants themself exert various kinds of defensive mechanisms to combat the ill effects of climate change. For example, plants with small leaves, benefit from aridity as part of their strategy for modifying the soil to water shortages and nutrient restrictions. Furthermore, low genetic diversity among significant crop species, together with ecological productivity limits, must be addressed in order to adapt crops to episodic drought spells in the coming days. A deeper understanding of the molecular and genetic underpinnings of the most important intrinsic adaptation responses to drought stress seems to be beneficial for gene engineering as well as gene-based expression investigations in plant systems under hostile environment. Recently, molecular markers and "omics" have opened a huge opportunity to identify and develop specific gene constructs governing plant adaptation to environmental stress.

Keywords: environment, aridity, drought, molecular markers, omics

1. Introduction

The ability to produce enough food for an endlessly expanding population is a key issue for mankind in the twenty-first century [1]. Currently, the scenario has been more difficult by the loss of arable farmland brought on by human habitation, the deterioration of the soil, and a range of environmental conditions, such as flooding, drought, salinity, temperature, and heavy metal pollution [2]. Eventually, accumulation of osmolytes at cellular level, modification of water flow, and scavenging of reactive oxygen species are some of the most frequent and well-documented adaptations that plants, which are sessile, have developed to recognize and respond to stress situations [3].

Environmental stress known as abiotic stress restricts plant growth and metabolism. Abiotic stressors are thought to diminish major food and cash crop yields and output by more than 50% [4]. Abiotic pressures may be divided into two categories: above- and below-ground abiotic stresses. Abiotic strains that are atmospherically produced come from the atmosphere, whereas abiotic stresses that are edaphic come from the soil [5]. In regions where climatic variability and precipitation patterns alter with extended periods of drought interspersed with spells of copious rainfall, abiotic stressors of atmospheric origin are prevalent [6]. On the other hand, anthropogenic activities such as the use of brackish water and sewage water for irrigation, sewage sludge for fertilisation, and inorganic chemicals for fumigation may result in abiotic pressures of edaphic origin. This issue is frequently made worse by inadequate waste management procedures, the weathering of local rocks, and subpar cultural practises that have rendered vast tracts of land unsuitable for agricultural development [7].

Aridity has a significant impact on community structure as well as ecological exploits, including primary productivity and nutrient cycling, by acting as a powerful environmental filter for plant survival, growth, and development [8]. For instance, plants with tiny leaves benefit from aridity as part of their strategy for adjusting the soil to water shortfalls and nutrient constraints. Plants adapt to their environment and develop an ideal phenotype, producing a set of adaptation tactics at both the collective and individual levels.

2. Relevance of plant-water kinship in agriculture

Water is the most abundant material in any living entities across the globe. The weight of water contained in a plant is usually four to five times the total weight of dry matter [9]. Inside a plant body, about 80–90% of cell mass is comprised of water. Plants absorb water from soil through their roots and other parts in the way of vascular system. Xylem tissues of plant vascular system play a crucial role in the movement of water containing essential elements from roots to the shoot. Water supply through cells by diffusion alone is not enough to maintain the hydration of a perspiring canopy plant. The necessity for a vascular system becomes more apparent while studying the hydraulic dynamic of a tree on a hot day, which requires a massive flow of water. Water transport through xylem is over a million times more efficient than water transport through plasmodesmata of parenchyma. Several theories have been proposed to explain the mechanism of movement of water into xylem against the concentration gradient. The cohesion-tension theory, proposed by Boehm, Dixon and Joly (1894) in the late 19th century is thought to be the most appropriate tenet to explain the mechanism of upward movement of water. According to this theory, the water evaporated from leaf surface establishes a tensile strength in the xylem, where the hydrogen bonds provide a continuous intermolecular attraction (cohesion) between the water molecules from the leaf to the root. Thus, the water column in the xylem lumen is driven out of a region with a higher water potential, *i.e.,* from the root and the stem, to a region with a lower water potential, as the leaves, and finally toward the air that can reach very low water potential. Once water reaches the xylem, it enters conducting elements of either conifer tracheid or angiosperm vessels, and flows upwards through the stem to the leaves. The conduit diameter of xylem gets smaller and tapered with plant height, indicating the widening aspect of xylem anatomy from apex to the base of plant. Plants that have an increased number of xylem conduits per cross-sectional area can maintain hydraulic conductance by reducing effects of path length [10].

Droughts can be classified as meteorological, agricultural, hydrological, or socioeconomic, according to the American Meteorological Society (1997). Precipitation deficits can be used to categories meteorological droughts, and these crises can develop in other categories of droughts. Agricultural drought focusing on precipitation shortages, discrepancies between actual and potential evapotranspiration, inadequate soil water, and lower reservoir levels. A lack of water in the hydrological system is referred to as a "hydrological drought," which is characterised by reduced river flow as well as declining dam, lake, and subsurface water levels on a basin-scale. Economic, social, and environmental harm brought on by many sorts of droughts refers to what is meant by socioeconomic droughts [11]. There is limited clarity over the metrics that better reflect the effects of drought on the environment and society.

Agriculture is a key activity of human being since it provides basic needs and water is a critical input for agriculture production. Several factors pose significant risk to farms leading to yield reduction like limited water condition. A limited water availability leading to drought, increased diseases and pest incidence and extreme weather events at local to regional scale. Limited water availability accounts for about 30–70% loss of productivity. It also results in abnormal metabolism that may reduce plant growth or cause the death of plant. Water stress is one of the most detrimental factors seriously affecting the growth and production of many plants mostly during the flowering phases. Under the exposure of severe water crisis, significant diminution in the major growth attributing characters including number of leaves, leaf area, stem length is very often in various plants. Furthermore, the crop yield and productivity are also found to be affected severely under water stress. The damaging effects on plants are associated with oxidative damage in the plant cells are commonly realised by elevated lipid peroxidation, reactive oxygen species (ROS) accumulation, and electrolyte leakage. Under usual conditions, ROS exist in plant organelles, mainly mitochondria, chloroplasts, and peroxisomes, while under stressful conditions such as drought, ROS levels increase resulting in lipid peroxidation and proteins degradation [12]. Also, biological yield and physiological characters such as stem length, number of leaves, leaf area, relative water content, and chlorophyll concentration as well as overall biological yield are decreased under stress condition in many plants [13]. Drought during blossoming is frequently associated with infertility [14], owing to a reduction in assimilating flow to the developing ear. Drought stress can significantly reduce production in important field crops by prolonging the anthesis period and delaying grain filling [15]. Numerous factors could explain the decline in yield, including decreased photosynthesis, inefficient flag leaf formation, uneven assimilate portioning, and a depleted pool of critical biosynthesis enzymes such as starch synthase, sucrose synthase, starch enzymes, and α-amylase.

3. Impact of water stress: physiological and biochemical alterations in plants

Agriculture output is gradually been threatened every year due to drought stress. Drought proves an obtrusive climatic factor for agriculture, livestock and climate. Climate change led to increased temperature and varied environmental conditions globally. So, we need plant varieties that are adapted to these environmental conditions specially drought stress. Water is crucial for plant survival and responsible for various biological, physical and biochemical activities of plant system. Influence of drought desperately hampers plant functioning and limits plant growth at various development stages. Water deficit conditions alter many metabolic activities in plants like reduced photosynthetic rate, increased reactive oxygen species (ROS) accumulation, and production of plant secondary metabolites etc.

3.1 Influence of water tension on plant adaptation at physiological level

The ever-changing nature of mercurial environment has forced the higher group of plants to develop a variety of intrinsic tactics at morphological, physiological (**Figure 1**), biochemical, and molecular levels for survival especially at limited water conditions. On the other hand, some plant species avoid water shortage circumstances by finishing their life cycle, for instance, before or after a drought period, while others showed adaptations to increase water absorption and minimise water loss to prevent its negative effects [15]. For example, *Phedimus aizoon* L., which was observed to respond the severity of drought stress by accelerated root system, thickened the waxy layer of leaf surface and closure of stomata for making sure of maximum water retention [16]. Under extreme arid conditions, the xerophyte *Zygophyllum xanthoxylum* is surprisingly found to accumulate ample amounts of Na^+ ions, coming from the soils they thrive on. The primary role of accumulated Na^+ in *Z. xanthoxylum* has been attributed to their ability to drastically reduce the osmotic potential of leaves, which enhances their ability to absorb water during drought spells [17]. *Reaumuria soongorica* shows specific characteristics during the process of adapting to desertification, such as an incredibly thick cuticle, hollow stomata, specialised leaf shape, deep root system, and efficient physiological mechanisms like a decreased transpiration rate, increased water use efficiency, and maintaining stem vigour to survive desiccation by leaf abscission [18].

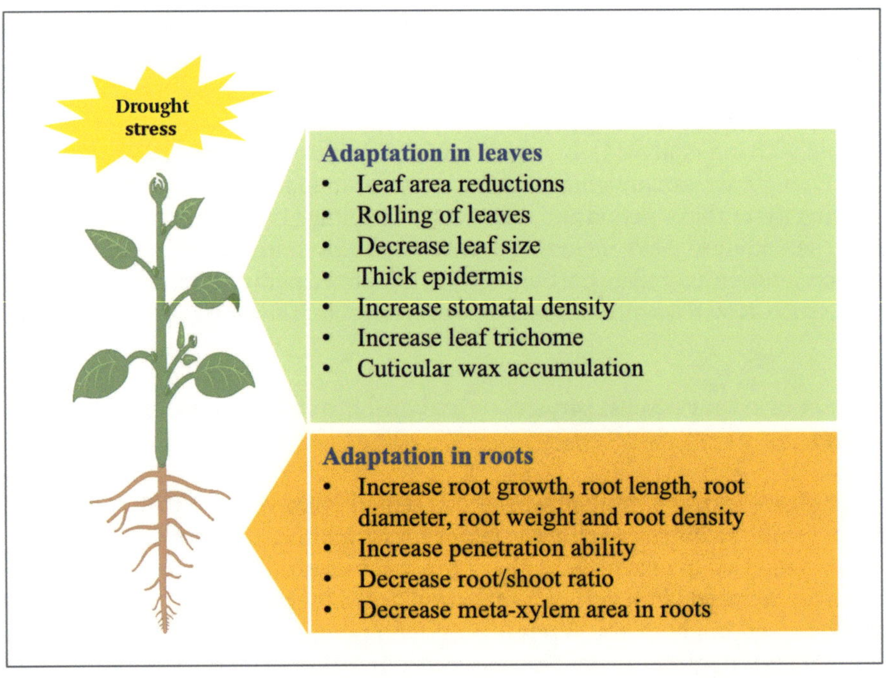

Figure 1.
Plausible alterations in plant physiognomy under drought stress.

3.1.1 Root modification

The soil provides nutrients and water to roots. As a result, the morphological and physiological traits of roots greatly influence the growth of shoots and overall production [19]. Plants attempt to extract water from deeper soil layers when there is a water shortage by strengthening their root architecture. In addition, roots are the primary organ that detects the presence of water, and control key aspects of plant growth and development [20]. In comparison with plants with shallow roots, those with deep root systems and perennial growth patterns demonstrated greater drought tolerance [21]. In addition to increasing the amount of soil that may be investigated for water and the surface area of roots in contact with moisture, roots with small diameters and long specific root lengths also boost hydraulic conductivity by lowering the apoplastic barrier to water entering the xylem. Additionally, decrease in root diameter also attribute the enhancement of water access and increases the productivity of plants under water stress. An examination of the root system of marigolds revealed a sharp decrease in the meta-xylem area (*Tagetes erecta* L.). Reducing the diameter of the meta-xylem vessels reduced embolism risk and improved water flow. Increased meta-xylem area is related to the flow of minerals and water and necessary for the growth of cortical parenchyma [22]. According to reports, the mechanism for drought tolerance in winter wheat, is supported by development of a deep root system, whereas a well-branched (albeit shallow) root system is found in spring wheat [23].

There are three alternative strategies to confer drought resistivity, *viz*., drought escape, drought avoidance, and drought tolerance. Each of these tactics could develop into a constitutive reaction that happens independently of environmental cues such as water deficit. Drought tolerance and drought avoidance are the major strategies of plants against water deficit stress. The ability of a plant to withstand a dry environment through a variety of physiological processes, such as osmotic adjustment using osmoprotectants, is known as drought tolerance [24]. The continuation of physiological functions including stomata regulation, and root system development even at the period of prolonged dry spell is known as drought avoidance. The ability to adjust short life cycle to avoid drought stress is known as drought resistance [25]. The root system plays a crucial part in the plant's response to drought stress and may be the first organ to detect it. Shorter roots are less suited to drought tolerance than longer roots. Drought stress results in a significant reduction in the number of roots, as shown by *Helichrysum petiolare* [26]. Drought tolerant adaptive characters of plant roots including long roots, high density, and intense root system. Long roots with a high density are necessary for plants to retain performance when water is scarce, especially when the water is deeper. Factually, more roots may come into contact with more water vapours in the soil, and a denser root system absorbs comparatively more water than thinner ones [27].

3.1.2 Leaf modification

The majority of photosynthetic products are produced primarily in the leaf, which is the main portion of the plant. When *Andrographis paniculate* was subjected to water stress, precocious leaf fall was found [28]. Reduced leaf area due to water stress results in less photosynthesis, which lowers crop output. In order to achieve stability between the water received by roots and the water status in different plant parts, leaf area was found to be decreased in *Petroselinum crispum* and *Stevia rabaudiana* at limited water conditions [29]. Reducing leaf area is a method for avoiding drought because it reduces the amount of water lost by transpiration. This reduction in leaf area is due to the suppression of leaf growth caused by a decline in cell division, which causes a loss in cell turgidity [30]. Reduced leaf area is probably a fundamental element of the drought resistance strategy used by eucalypts, and it might be more beneficial to survival than any physiological changes that have been observed [31].

The decline in leaf water potential is typically followed by the rolling of the leaves. Reduced leaf rolling, which occurs in plants with high osmotic adjustment, is thought to indicate that the plant is avoiding desiccation to a larger extent through a deep root system [32].

In addition, thick epidermis with large epidermal cells in plants also comes under the potential strategy of plant drought tolerance. Epidermal tissue thickness offers higher resistance of plants to water loss from root surface under arid climate [33].

With the application of the drought hardening treatment, the stomatal density of potato seedling leaves dramatically increased while the leaf area, stomatal size, and stomatal aperture decreased. These changes led to reduced leaf transpiration rate and improved water utilisation efficiency (WUE). The drought resistance of the potato seedlings that had undergone drought hardening was also enhanced by the alterations in leaf microstructure [34].

An intensive study on leaf trichomes in *Caragana korshinskii* has revealed that leaf trichomes are important structures on epidermis which uptake the dew from outer environment that assist in sustaining the leaf hydraulic assimilation system and mitigate the adverse effects of drought stress [35]. The outermost layer of defence against abiotic stress on plants is called cuticular wax. It was found that compared to healthy plants, sunflower genotypes exposed to drought stress had increased wax loads [36].

3.2 Influence of water tension on plant adaptation at biochemical level

3.2.1 Photosynthesis

A severe drought results in decrease or suppression of photosynthesis. Increased stomatal closure, reduced leaf area, and consequent reduced leaf cooling by evapotranspiration leading to damages to the photosynthetic apparatus contribute as the major obstacles for photosynthesis [37]. Decline in CO_2 conductance via reduced stomatal activity enhances diffusive resistance and other vital metabolic processes [38]. Loss of CO_2 uptake, affect Rubisco activity and decrease the function of nitrate reductase and sucrose phosphate synthase and the ability for ribulose bisphosphate (RuBP) production [39]. The closing of stomata, restriction of gas exchange, degraded photosynthetic apparatus, primarily PSI and PSII, and increased metabolite fluxes are all factors that also contribute to reduced photosynthesis [40]. Drought induced water loss affects the activity of photosynthesis-related enzymes, causing the photosynthetic device to malfunction and resulting in the poor execution of metabolic processes [14]. Reduction in photosynthesis attributed to increased metabolite fluxes result in the production reactive oxygen species, which impede cell growth by causing oxidative stress [41]. Extreme water limitations substantially hinder the rate of CO_2 uptake and the photosynthetic system in cedar seedlings (*Cedrus atlantica* and *Cedrus libani*). The chlorophyll content, net photosynthesis, potential yield of the photochemical reaction of PSII and stomatal conductance of *Atractylodes lancea* shown persistent negative trends as the length of drought stress treatment increased [42].

3.2.2 Mineral nutrition

Water deficit situations usually lessen the ion content in various plant tissues by reducing the overall soil nutrient accessibility and root nutrient translocation [43]. Water stress conditions decreased plant potassium (K) uptake [44]. Reduced K mobility, declined transpiration rate and weakened action of root membrane transporters [44, 45]. Decrease in K level in leaves due to disrupted stomatal dynamism as well as irregular guard cell turgidity, also restricts the rate of photosynthesis and, backpedal the plant biomass production [46]. K transporters were inhibited by water stress conditions [47] and inner K channels were stimulated by a protein kinase, CIPK23, which in turn cooperates with calcium sensors (calcineurin B). This K channel was inhibited in roots but activated in leaves of grapevine [48]. K level decreased in *Ocimum basilicum* and *Ocimum americanum* plants subjected to limited water availability [49].

Leaf nitrogen (N) content did not change under drought-stress in *Mentha piperita*, *Salvia lavandulifolia*, *Salvia sclarea* and *Thymus capitatus*, whereas, in *Lavandula latifolia* and *Thymus mastichina* plants, reduced N level were observed. While leaf phosphorus (P) level reduced in all species except *S. sclarea* whose concentration remained the same [50]. Reduced N level and decline in K level in *Thymus daenensis* was considered as the main responsible factor for photosynthesis decline and leaf senescence under water deficit conditions [51]. Water deficit conditions increased the accumulation of manganese (Mn), molybdenum (Mo), P, K, copper (Cu), calcium (Ca) and zinc (Zn) in soybean [52].

3.2.3 Antioxidant defence system

Plants defensive system prevents the unwanted exposure of extraneous physical and biological agents which harm the plant body. In this context, a prompt, powerful and efficient antioxidant system is of pivotal importance to provide drought tolerance [53]. This system involves enzymatic and non-enzymatic detoxification moieties, which lessen and repair injury triggered by ROS. Antioxidant defence system helps in ROS scavenging that decreases electrolyte leakage and lipid peroxidation, therefore maintaining the vitality and integrity of organelles and cell membrane [54].

It is well established that drought induces oxidative stress by generating ROS, for instance $O_2^{\bullet}$, hydroxyl radicals ($OH^{\bullet}$), singlet oxygen ($^1O_2$) and H_2O_2 [55]. Numerous studies conducted under water stress conditions found enhanced activities of pivotal antioxidant enzymes, namely CAT, SOD, POD and APX [56]. Usually, an enhanced antioxidant enzymes activity is observed in stress tolerant genotypes as compared to non-tolerant plants.

Antioxidant enzymes like superoxide dismutase (SOD), peroxidase (POD) and catalase (CAT) significantly involve in the production of antioxidants such as $O_2{\bullet}$- and H_2O_2 [57]. Ascorbate peroxidase (APX) also participates as ROS scavenger. APX mainly occurs in the chloroplast and cytoplasm and is a crucial enzyme for scavenging H_2O_2 in chloroplasts which convert H_2O_2 to H_2O), and its activity is usually elevated under stress conditions. APX mainly occurs in the chloroplast and cytoplasm and is a crucial enzyme for scavenging H_2O_2 in chloroplasts [58].

Enzymatic activities of SOD, CAT and POD were stimulated by limited water availability in *Vicia faba* [59]. The amount of enzymatic and non-enzymatic antioxidants was improved in drought tolerant plants under mild and moderate water deficit conditions. CAT, SOD, POD and APX activities indicating that improved functioning of these enzymes helps to lower the level of ROS and mitigate the drought generated oxidative stress [60]. Water deficit boosted the levels of SOD and POD levels of these enzymes which stimulate tolerance against drought stress and are vital to reduce its adverse effects [61].

3.2.4 Secondary metabolites

Plants produce some chemical compounds in response to various environmental stresses, called secondary metabolites [62]. Biosynthesis of secondary metabolites (SMs) is regulated by environmental factors, such as temperature, light regime and nutrient availability. In this context, the drought stress signals induce systemic SM biosynthesis such as terpenes, alkaloids, and phenolic complexes to protect the plant system from oxidative stress [63]. On the order hand, high temperatures can also induce changes in SM biosynthesis. For example, heat stress has shown that isoprene levels increase; this biosynthesis is energetically costly for the plant, but these SM protect the cell membrane against oxidative stress, showing physiological benefits that far outweigh their energetic cost. Improved production of secondary metabolites is usually observed under water deficit conditions, which is caused by reduction in biomass formation and destination of assimilated CO_2 to C-based secondary metabolites to avoid sugar-promoted feedback of photosynthesis.

4. Molecular symphony of plant adaptation: innate shield to aridity

The intrinsic ability of plant system to respond against drought stress involves a complex cascade of highly regulated genes and signal transduction pathways. Under drought prone conditions, competent stimuli are perceived and captured by uncharacterized membrane sensors, and the signals are then passed down through multiple signal transduction pathways, resulting in the expression of drought-responsive genes and drought adaptation. Secondary messengers (such as Ca^{2+}, ROS, phosphoglycerol, ABA, and diacylglycerol) and transcriptional regulators all play important roles in signalling pathways.

Drought stress increase ABA accumulation in plants, and exogenous ABA application, such as gene induction, can have similar effects to osmotic stress. According to Mittler and Blumwald (2015), drought causes the production of ABA in roots, which is then transferred to the shoots and causes stomatal closure, ultimately limiting development [64]. Additionally, ABA is produced in leaf cells and distributed throughout the plant. According to recent data, xylem/apoplastic pH affects ABA compartmentation, which in turn affects the quantity of ABA that reaches stomata. As a result of less ABA being removed from the xylem and leaf apoplast to the symplast in drought-stressed plants (a process known as alkaline trapping of ABA), more ABA reaches the guard cells, allowing for the modulation of stomatal aperture in response to various environmental factors.

Transcription factors are early genes that are activated within minutes of being stressed. Some of the gene families including RD29A contains both ABRE and DRE/CRT elements [65]. The RD29A gene has served as a model for both ABA-dependent and ABA-independent gene regulation. Although ABA does not activate the DRE element, it is required for the DRE to be fully activated by osmotic stress.

Both cis-acting and trans-acting regulatory elements involved in drought-induced ABA-independent/ ABA-responsive gene expression have been thoroughly

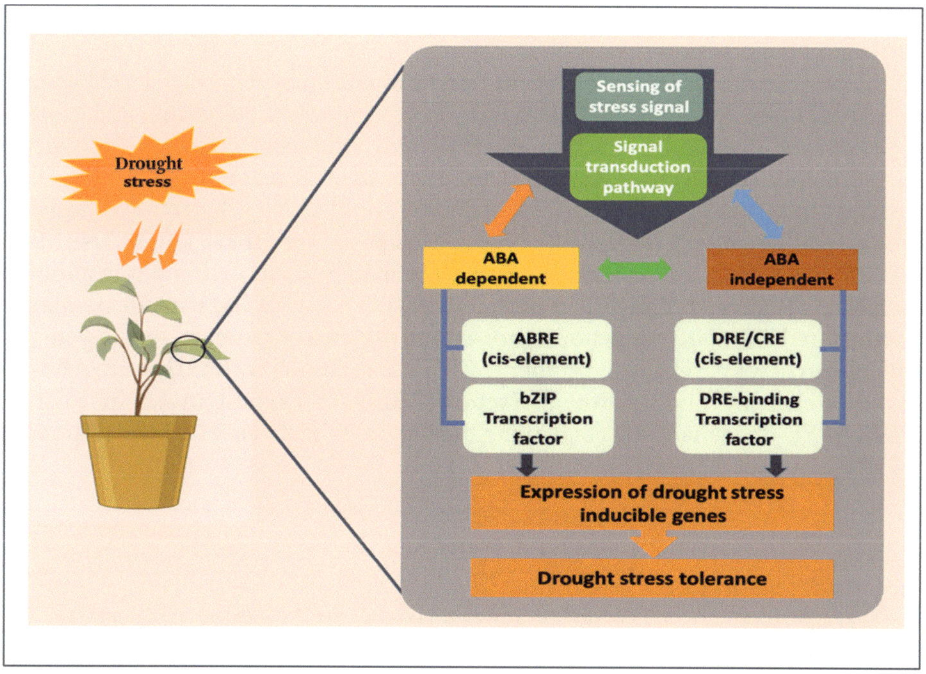

Figure 2.
Two intrinsic immune strategies in plant system for survival against aridity. ABA: Abscisic acid; ABRE: ABA-responsive element; DRE/CRT: Dehydration-responsive element/C-repeat; bZIP: Basic-leucine zipper.

investigated at the molecular level [66]. Several drought-inducible genes, on the other hand, do not respond to ABA treatment, implying the existence of an ABA-independent pathway during the dehydration stress response. Exogenous ABA treatment increases the expression of many osmotic stress inducible genes. ABRE is a key cis-acting element in ABA responsive gene expression. Two ABRE motifs are critical cis-acting elements that regulate RD29B-mediated ABA responsive expression [67]. ABA up-regulates three members of the AREB/ABF subfamily, AREB1, AREB2, and ABF3, and their full activation requires ABA. The triple mutant areb1 areb2 abf3 exhibits increased ABA resistance and decreased drought tolerance, indicating that the three factors co-ordinately govern ABRE-dependent gene expression under water stress conditions. Involvement of ABA in drought responsive system has been well depicted in the **Figure 2**.

Some ROS genes have been used to create drought-tolerant plants. The formation of ROS, also referred to as the "oxidative burst," is a primary defence response of plants to water stress and serves as a secondary messenger to start additional defence responses in plants [68]. Overexpression of a pea manganese superoxide dismutase (*MnSOD*) gene in rice chloroplasts under the control of an oxidative stress-inducible promoter SWPA2 improved transgenic rice drought tolerance. Cytosolic APX1 has been shown to play an important role in the response to a combination of drought and heat stress.

Multiple mechanisms increase ROS generation when there is a drought stress. The Mehler process leaks more electrons to O_2 when photosynthesis is occurring. One of the main risks to the chloroplast during a drought is the Fenton reaction's creation of the hydroxyl radical in the thylakoids. Since it has the strongest oxidising potential and the shortest half-life, the hydroxyl radical is the ROS that reacts with the majority of biological molecules.

LEA proteins are expressed at specific stages of late embryonic development and play critical roles in desiccation tolerance by capturing water, stabilising and protecting protein and membrane structure and function, and acting as molecular chaperons and hydrophilic solutes to protect cells from water stress damage [69].

Many transcription factor families, including APETALA2/Ethylene-responsive element binding protein (AP2/EREBP), basic leucine zipper (bZIP), MYB, NAM-ATAF1/2-CUC2 (NAC), and zinc finger, have been implicated in drought responses (as shown in **Figure 3**). Zinc finger proteins (bZIPs), a big family with 75 members identified in the *Arabidopsis* genome, are among the transcription factors dependent on ABA. Two basic leucine zipper (bZIP) transcription factors that are ABA-responsive element-binding proteins/factors (AREBs/ABFs) best known for their roles in ABRE-dependent ABA signalling during drought stress. The ABA-responsive

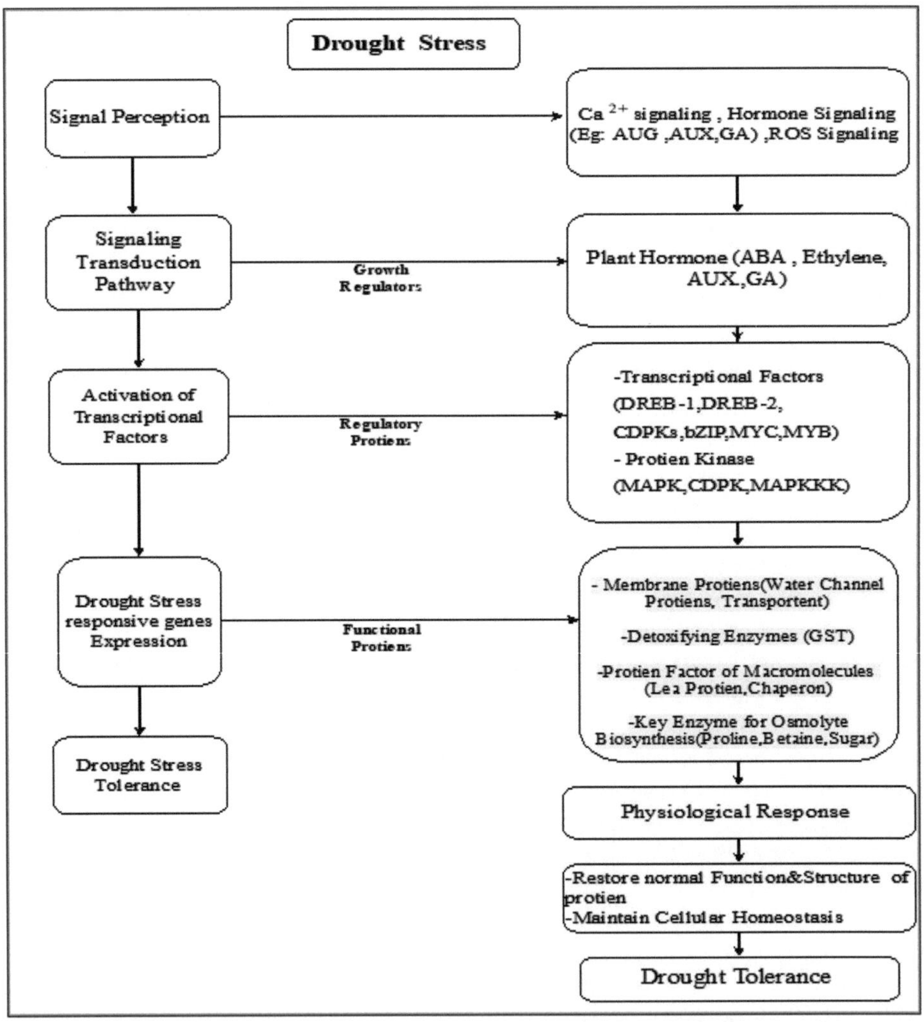

Figure 3.
Schematic diagram showing genetic cross-talk as an important part of drought responsive system in plants.

elements-binding (AREB) proteins react to drought at the transcriptional and post-transcriptional level, enhancing tolerance to drought stress. AREB/ABF, bind to ABRE and activate ABA-dependent gene expression [67]. The AREB/ABF proteins require an ABA-mediated signal to be activated, as evidenced by their decreased activity in *Arabidopsis* ABA deficient aba2 and ABA insensitive eabi1 mutants and increased activity in *Arabidopsis* ABA hypersensitive era1 mutant [66]. Several rice bZIP proteins, including OsbZIP23 and the constitutive active form of OsbZIP46, have also been identified as having a high potential for improving rice drought resistance [70, 71].

In transgenic petunia, constitutive over-expression of a Cys2/His2 (C2H2)-type zinc finger protein encoding the ZPT2-3 gene improved tolerance to dehydration stress. DST, another zinc finer protein, has been shown to act as a negative regulator of drought and salt tolerance in rice by controlling the genes involved in H_2O_2-mediated stomatal movement. *ATGPX3*, a gene encoding an *Arabidopsis thaliana* glutathione peroxidase, was discovered to function as a scavenger and an oxidative signal transducer in ABA and drought stress signalling, as well as a key player in H_2O_2 homeostasis [72]. Several other genes, including *OsSKIPa* and *OsSRO1c*, have been shown to modulate drought resistance in plants by controlling ROS metabolism and regulating ROS homeostasis.

Numerous factors including heat-shock proteins, and other key enzymes involved in protein folding make up the most prevalent functional group of proteins responding to drought. Additionally, in order to generate drought-tolerant crop plants, aquaporin proteins could be used as possible targets. In *Arabidopsis*, constitutive over-expression of the aquaporin gene *GoPIP1* enhanced the rosette/root ratio while lowering drought resistance due to stunted development. It was discovered that the expression of stress-responsive genes, particularly genes of a large set of antioxidant enzymes that directly affect water stress-related traits in rice, was regulated by the plant-specific protein OsGRAS23.

Numerous candidate genes identified through mutant screening or expression profiling studies have been studied further for their roles in drought response. Regulatory proteins have been shown to play critical roles in plant responses to drought stress. Protein phosphorylation and dephosphorylation are common events in plants caused by drought stress. Several kinases have been implicated in drought response, including calcium dependent protein kinases (CDPKs), CBL (calcineurin B-like) interacting protein kinase (CIPK), mitogen-activated protein kinases (MAPKs), and sucrose nonfermenting protein (SNF1)-related kinase 2 (SnRK2). In response to drought stress, the Arabidopsis CDPK gene CPK10 was found to mediate stomatal movement via the ABA and Ca^{2+} signalling pathways [73].

5. Conclusion

The frequency and severity of agricultural aridity are predicted to increase in the near future due to a warming environment. Under intermittent drought situations, it will be crucial to provide sustainable agricultural production so that plants can retain physiological activities at low plant water status and swiftly recover once the stress is eliminated. In this scenario, selection of individuals with better water use efficiency, stronger antioxidant defences, and ability to produce important osmolytes as well as secondary metabolites seems potential approach to minimise yield loss under limited water conditions. Currently, the use of genetically modified agricultural plants to introduce and/or overexpress candidate genes appears to be a promising alternative for accelerating the breeding of improved adaptable and high-yielding crop genotypes. The introduction of genomic technology and gene mapping methods like as genome-wide association studies (GWAS) and precision genome editing with the CRISPR/Cas9 system has aided in the development of alleles that can increase plant yield and performance under a variety of conditions. The sincere research of drought response networks that may be targeted by diverse strategies has currently been possible by molecular studies that combine tissue- or cell-specific promoters with live imaging methods for real-time monitoring of cellular processes. In addition, trans- and multidisciplinary research is urgently required to develop pertinent answers for all the environmental issues affecting agricultural yields and guaranteeing food security. Together, research projects targeted at revealing the physiology of plant responses to water scarcity in model systems and employing innovative discoveries to agriculture are believed to find out some effective avenue to deal with aridity.

Author details

Kinjal Mondal[1*], Shani Raj[2], Kalpna Thakur[3], Anjali Verma[4], Neerja Kharwal[5], Animesh Chowdhury[6], Supratim Sadhu[7], Mala Ram[1], Pooja Bishnoi[1], Sukanya Dutta[8], Ayush G Jain[9] and Saroj Choudhary[1]

1 Department of Molecular Biology and Biotechnology, Rajasthan College of Agriculture, Maharana Pratap University of Agriculture and Technology, Udaipur, Rajasthan, India

2 Department of Botany, Mohanlal Sukhadia University, Udaipur, Rajasthan, India

3 Department of Biotechnology, College of Horticulture and Forestry, Dr. Yashwant Singh Parmar University of Horticulture and Forestry, Mandi, Himachal Pradesh, India

4 Division of Plant Biotechnology, Faculty of Horticulture, Sher-e-Kashmir University of Agricultural Sciences and Technology, Srinagar, India

5 Department of Chemistry and Biochemistry, College of Basic Sciences, CSK Himachal Pradesh Agriculture University, Palampur, Himachal Pradesh, India

6 Department of Agronomy, Palli Siksha Bhavana, Visva Bharati University, Sriniketan, West Bengal, India

7 Department of Genetics and Plant Breeding, Uttar Banga Krishi Viswavidyalaya, Pundibari, Cooch Behar, West Bengal, India

8 Department of Agronomy, College of Agriculture, Bidhan Chandra Krishi Viswavidyalaya, Mohanpur, West Bengal, India

9 Department of Environmental Biotechnology, Gujarat Biotechnology University, Gandhinagar, Gujarat, India

*Address all correspondence to: kinjal.mondal1234@gmail.com

References

[1] Silva JV, Giller KE. Grand challenges for the 21st century: What crop models can and can't (yet) do. The Journal of Agricultural Science. 2020;**158**:794-805

[2] Ojuederie O, Olanrewaju O, Babalola O. Plant growth promoting rhizobacterial mitigation of drought stress in crop plants: Implications for sustainable agriculture. Agronomy. 2019;**9**:712

[3] Fadiji AE, Babalola OO, Santoyo G, Perazzolli M. The potential role of microbial biostimulants in the amelioration of climate change-associated abiotic stresses on crops. Frontiers in Microbiology. 2022;**12**:829099

[4] Cai Z, Wang X, Bhadra S, Gao Q. Distinct factors drive the assembly of quinoa-associated microbiomes along elevation. Plant and Soil. 2020;**448**:55-69

[5] Reynaert S, De Boeck HJ, Verbruggen E, Verlinden M, Flowers N, Nijs I. Risk of short-term biodiversity loss under more persistent precipitation regimes. Global Change Biology. 2021;**27**:1614-1626

[6] Abdul Rahman NSN, Hamid NWA, Nadarajah K. Effects of abiotic stress on soil microbiome. International Journal of Molecular Sciences. 2021;**22**:9036

[7] Igiehon NO, Babalola OO, Cheseto X, Torto B. Effects of rhizobia and arbuscular mycorrhizal fungi on yield, size distribution and fatty acid of soybean seeds grown under drought stress. Microbiological Research. 2021;**242**:126640

[8] Maracahipes L, Carlucci MB, Lenza E, Marimon BS, Marimon BH Jr, Guimaraes FA, et al. How-to live-in contrasting habitats? Acquisitive and conservative strategies emerge at inter- and intraspecific levels in savanna and forest woody plants. Perspectives in Plant Ecology, Evolution and Systematics. 2018;**34**:17-25

[9] Kumar R, Singh VP, Jhajharia D, Mirabbasi R, editors. Applied Agricultural Practices for Mitigating Climate Change. Vol. 2. CRC Press; 2019. p. 191

[10] Qaderi MM, Martel AB, Dixon SL. Environmental factors influence plant vascular system and water regulation. Plants. 2019;**8**(3):65

[11] Bae H, Ji H, Lim YJ, Ryu Y, Kim MH, Kim BJ. Characteristics of drought propagation in South Korea: Relationship between meteorological, agricultural, and hydrological droughts. Natural Hazards. 2019;**99**(1):1-16

[12] Huang S, Van Aken O, Schwarzländer M, Belt K, Millar AH. The roles of mitochondrial reactive oxygen species in cellular signaling and stress response in plants. Plant Physiology. 2016;**171**(3):1551-1559

[13] Hafez Y, Attia K, Alamery S, Ghazy A, Al-Doss A, Ibrahim E, et al. Beneficial effects of biochar and chitosan on antioxidative capacity, osmolytes accumulation, and anatomical characters of water-stressed barley plants. Agronomy. 2020;**10**(5):630

[14] Zargar SM, Gupta N, Nazir M, Mahajan R, Malik FA, Sofi NR, et al. Impact of drought on photosynthesis: Molecular perspective. Plant Gene. 2017;**11**:154-159

[15] Deka D, Singh AK, Singh AK. Effect of drought stress on crop plants with special reference to drought avoidance and tolerance mechanisms: A review. International Journal of Current Microbiology and Applied Sciences. 2018;7:2703-2721

[16] Liu Y, He Z, Xie Y, Su L, Zhang R, Wang H, et al. Drought resistance mechanisms of Phedimus aizoon L. Scientific Reports. 2021;**11**:13600

[17] Xi JJ, Chen HY, Bai WP, Yang RC, Yang PZ, Chen RJ, et al. Sodium-related adaptations to drought: New insights from the xerophyte plant Zygophyllum xanthoxylum. Frontiers in Plant Science. 2018;**9**:1678

[18] Shi W, Liu C, Shu Y, Feng C, Lei Z, Zhang Z. Synergistic effect of rice husk addition on hydrothermal treatment of sewage sludge: Fate and environmental risk of heavy metals. Bioresource Technology. 2013;**149**:496-502

[19] Ghosh D, Xu J. Abiotic stress responses in plant roots: A proteomics perspective. Frontiers in Plant Science. 2014;5:6

[20] Salazar C, Hernandez C, Pino MT. Plant water stress: Associations between ethylene and abscisic acid response. Chilean Journal of Agricultural Research. 2015;**75**:71-79

[21] Chowdhury JA, Karim MA, Khaliq QA, Ahmed AU, Khan MSA. Effect of drought stress on gas exchange characteristics of four soybean genotypes. Bangladesh Journal of Agricultural Research. 2016;**41**:195-205

[22] Zulfiqar F, Younis A, Riaz A, Mansoor F, Hameed M, Akram NA, et al. MORPHO-anatomical adaptations of two Tagetes erecta L. Cultivars with contrasting response to drought stress. Pakistan Journal of Botany. 2020;**52**(3):801-810

[23] Djanaguiraman M, Prasad PVV, Kumari J, Rengel Z. Root length and root lipid composition contribute to drought tolerance of winter and spring wheat. Plant and Soil. 2019;**439**:57-73

[24] Khan MN, Zhang J, Luo T, Liu J, Rizwan M, Fahad S, et al. Seed priming with melatonin coping drought stress in rapeseed by regulating reactive oxygen species detoxification: Antioxidant defense system, osmotic adjustment, stomatal traits and chloroplast ultrastructure perseveration. Industrial Crops and Products. 2019;**140**:111597

[25] Ilyas RA, Sapuan SM, Atikah MSN, Asyraf MRM, Rafiqah SA, Aisyah HA, et al. Effect of hydrolysis time on the morphological, physical, chemical, and thermal behaviour of sugar palm nanocrystalline cellulose (Arenga pinnata (Wurmb.) Merr). Textile Research Journal. 2021;**91**(1-2, 152):-167

[26] Caser M, D'Angiolillo F, Chitarra W, Lovisolo C, Ruffoni B, Pistelli L, et al. Water deficit regimes trigger changes in valuable physiological and phytochemical parameters in Helichrysum petiolare Hilliard & BL Burtt. Industrial Crops and Products. 2016;**83**:680-692

[27] Abdelraheem A, Esmaeili N, O'Connell M, Zhang J. Progress and perspective on drought and salt stress tolerance in cotton. Industrial Crops and Products. 2019;**130**:118-129

[28] Bhargavi B, Kalpana K, Reddy JK. Influence of water stress on morphological and physiological changes in *Andrographis paniculata*. International Journal of Pure and Applied Bioscience. 2017;**5**:1550-1556

[29] Srivastava S, Srivastava M. Morphological changes and antioxidant activity of Stevia rebaudiana under water stress. American Journal of Plant Sciences. 2014;**5**:3417

[30] Bangar P, Chaudhury A, Tiwari B, Kumar S, Kumari R, Bhat KV. Morphophysiological and biochemical response of mungbean [Vigna radiata (L.) Wilczek] varieties at different developmental stages under drought stress. Turkish Journal of Biology. 2019;**43**:58-69

[31] Pritzkow C, Szota C, Williamson V, Arndt SK. Previous drought exposure leads to greater drought resistance in eucalypts through changes in morphology rather than physiology. Tree Physiology. 2020;**41**:1186-1198

[32] Morka EA. Physiological Indices for Drought Tolerance in Stay-green Sorghum (Sorghum bicolor L. Moench) Accessions. Master of Science (Plant Biology and Biodiversity Management). Ethiopia: Addis Ababa University, Addis Ababa; 2015

[33] Riaz A, Tariq U, Qasim M, Shaheen MR, Iqbal A, Younis A. Effect of water stress on growth and dry matter partitioning of Conocarpus erectus. Acta Horticulturae. 2016;**1112**:163-172

[34] Zhang S, Xu X, Sun Y, Jhang J, Li C. Influence of drought hardening on the resistance physiology of potato seedlings under drought stress. Journal of Integrative Agriculture. 2018;**17**:336-347

[35] Waseem M, Nie Z, Yao G, Hasan M, Xiang Y, Fang X. Dew absorption by leaf trichomes in Caragana korshinskii: An alternative water acquisition strategy for withstanding drought in arid environments. Physiologia Plantarum. 2021;**172**:528-239

[36] Ahmad HM, Wang X, Rahman M, Fiaz S, Azeem F, Shaheen T. Morphological and physiological response of helianthus annuus L. to drought stress and correlation of wax contents for drought tolerance traits. Arabian Journal for Science and Engineering. 2021;**47**:6747-6761

[37] Bhargava S, Sawant K. Drought stress adaptation: Metabolic adjustment and regulation of gene expression. Plant Breeding. 2013;**132**:21-32

[38] Singh J, Thakur JK. Photosynthesis and abiotic stress in plants. In: Biotic and Abiotic Stress Tolerance in Plants. Singapore: Springer; 2018. pp. 27-46

[39] Bansal S, Thakur A, Singh S, Bakshi M, Bansal S. Changes in crop physiology under drought stress: A review. Journal of Pharmacognosy and Phytochemistry. 2019;**8**(4):1251-1253

[40] Siddiqui SA, Khatri K, Patel D, Rathore MS. Photosynthetic gas exchange and chlorophyll a fluorescence in *Salicornia brachiata* (Roxb.) under osmotic stress. Journal of Plant Growth Regulation. 2022;**41**:429-444

[41] Zhao F, Zeng N, Asrar G, Friedlingstein P, Ito A, Jain A, et al. Role of CO_2, climate and land use in regulating the seasonal amplitude increase of carbon fluxes in terrestrial ecosystems: A multimodel analysis. Biogeosciences. 2016;**13**(17):5121-5137

[42] Zhang A, Liu M, Gu W, Chen Z, Gu Y, Pei L, et al. Effect of drought on photosynthesis, total antioxidant capacity, bioactive component accumulation, and the transcriptome of Atractylodes lancea. BMC Plant Biology. 2021;**21**(1):1-14

[43] Kheradmand MA, Fahraji SS, Fatahi E, Raoofi MM. Effect of

water stress on oil yield and some characteristics of Brassica napus. International Research Journal of Basic and Applied Science. 2014;**8**:1447-1453

[44] Hu Y, Schmidhalter U. Drought and salinity: A comparison of their effects on mineral nutrition of plants. Journal of Plant Nutrition and Soil Science. 2005;**168**:541-549

[45] Hu L, Wang Z, Huang B. Effects of cytokinin and potassium on stomatal and photosynthetic recovery of Kentucky bluegrass from drought stress. Crop Science. 2013;**53**:221-231

[46] Sarani M, Namrudi M, Hashemi SM, Raoofi MM. The effect of drought stress on chlorophyll content, root growth, glucosinolate and proline in crop plants. International Journal of Farming and Allied Sciences. 2014;**3**:994-997

[47] Li Y, Sun C, Huang Z, Pan J, Wang L, Fan X. Mechanisms of progressive water deficit tolerance and growth recovery of Chinese maize foundation genotypes Huangzao 4 and Chang 7-2, which are proposed on the basis of comparison of physiological and transcriptomic responses. Plant & Cell Physiology. 2009;**50**:2092-2111

[48] Cuéllar T, Pascaud F, Verdeil JL, Torregrosa L, Adam-Blondon AF, Thibaud JB, et al. A grapevine Shaker inward K (+) channel activated by the calcineurin B-like calcium sensor1-protein kinase CIPK23 network is expressed in grape berries under drought stress conditions. The Plant Journal. 2010;**61**:58-69

[49] Khalid KA. Influence of water stress on growth, essential oil, and chemical composition of herbs (Ocimum sp.). International Agrophysics. 2006;**20**:289-296

[50] García-Caparrós P, Romero MJ, Llanderal A, Cermeño P, Lao MT, Segura ML. Effects of drought stress on biomass, essential oil content, nutritional parameters, and costs of production in six Lamiaceae species. Water. 2019;**11**:573

[51] Bahreininejad B, Razmjou J, Mirza M. Influence of water stress on morpho-physiological and phytochemical traits in Thymus daenensis. International Journal of Plant Production. 2013;**7**:151-166

[52] Samarah N, Mullen R, Cianzio S. Size distribution and mineral nutrients of soybean seeds in response to drought stress. Journal of Plant Nutrition. 2004;**27**:815-835

[53] Hussain S, Rao MJ, Anjum MA, Ejaz S, Zakir I, Ali MA, et al. Oxidative stress and antioxidant defense in plants under drought conditions. In: Hasanuzzaman M, Hakeem KR, Nahar K, Alharby HF, editors. Plant Abiotic Stress Tolerance: Agronomic, Molecular and Biotechnological Approaches. Cham, Switzerland: Springer International Publishing; 2019. pp. 207-219

[54] Gharibi S, Tabatabaei BES, Saeidi G, Goli SAH. Effect of drought stress on total phenolic, lipid peroxidation, and antioxidant activity of Achillea species. Applied Biochemistry and Biotechnology. 2016;**178**:796-809

[55] Impa SM, Nadaradjan S, Jagadish SVK. Drought stress induced reactive oxygen species and anti-oxidants in plants. In: Abiotic Stress Responses in Plants. New York, NY, USA: Springer; 2012. pp. 131-147

[56] Cao Y, Luo Q, Tian Y, Meng F. Physiological and proteomic analyses of the drought stress response in Amygdalus Mira (Koehne) Yü et Lu roots. BMC Plant Biology. 2017;**17**:53

[57] Shi S, Fan M, Iwama K, Li F, Zhang Z, Jia L. Physiological basis of drought tolerance in potato grown under long-term water deficiency. International Journal of Plant Production. 2015;**9**:305-320

[58] Caverzan A, Casassola A, Brammer SP. Reactive oxygen species and antioxidant enzymes involved in plant tolerance to stress. In: Shanker A, Shanker C, editors. Abiotic and Biotic Stress in Plants-Recent Advances and Future Perspectives. London, UK: InTech Open; 2016. pp. 463-480

[59] Abid G, M'hamdi M, Mingeot D, Aouida M, Aroua I, Muhovski Y, et al. Effect of drought stress on chlorophyll fluorescence, antioxidant enzyme activities and gene expression patterns in faba bean (*Vicia faba* L.). Archives of Agronomy and Soil Science. 2017;**63**:536-552

[60] Gao S, Wang Y, Yu S, Huang Y, Liu H, Chen W, et al. Effects of drought stress on growth, physiology and secondary metabolites of Two Adonis species in Northeast China. Scientia Horticulturae. 2020;**259**:108795

[61] Gurumurthy S, Sarkar B, Vanaja M, Lakshmi J, Yadav S, Maheswari M. Morpho-physiological and biochemical changes in black gram (Vigna mungo L. Hepper) genotypes under drought stress at flowering stage. Acta Physiologiae Plantarum. 2019;**41**:42

[62] Sharma A, Shahzad B, Rehman A, Bhardwaj R, Landi M, Zheng B. Response of phenylpropanoid pathway and the role of polyphenols in plants under abiotic stress. Molecules. 2019;**24**:2452

[63] Anjum SA, Xie XY, Wang LC, Saleem MF, Man C, Lei W. Morphological, physiological and biochemical responses of plants to drought stress. African Journal of Agricultural Research. 2011;**6**:2026-2032

[64] Mittler R, Blumwald E. The roles of ROS and ABA in systemic acquired acclimation. Plant Cell. 2015;**27**(1):64-70

[65] Mahajan S, Tuteja N. Cold, salinity and drought stresses: An overview. Archives of Biochemistry and Biophysics. 2005;**444**:139-158

[66] Shinozaki K, Yamaguchi-Shinozaki K. Gene networks involved in drought stress response and tolerance. Journal of Experimental Botany. 2007;**58**:221-227

[67] Uno Y, Furihata T, Abe H, Yoshida R, Shinozaki K, Yamaguchi-Shinozaki K. Arabidopsis basic leucine zipper transcription factors involved in an abscisic acid-dependent signal transduction pathway under drought and high-salinity conditions. Proceedings of National Academy of Sciences USA. 2000;**97**(21):11632-11637

[68] Miller G, Suzuki N, Yilmaz SU, Mittler R. Reactive oxygen species homeostasis and signalling during drought and salinity stress. Plant Cell and Environment. 2010;**33**:453-467

[69] Hand SC, Menze MA, Toner M, Boswell L, Moore D. LEA proteins during water stress: Not just for plants anymore. Annual Review of Physiology. 2011;**73**:115-134

[70] Xiang Y, Tang N, Du H, Ye H, Xiong L. Characterization of OsbZIP23 as a key player of the basic leucine zipper transcription factor family for conferring abscisic acid sensitivity and salinity and drought tolerance in rice. Plant Physiology. 2008;**148**:1938-1952

[71] Tang N, Zhang H, Li X, Xiao J, Xiong L. Constitutive activation of

transcription factor OsbZIP46 improves drought tolerance in rice. Plant Physiology. 2012;**158**:1755-1768

[72] Miao Y, Lv D, Wang P, Wang XC, Chen J, Miao C, et al. An Arabidopsis glutathione peroxidase functions as both a redox transducer and a scavenger in abscisic acid and drought stress responses. Plant Cell. 2006;**18**:2749-2766

[73] Zou JJ, Wei FJ, Wang C, Wu JJ, Ratnasekera D, Liu WX, et al. Arabidopsis calcium-dependent protein kinase CPK10 functions in abscisic acid- and Ca^{2+}-mediated stomatal regulation in response to drought stress. Plant Physiology. 2010;**154**:1232-1243

Section 3

Stress Mitigation: Plant Management

Chapter 9

Improvement of Abiotic Stress Tolerance in Plants with the Application of Nanoparticles

Saba Nawaz, Iqra Maqsood, Fatima Batool, Zainab Y. Sandhu, Sameera Hassan, Faheem Akram and Bushra Rashid

Abstract

Plants are under the threat of climatic changes and there is a reduction in productivity and deterioration in quality. The application of nanoparticles is one of the recent approaches to improve plant yield and quality traits. A number of nanoparticles, such as zinc nanoparticles (ZnO NPs), iron nanoparticles (Fe_2O_3 NPs), silicon nanoparticles (SiO_2 NPs), cerium nanoparticles (CeO_2 NPs), silver nanoparticles (Ag NPs), titanium dioxide nanoparticles (TiO_2 NPs), and carbon nanoparticles (C NPs), have been reported in different plant species to play a role to improve the plant physiology and metabolic pathways under environmental stresses. Crop plants readily absorb the nanoparticles through the cellular machinery of different tissues and organs to take part in metabolic and growth processes. Nanoparticles promote the activity of a range of antioxidant enzymes, including catalase (CAT), peroxidase (POD), and superoxide dismutase (SOD), in plant species, which in turn improve the growth and development under stressful conditions. The present review focuses on the mode of action and signaling of nanoparticles to the plant systems and their positive impact on growth, development, and ROS scavenging potential. The appropriate elucidation on mechanisms of nanoparticles in plants leads to better growth and yields under stress conditions, which will ultimately lead to increased agricultural production.

Keywords: agriculture, climatic challenges, crop development, food security, nanotechnology

1. Introduction

Population of the globe is rising and is predicted to reach almost 9.6 billion by the year 2050. Sustained growth of 70–100% in global agricultural and food production is essential to feed the growing population [1]. The area under cultivation may shrink over time due to the increasing nonagricultural uses of the land and urbanization, making it difficult to increase agricultural production [2]. Plants are constantly exposed to environmental changes during their life cycle. Deteriorating soil health conditions inevitably have a detrimental impact on plant development and productivity [3]. Billions of dollars worth of crops are being destroyed each year due to abiotic stresses, such as salinity and drought. Prior to the advent of efficient selection techniques, conventional breeding was used to maintain agricultural productivity, but its effectiveness was constrained by the diversity of stress tolerance traits [2]. Determining innovative solutions to reduce abiotic stress challenges and maintain food security is therefore urgently required under these deteriorating environmental conditions [1] as most promising approach currently available is nanotechnology.

To improve abiotic stress tolerance in agricultural biosystems, Eric Drexler initially coined the term "Nanotechnology" [4]. It deals with the study of nanostructures that possess diverse physicochemical properties and biochemical activities that are dependent on their surface-to-volume ratio [5]. Different physical, chemical, and biological processes can be used to manufacture nanoparticles (NPs), and they can interact with plants in a variety of ways [6]. Crop plants readily absorb NPs, which can enter the cells and play crucial roles in metabolic and growth processes [7]. There is a surge in the use of nanobiotechnology tools in agricultural production that has the potential to boost plant metabolism since NPs promote plant growth, development, and yield to withstand environmental stresses [8]. Additionally, it has been observed that NPs promote the activity of a range of antioxidant enzymes, including catalase (CAT), peroxidase (POD), and superoxide dismutase (SOD) [9]. Extensive research revealed that NPs are crucial for plants dealing with abiotic stress conditions [10].

Nanobiotechnology will improve plant functions that will help them cope with environmental challenges [11]. Therefore, this technology is strongly encouraged due to the rising global food demand as well as the potential for positive effects on the economy and ecology [12]. Despite its extraordinary potential in the enhancement of agricultural productivity and improvement in abiotic stress tolerance, the wider application of nanotechnology at the field level is limited in agriculture. In this review article, updates on the positive effects of nanotechnology for the improvement of abiotic stresses in crops have been discussed in detail.

2. Role of nanoparticles in salinity stress

Salinity has a negative impact on crops in various physiological and biochemical processes that decrease crop production drastically [13]. Water scarcity in the soil causes low osmotic potential and ionic toxicity of Cl^- and Na^+ in plant cells [14]. It is also seen that salt stress results in decreased concentration of photosynthetic pigments, reduced stomatal flow, lack of efficiency of photosystem II, and increased production of ROS.

2.1 Zinc nanoparticles (ZnO NPs)

Salt stress causes chlorophyll concentration leads to membrane disintegration and the rate of photosynthesis is significantly decreased. It also causes injury in thylakoid and grana that results in limited starch content [15]. Lupine (*Luminus termis*) was protected in saline conditions by priming with ZnO NPs, which enhanced the photosynthetic pigments, antioxidant responses, and growth [16]. ZnO NPs treatment also increased the antioxidant enzymes in *Zea mays* [17]. When *Abelmoschus esculentus* was treated with ZnO NPs, it increased the SOD and CAT activities and photosynthetic pigment [18]. Canola (*Brassica napus*) plant treated with ZnO NPs, alleviated the harmful effects of salt by upregulating the osmolyte biosynthesis, ionic regulation, and antioxidant system under saline conditions [19]. *Gossypium hirsutum* plants treated with ZnO NPs enhanced the contents of carotenoids, chlorophyll a and b, and total chlorophyll under salt stress conditions [20, 21]. When salt-stressed citrus plants (*Citrus reticulata*) were treated with ZnO NPs, it results in decreased accumulation of total soluble sugars and proline contents that help in the osmoregulation of plants and maintain the growth rate of treated plants [22].

2.2 Iron nanoparticles (Fe_2O_3 NPs)

Iron oxide nanoparticles are a rich source of Fe for plants. When peppermint was exposed to Fe_2O_3 NPs, proline content, and lipid peroxidation were decreased significantly in saline soil. Antioxidant enzyme activities (guaiacol peroxidase, CAT, and SOD) declined in plants. They also increased the potassium, zinc, calcium, iron, leaf dry and fresh weight, and phosphorus [23]. Grape softwood showed a prominent increase in protein content and reduced production of hydrogen peroxide, proline, and antioxidant enzyme activities when treated with potassium silicate and Fe_2O_3 NPs [24]. Under salt stress, the application of Fe_2O_3 NPs on *Helianthus annuus* increased the activities of POD and CAT [25]. Ajowan (*Trachyspermum ammi*) was treated with Fe_2O_3 NPs under saline conditions increased antioxidant activities, osmolyte synthesis, and maintained Na+/ K+ ratio. These adaptations help plants to improve leaf pigments, seed yield, membrane stability, and shoot and root growth [9].

2.3 Silicon nanoparticles (SiO_2 NPs)

SiO_2 NPs are used to help plants by forming a layer in cell walls and maintaining yield. In squash and tomato plants, the antioxidant system is enhanced and seed germination increases due to SiO_2 NPs under salt stress [26]. In Basil plants, silica nanoparticles have shown promising results related to morphological and physiological traits under salt stress [27]. SiO_2 NPs increased the seedling growth of lentils and seed germination and improved the defense mechanism of plants in saline conditions [28]. They help plants to cope up with salt stress by increasing the fresh weight in maize [29]. Under salt stress, the application of SiO_2 NPs on soybean decreased toxic ROS production and Na + level in leaves [30]. Wheat cultivars treated with SiO_2 NPs improved biological antioxidant levels and seedling growth under salt stress [31]. Application of SiO_2 NPs on the strawberry plant in saline conditions increased the photosynthetic pigment and maintained the carotenoid and chlorophyll content, decreasing the effect on epicuticular wax [32].

2.4 Cerium nanoparticles (CeO_2 NPs)

They can be used as fertilizer to stimulate the growth of roots, enhance the antioxidant enzyme activities, and to prevent membrane leakage and peroxidation [33]. Moreover, CeO_2 NPs help to preserve cell wall and chloroplast structure [34]. Activation of CeO_2 NPs as antioxidants depends upon the pH of surroundings, sub-cellular localization, surface charge, concentration, and particle size. CeO_2 NPs increased the growth in *Dracocephalum Moldavica* (a herbaceous plant also called Moldavian balm), by regulating nonenzymatic and enzymatic defense mechanisms under saline conditions [35]. *Brassica Napus* plants treated with CeO_2 NPs have efficient chloroplast and biomass under salt stress [36]. Anatomical changes, such as low accumulation of Na + in roots and high Na + flow toward shoots have also been reported [37].

2.5 Silver nanoparticles Ag NPs

Silver nanoparticles enhanced the sodium, potassium, and chloride to regulate the osmolality level in treated plants under salt stress conditions. The stability of Ag NPs can be easily controlled in aquatic environments as compared to soil conditions [38]. Priming of seeds was carried out with Ag NPs to enhance the seed germination in wheat and the development of tomato plants [39]. The combined effect of Ag NPs with NaCl reduced the thiobarbituric acid reactive substances, electrolyte leakage, and hydrogen peroxide to control the oxidative damage in plants that is linked with the overproduction of ROS [40]. *Triticum aestivum* treated with Ag NPs increased the fresh and dry biomass under saline conditions [41]. Seeds of *Pennisetum glaucum* treated with Ag NPs improved the growth, proline, and relative water content (RWC) and decreased the oxidative damage by increasing the antioxidant enzyme activities under saline conditions [42].

2.6 Titanium dioxide nanoparticles (TiO_2 NPs)

Titanium is a transition element and the 9th most abundant element that contributes 0.33% of the earth's outer layer [43]. It improves photosynthesis and chlorophyll pigments in plants by altering antioxidant enzyme activities [44]. Titanium has an integral role in plants' tolerance under stressful conditions [45]. *Dracocephalum moldavica* plants were treated with TiO_2 NPs improved plant growth, and proline content and increased enzymatic activities, soluble sugars, and amino acids under salinity stress [46].

Type of NPs	Plant species	Physiological responses of plants under salt stress	References
Zinc NPs (ZNO NPs)	*Luminus termis*	Enhanced photosynthetic pigments, antioxidant responses, and growth.	[16]
	Zea mays	Increased antioxidant enzymes and improved salinity tolerance.	[17]
	Abelmoschus esculentus	Increased the activities of CAT and SOD and photosynthetic pigments and decreased the accumulation of soluble sugar and proline content.	[18]
	Brassica napus	Ionic regulation, osmolyte synthesis, and upregulated oxidative defense system.	[19]
	Gossypium hirsutum	Enhanced content of carotenoids and chlorophyll a & b.	[20]
	Citrus reticulata	Decreased accumulation of total soluble sugars and proline contents, osmoprotectants.	[22]
Iron NPs (Fe_2O_3 NPs)	*Mentha piperita*	Decreasing lipid peroxidation and increase in leaf dry and fresh weight, potassium, calcium, iron, phosphorus, and zinc contents.	[23]
	Grape softwood	Increased protein content and reduced production of hydrogen peroxide, proline, and antioxidant enzymes.	[24]
	Helianthus annuus	Increased antioxidant activities.	[25]
	Trachyspermum ammi	Improved osmolyte synthesis, antioxidant activities, and improved $Na+/K+$ ratio.	[9]

Type of NPs	Plant species	Physiological responses of plants under salt stress	References
Silicon NPs (SiO_2 NPs)	Tomato and squash	Increased ROS and seed germination. Maintained vitamin C concentration and chlorophyll content.	[26]
	Ocimum basilicum	Increasing proline accumulation, antioxidant responses, leaf dry and fresh weight, and chlorophyll content.	[27]
	Zea mays	Boosting defense mechanism by ROS.	[29]
	Glycine max	Decreased ROS, lipid peroxidation, and Na^+, increased antioxidant activities and K^+ content.	[30]
	Triticum aestivum	Improved chlorophyll content and seed germination.	[31]
	Fragaria ananassa	Maintained carotenoid and chlorophyll content, and decreased the effect on epicuticular wax.	[32]
Chitosan	*Zea mays*	Increasing chlorophyll content, growth, photosystem II, and enhanced nitric oxide bioactivity.	[53]
Cerium NPs (CeO_2 NPs)	*Glycine max*	Rubisco carboxylase activity stimulates growth and increased photosynthesis.	[33]
	Dracocephalum Moldavica	Increased enzymatic and nonenzymatic defense system.	[35]
	Brassica napus	Efficient chloroplast and increase in biomass.	[36]
Silver NPs (Ag NPs)	*Solanum lycopersicum*	Increased germination with seed priming.	[39]
	Triticum aestivum	Enhanced dry and fresh biomass of plants. Increased the activity of CAT and decreased the activity of POD, and increased the accumulation of soluble sugar and proline content.	[41]
	Pennisetum glaucum	Improved proline and RWC, decreased oxidative damage by increasing the antioxidant enzyme activities.	[42]
Titanium dioxide NPs (TiO_2NPs)	*Dracocephalum moldavica*	Increased seedling growth, dry and fresh weight, root and shoot length. Increased antioxidant activity.	[46]
Carbon NPs (C NPs)	*Brassica napus*	Reestablishing ion homeostasis and redox balance.	[49]
	Ocimum basilicum	Increased enzymatic and nonenzymatic defense system, increased carotenoid, and chlorophyll content.	[50]
	Sophora alopecuroides	Increased proline content in roots and leaves, photosystem II activity, soluble sugar in leaves, and membrane integrity was maintained.	[51]
	Triticum aetivum	Enhanced phosphorus and potassium contents in root and phosphorus content in the shoot. Increased activity of antioxidant enzymes. Improved chlorophyll content, free ascorbic acid, amino acid, and soluble sugars.	[52]

Table 1.
Effect of nanoparticles on plants under salinity stress.

2.7 Carbon nanoparticles (C NPs)

Carbon nanoparticles have distinctive properties, including morphologically small surface area and better chemical reactivity [47]. They help in increasing seed germination and crop production under salt stress [48]. Application of C NPs on *Brassica napus* increased the NaCl stress tolerance in the plant by reestablishing ion homeostasis and redox balance [49]. When *Ocimum basilicum* was treated with C NPs increased enzymatic and nonenzymatic defense systems, and increased carotenoid and chlorophyll content [50]. Proline content in roots and leaves, photosystem II activity, soluble sugar in leaves were increased, and membrane integrity was maintained due to an increase in unsaturated fatty acid in *Sophora alopecuroides* C NPs treated plants under salt stress [51]. Application of C NPs on *Triticum aetivum* plants under saline conditions enhanced phosphorus and potassium content in roots and phosphorus content in shoot and increased the activity of antioxidant enzymes and improved the chlorophyll content, free ascorbic acid, amino acid, and soluble sugars [52]. **Table 1** consists of the reports of different scientists who studied the effects of the treatment of different nanoparticles on different plant species and their responses to physiological levels under salt stress.

3. Role of NPS in drought tolerance

Water crisis is one of the several issues that have been afflicted by climate change and global warming. Water is crucial for plant vitality as it has a role in the transportation of nutrients. Therefore, water deficit results in drought stress, which harshly affects the survival of plants and reduces agricultural production [54]. Therefore, a key solution in relation to sustainable agriculture is the identification of resistant crop varieties or improving drought tolerance in plants. Crop management and coping with various environmental challenges are possible by the new features of nanotechnology. The negative impacts of a restricted water supply on agriculture have been attempted to be reduced utilizing nano-materials. Farmers may be able to identify the effects of stress on plants at an early stage by using nano-sensors in global positioning systems that produce satellite photographs of fields [55]. Crop production in drought-prone locations may rise if soils have been given better water-retention capabilities [56]. Using nanoparticle-based plant modifications, conventional technologies may be used to improve crop plants by increasing the capacity of food crops to retain water, and nanoparticles improve the effectiveness of water consumption in the plants [57].

3.1 Silicon nanoparticles (SNPs)

Only a few studies have documented the biological activity of silica, an element that makes up a major portion of the Earth's crust and is found as silicon [58]. The tolerance of Hawthorn (*Crataegus* sp.) plants to drought stress is increased by applying various concentrations of silica nanoparticles [59]. The findings indicated that pretreating SNPs had a favorable impact on the photosynthetic metrics, RWC, malondialdehyde, membrane electrolyte leakage, as well as the levels of carbohydrate and proline. Two *Sorghum bicolor* cultivars treated with silicon demonstrated enhanced drought tolerance by reducing their shoot-to-root ratio, which may have indicated enhanced root development and retention of photosynthetic rate. This suggests that increasing plant water uptake efficiency will increase resistance to drought [60]. Using sodium silicate at 1.0 mM enabled the reduction of the effects of drought stress on wheat [61]. Although the precise process is unknown but silicon helps stressed plants to boost shoot growth, preserve RWC, and chlorophyll content, and reduce the membrane lipid peroxidation.

3.2 Zinc nanoparticles (ZnO NPs)

Application of Zn plays a role in increasing the radical growth and seed viability along with the establishment of germinated seeds, especially in Zinc deficient areas. Soybean seeds subjected to water stress showed that nano-zinc oxide has improved seed germination [62]. A composite of ZnO, B_2O_3, and CuO NPs reduced the effects of drought stress on soybean [63] and increased grain production and shoot growth by 36 and 33%, respectively, and improved nitrate and phosphorus uptake. By boosting the activity of the antioxidant enzymes SOD and CAT in wheat, ZnO NPs improved drought resistance. Zn and Cu NPs also increased the antioxidative enzyme activity, decreased lipid peroxidation, stabilized the photosynthetic pigments with increased RWC, and enhanced drought tolerance in wheat [64]. CuO and ZnO NPs modified the root morphology of plants colonized by *Triticum aestivum*, a beneficial pseudomonad, altering the plants to withstand drought [65]. CuO NPs boosted the production of lateral roots in wheat seedlings, and ZnO NPs stimulated the growth of extended root hairs proximal to the root tip. Drought stress severely affected eggplant development and production [66]. Exogenous ZnO NPs with 50 and 100 ppm, boosted the RWC and membrane stability index, improved stem and leaf morphology and better photosynthesis in water-stressed eggplant and yield rose by 12.2 and 22.6%, respectively.

3.3 Iron nanoparticles (Fe_2O_3 NPs)

Iron is a key micronutrient that is essential for plant growth and development and its reduction causes chlorosis in plants [67]. Thus, iron absorption in plants experiencing drought stress may play a crucial role in their ability to withstand drought [68]. Fe_2O_3 NPs applied topically to *Carthamus tinctorius* leaves, reduced the negative effects of drought stress while simultaneously promoting yield, growth, and development metrics [69]. Plants under drought stress have a substantial impact of Fe_2O_3 NPs on the number and weight of seeds, number of bolls yield, and oil percentage in cotton. Drought stress in sunflower was significantly reduced by the application of maghemite nanoparticles (a member of iron oxide) [70]. The activity of zerovalent (nZVI) iron nanoparticles is hypothesized to activate the proton pump ATPase (H+ -ATPase) of the plasma membrane in leaves of *Arabidopsis thaliana* plants, which in turn contributes to their ability to withstand drought by increasing the stomatal aperture [71]. They retained the drought sensitivity and boosted CO_2 uptake, the rate of the stomatal opening was accelerated, which in turn increased the plant biomass and chlorophyll content [72].

3.4 Titanium nanoparticles (titanium dioxide TiO_2 & anatase titanium dioxide $AnTiO_2$)

Wheat seed gluten and starch contents have responded favorably for using TiO_2 NPs foliar applications and improved plant height, seed and ear numbers, ear weight, gluten and starch content, yield, biomass, and harvest index under drought stress [73]. Enhanced photosynthesis and increased the maize plant's capacity to absorb light under water deficit [74]. Applying nanoparticles and Gibberellic acid to basil plants effectively enhanced the rate of photosynthesis, enhancing their tolerance to drought stress [75]. Various doses and sizes (10–25 nm) of $AnTiO_2$NP on the flax plant under water-scarce conditions responded favorably for growth, development, hydrogen peroxide (H_2O_2), malondialdehyde content, seed oil production, protein content, and photosynthetic pigments [76]. The effects of TiO_2 NP on onion seedlings have also been documented as they boosted SOD activity. TiO_2 NP concentrations of 40 and 50 mg/ml, lowered the CAT and POD activities and secretion of amylase was reduced. However, seed germination and seedling growth were reduced at higher concentration of TiO_2 NPs, whereas the effect was enhanced at lower concentration [77].

Type of NPs	Plant species	Physiological responses of plants under drought stress	References
Silicon NPs (SiO_2NPs)	*Crataegus* sp.	Positive effects on photosynthesis, malondialdehyde, RWC, membrane electrolyte leakage, chlorophyll, carotenoid, carbohydrate, and proline content.	[59]
	Sorghum bicolor	Reduced shoot-to-root ratio, improved root growth and the maintenance of photosynthetic rate, augmentation of water uptake efficiency, increase in leaf area index and leaf weight.	[40]
	Triticum aestivum	Improved shoot growth, increased the leaf chlorophyll content, maintained leaf water potential in stressed plants, reduced membrane lipid peroxidation.	[61]
Zinc NPs (ZnO NPs)	*Glycine max*	Increased germination, radical growth and seed viability, shoot growth and grain yield, and uptake of N and P.	[62, 63]
	Triticum aestivum	Increased antioxidant enzymes, reduced lipid peroxidation, stabilized the photosynthetic pigments, proliferation of elongated root hairs, and increased water stress tolerant gene expression.	[64]
	Solanum melongena	Increased RWC and MSI, improved stem and leaf anatomy, photosynthetic efficiency, growth, and yield.	[66]
Iron NPs (Fe_2O_3 NPs)	*Carthamus tinctorius*	Improved crop yield at the flowering stage and enhanced oil percentage.	[69]
	Arabidopsis thaliana	Enhanced plasma membrane proton pump ATPase activity, plant biomass, chlorophyll content, and CO_2 assimilation.	[71, 72]
Titanium NPs (Titanium dioxide TiO_2	*Triticum aestivum*	Enhancement in seed gluten and starch contents, plant height, seed number, and biomass.	[73]
	Zea mays	Improved sunlight absorbance, synthesis of photosynthetic pigments, and photosynthesis.	[74]
	Ocimum basilicum	Increased plant drought resistance and improve photosynthetic mechanism.	[75]
	Linum usitatissimum	Increased chlorophyll and carotenoids, enhancing flax development and yield, declining malondialdehyde and H_2O_2 content.	[76]
	Allium cepa	Increased SOD activity, seed germination, and seedling growth.	[77]
Silver NPs (Ag NPs)	*Lens culinaris*	Increased germination rate, root length, root fresh, and dry weight.	[68]
Cerium Oxide NPs (CeO_2 NPs)	*Sorghum bicolor*	Enhanced catalytic scavenging of ROS, reduced O_2^- (41%), H_2O_2 (36%), increased yield, pollen germination, and CO_2 absorption.	[80]

Table 2.
Effects of nanoparticles on plants under drought stress.

3.5 Silver nanoparticles (Ag NPs)

Developing drought-tolerant cultivars and reducing drought stress is essential for maintaining food security [78]. Only a small amount of historical literature focused on the interaction between drought and Ag NPs. Diminishing the detrimental impacts of drought stress on *Lens culinaris* has been credited to the application of Ag NPs [68]. The study indicated that varying PEG and Ag NPs concentrations had a significant impact on seed germination, root length, and biomass. Hence, Ag NPs could help to reduce water stress that affects plant development and productivity [79].

3.6 Cerium oxide nanoparticles (CeO_2 NPs)

Cerium oxide NPs could help *Sorghum bicolor* to cope with drought stress by inducing oxidative stress by catalytically neutralizing ROS and maintaining the photosynthetic activity and grain yield [80]. The foliar spray of nanoceria dramatically decreased free radicals (H_2O_2 and superoxide) under drought conditions, for instance, by 36 and 41%, respectively. ROS-scavenging enzymes were shown to be more active in plants exposed to drought: POD (54%), SOD (94%), and CAT (117%). The rate of pollen germination, carbon absorption, and seed output per plant was improved by 38, 31, and 31%, respectively. Different reports summarizing the effects of a number of nanoparticles on different plant species and their physiological responses under drought stress are discussed in **Table 2**.

4. Nanoparticles and other environmental stresses

Titanium oxide nanoparticles play a substantial role in the mitigation of light stress in crops as they catalyze the oxidation-reduction reaction, which then forms superoxide anion radicals and hydroxides. Oxidative stress is induced by ultraviolet (UV) light and has a negative impression on the growth of the plant. UV-B produces H_2O_2 and superoxide radicals and enhanced the leakage of electrolytes and lipid peroxidation, which leads to reduce the photosynthesis rate and normal leaf structure is also deteriorated [81]. In wheat plants, Silicon NPs increase antioxidant activities for the regulation of oxidative stress after UV-B exposure [82]. Herbicides are used to control weeds in agroecosystems. A methyl viologen herbicide, Paraquat is used extensively to control weeds in rice. Multiwall carbon nanotubes can modulate the toxicity of Paraquat [50], which promotes lateral root growth and photosynthesis in *Arabidopsis* and protect against the toxicity of Paraquat by lowering its bioavailability and promoting the oxidative-stress-related protein expression and photosynthesis. Therefore, the NPs can modulate abiotic stress-induced responses in plant growth at different levels. However, their physiochemical, electrical, optical, and biological properties are crucial [83]. Plants' tolerance to low temperature in green beans is increased by the exogenous application of Ag NPs [84] as they are used to reduce the oxidative stress in wheat.

5. Effect of nanoparticles on antioxidant and molecular mechanism of plants

Nanoparticles have an impact on plants' antioxidant system at the molecular level as they increase the capability of plants to tolerate oxidative stress. When *Brassica*

Type of nanoparticles	Plant species	Impact on antioxidant system and ROS	References
Zinc NPs (ZnO NPs)	*Solanum lycopersicum*	Decreased malondialdehyde content and increased SOD, CAT, POD, and APX activities under salinity stress.	[91]
Silver NPs (Ag NPs)	*Pennisetum glaucum*	Significant increase in proline content, antioxidant enzyme activities, flavonoid contents, and phenolics.	[42]
Titanium NPs (Titanium dioxide TiO_2	*Spinacia oleracea*	Enhance antioxidant stress tolerance.	[92]
	Glycine max	Enhance antioxidant enzyme activities contribute to reduction in hydrogen peroxide and malondialdehyde content under salinity.	[45]
Silicon NPs (SiO_2NPs)	*Coriandrum sativum*	Reduce the detrimental impact of Pb under lead stress by altering vitamin C, antioxidant enzyme activation, and flavonoids and increase plant capabilities to endure abiotic stresses.	[93]
Cerium oxide NPs (CeO_2 NPs)	*Phaseolus vulgaris*	Scavenge ROS in isolated chloroplasts protect plant photosynthesis from detrimental effects Of ROS accumulation during abiotic stresses.	[87]

Table 3.
Effect of nanoparticles on plants for ROS and antioxidant system.

juncea plant was treated with silver NPs, antioxidant enzyme activities (CAT, ascorbate peroxidase APX and guaiacol peroxidase) were increased which reduced the ROS [78]. The antioxidant system of *Spirodela polyrhiza* plants was activated when copper NPs were used to induce the activity of CAT, POD, and SOD. Moreover, ROS level also increased remarkably because of malondialdehyde and glutathione [85]. When seedlings of *Brassica juncea* were treated with gold NPs, the activity of antioxidant enzymes (guaiacol peroxidase, CAT, glutathione reductase, and APX) was significantly increased, in addition to the accumulation of the higher amount of proline and hydrogen peroxide. The activity of glutathione reductase was maximum at 200 ppm and the activity of other antioxidant enzymes, such as APX and guaiacol peroxidase, were also increased at 400 ppm of gold NPs treated plants [86]. When roots of kidney beans were exposed to CeO_2 NPs for a longer time, then antioxidant enzymes' activities were reduced and soluble protein was increased. While leaves treated with CeO_2 NPs showed increased activity of guaiacol peroxidase [87]. Plants exposed to ZnO NPs increase the Zn and SOD antioxidant enzyme minimizing the effect of oxidative stress [88].

The molecular mechanism of plants can be studied by using the model plant species. *Arabidopsis thaliana* treated with AgNPs gene expression analysis done by RT-PCR and cDNA microarray analyzed for transcriptome behavior [89] showed 281 upregulated genes associated with metal and oxidative stress and 80 downregulated genes associated with hormonal stimuli and plant defense system. The effect of AgNPs on rice has also been studied and some responsive proteins were associated with transcription, oxidative stress, protein degradation, cell division, calcium signaling and regulation, and apoptosis [90]. Hence, the effect of different nanoparticles on different plant species for the functioning of ROS and antioxidant enzymes has been briefed in **Table 3**.

6. Mechanism of nanoparticle (NPs) absorption in plants

Absorption and translocation of NPs in plants are one of the latest disciplines of study. The most commonly used NPs to enhance abiotic stress tolerance in plants are metal-based (MB) and carbon-based (CB). Among MB, the most widely studied nanomaterials are metal and metal oxides, such as copper, silver, titanium, iron, and zinc; while the most explored CB nanomaterials are carbon nanotubes (CNTs), fullerene (C_{70}), and fullerol ($C_{60}(OH)_{20}$) [94]. The impact of NPs on plants is determined by a number of variables, including availability, uptake, translocation, and accumulation. The plant cell wall restricts the entry of foreign elements; therefore, effective techniques are needed to introduce advantageous NPs and make them available to plants.

Different factors like size, chemical content, and plant species affect the entry of NPs, which is further influenced by their stability, transport and absorption, toxicity, and accumulation [95]. Particle size, surface charge, and the hydrophobicity of the plant surface all play important roles in their absorption [96]. Additionally, the absorption rate and translocation in plants are directly correlated to the structure of the nanomaterial utilized [97]. All of these elements highlight the requirement for developing and enhancing laboratory techniques to comprehend the NPs physicochemical qualities [98] as they undergo biotransformation in the soil, which has a direct impact on their toxicity and bioavailability. Foliar spraying or incubating isolated cells, roots, pollen, seeds, and protoplast with NPs, direct injection, irrigation of plants with NPs, delivery by biolistic, and hydroponic treatment have all been employed in previous research to make NPs available to the plant cells [99]. Bioaccumulation defines the uptake of NPs by plant roots and travels through apoplastic and symplastic routes to the cortex and pericycle [100].

Nanoparticles entered through the stomata, cuticle, stigma, trichomes, wounds, and lenticels and move through the phloem. They reach the xylem and phloem through the root tip meristem, where the Casparian strips continue to the shoot but have not fully developed. Endocytosis allows NPs to enter cells even when the cell wall, cell membrane, and Casparian strips block their uptake and transport. Additionally, transporters like aquaporins and carrier proteins facilitate their easier entry into cells [101]. The capacity of roots to absorb nutrients can change if NPs accumulate on the surface of the roots. Parenchymatic intercellular gaps in seeds enable NPs to be directly absorbed before being diffused into the cotyledon. Stomata allow for the internalization of NPs larger than 10 nm, which are then delivered to the plant's vascular system *via* apoplastic and symplastic pathways. Once internalized, NPs move through vascular systems carried by phloem alongside sugar flow, move-in both the directions, and eventually build up in organs that could serve as sap-sinks [102]. The apoplastic pathway has been extensively described to enhance the transfer of water nutrients and nonessential metal complexes. Leaf shape and chemical composition of surface waxes limits the entry of NPs through leaf [103]. Hence, to ensure the NPs' effective absorption in plants, it is crucial to consider their size, concentration, and physiological environment.

7. Mechanism of translocation and accumulation of NPs in plants

Mechanism of NPs translocation in different plant cells and organelles has been clarified [96]. Plant cell wall serves as a barrier that manages NPs uptake and establishes solubilization needed to enable their translocation. NPs with a size range of 40–50 nm can easily pass through the cell wall [104]. Composition of NPs affects their mobility through cell membrane or cell wall and also encourages their adsorption to radical exudates. Positively charged NPs have better adherence to cell walls. Their coating and morphology have a big impact on how they behave inside plants and the rhizosphere [105]. After penetration through cells, they go through the shoots [101] and roots are transmitted to various aerial tissues and the seeds [6]. Gold NPs were only collected in the shoots of *Oryza sativa* when used with *Cucurbita pepo*, *Raphanus raphanistrum*, and other plant species. Positively charged gold nanoparticles tend to be quickly absorbed by plant roots [106]. The entry of NPs into the cell is facilitated by capillary action and osmotic pressure [95]. Membrane proteins of NPs, including as receptors and transporters, are altered as a result of their interaction with the outermost layer.

Negatively charged gold nanoparticles are easily translocated from plant roots to shoots. The most stable are SiO_2 and TiO_2, as they remain present in plant tissues after their uptake. When *Zea mays* is exposed to ZnO NPs hydroponically, most of them are accumulated in its roots and shoots. It is explained by the maximal NPs dissolution in the rhizosphere, which produces the zinc ions and enhances its absorption and translocation in the plant [107]. Soil-grown wheat has also been observed for this perseverance.

Different processes have been identified by the translocation of CeO_2 and ZnO NPs into *Glycine max* [108]. CuO NPs have been shown to be capable of moving from *Zea mays* roots to shoots and vice versa. TiO_2 NPs with a diameter of 140 nm or larger may translocate in *Triticum aestivum* roots [103]. The data of different reports about accumulation of different NPs in different plant tissues have been summarized in **Table 4**.

Type of nanoparticles	**Plant species**	**Conc (mg/l *mg/kg)**	**Accumulation (mg/kg)**		**References**
			Roots	**Shoots**	
Copper NPs	*Oryza sativa*	1000	1544.1	17.27	[109]
	Lactuca sativa	250	3773	—	[110]
	Vigna radiate	125	—	18.46	[111]
	Brassica juncea	1500	190.4	—	[112]
	Cajanus cajan	20	5.82	19.06	[113]
	Phaseolus vulgaris	100	800	—	[114]
Silver NPs	*Glycin max*	4000	2102	11.35	[108]
	Oryza sativa	1000	20	5	[115]
	Solanum lycopersicum	250	—	50	[116]
Zinc NPs	*Solanum lycopersicum*	1000	—	250	[117]
	Zea mays	100	10	30	[107]
Titanium NPs	*Solanum lycopersicum*	1000	—	250	[117]
Magnesium NPs	*Zea mays*	1000	103	131	[118]

Table 4.
Accumulation of nanoparticles in different plant species' tissues.

8. Conclusion and prospective

Various abiotic factors are damaging the plants in terms of growth and development leading to a reduction in yield and deteriorating the quality. Applications of NPs have been proven very protective as they stimulate germination, growth, and improve yield. They increase tolerance in plants as they enhance the uptake of water and nutrients. They are capable to metabolize starch reserves in plant cells. They stimulate the process of photosynthesis and alter levels of phytochromes and modulate oxidative stress.

Although various studies suggest the beneficial roles of NPs in plants but the molecular basis of the actual mechanism is still unknown. Further elucidation on this mechanism may generate the smart NPs for the production of crops sustainable to the environment. In addition, their interaction with signaling molecules is also required to be explored. The economic stability of the use of NPs in agriculture is important as silver and gold nanoparticles costs very expensive. Understanding their mode of action, toxicity limits, signaling, and translocation; hence cheaper NPs may be used as an alternative.

Conflict of interest

The authors declare no conflict of interest.

Author details

Saba Nawaz[1†], Iqra Maqsood[1†], Fatima Batool[1], Zainab Y. Sandhu[2], Sameera Hassan[1], Faheem Akram[1] and Bushra Rashid[1*]

1 Centre of Excellence in Molecular Biology, University of the Punjab Lahore, Lahore, Pakistan

2 Montclair State University, Montclair, NJ, USA

*Address all correspondence to: bushra.cemb@pu.edu.pk

† Both the authors contributed equally and share the first authorship.

References

[1] Zhao L, Cui J, Cai Y, Yang S, Liu J, Wang W, et al. Comparative transcriptome analysis of two contrasting soybean varieties in response to aluminum toxicity. International Journal of Molecular Sciences. 2020;**21**(12):4316. DOI: 10.3390/ijms21124316

[2] Das S, Ho A, Kim PJ. Role of microbes in climate smart agriculture. Frontiers in Microbiology. 2019;**26**(10):2756. DOI: 10.3389/fmicb.2019.02756

[3] Bhatt D, Nath M, Sharma M, Bhatt MD, Bisht DS, Butani NV. Role of growth regulators and phytohormones in overcoming environmental stress. In: Roychoudaury A, Tripathi Dk, editors. Protective Chemical Agents in the Amelioration of Plant Abiotic Stress: Biochemical and Molecular Perspectives. 1st ed. Chichester: Wiley; 2020. p. 254-279. DOI: 10.1002/9781119552154.ch11

[4] Khan Z, Upadhyaya H. Impact of nanoparticles on abiotic stress responses in plants: An overview. In: Tripathi DK, Ahmad P, Sharma S, Chauhan DK, Dubey NK, editors. Nanomaterials in Plants, Algae and Microorganisms. 1st ed. Academic Press; 2019;2:305-322. DOI: 10.1016/B978-0-12-811488-9.00015-9

[5] Anastasiadis F, Manikas I, Apostolidou I, Wahbeh S. The role of traceability in end-to-end circular Agri-food supply chains. Industrial Marketing Management. 2022;**104**:196-211. DOI: 10.1016/j.indmarman.2022.04.021

[6] Tripathi DK, Singh S, Singh VP, Prasad SM, Dubey NK, Chauhan DK. Silicon nanoparticles more effectively alleviated UV-B stress than silicon in wheat (Triticum aestivum) seedlings. Plant Physiology and Biochemistry. 2017;**110**:70-81. DOI: 10.1016/j.plaphy.2016.06.026

[7] Javed T, Shabbir R, Hussain S, Naseer MA, Ejaz I, Ali MM, et al. Nanotechnology for endorsing abiotic stresses: A review on the role of nanoparticles and nanocompositions. Functional Plant Biology. 2022. DOI: 10.1071/FP22092

[8] Kandhol N, Jain M, Tripathi DK. Nanoparticles as potential hallmarks of drought stress tolerance in plants. Physiologia Plantarum. 2022;**174**(2):e13665. DOI: 10.1111/ppl.13665

[9] Abdoli S, Ghassemi-Golezani K, Alizadeh-Salteh S. Responses of ajowan (Trachyspermum ammi L.) to exogenous salicylic acid and iron oxide nanoparticles under salt stress. Environmental Science and Pollution Research. 2020;**27**(29):36939-36953. DOI: 10.1007/s11356-020-09453-1

[10] Tayyab QM, Almas MH, Jilani G, Razzaq A. Nanoparticles and plant growth dynamics: A review. Journal of Applied Agriculture and Biotechnology. 2016;**1**(2):14-22

[11] Lowry GV, Avellan A, Gilbertson LM. Opportunities and challenges for nanotechnology in the Agri-tech revolution. Nature Nanotechnology. 2019;**14**(6):517-522. DOI: 10.1038/s41565-019-0461-7

[12] Fincheira P, Tortella G, Duran N, Seabra AB, Rubilar O. Current applications of nanotechnology to develop plant growth inducer agents as an innovation strategy. Critical Reviews in Biotechnology. 2020;**40**(1):15-30. DOI: 10.1080/07388551.2019.1681931

[13] Shrivastava P, Kumar R. Soil salinity: A serious environmental issue and plant growth promoting bacteria as one of the tools for its alleviation. Saudi Journal of Biological Sciences. 2015;**22**(2):123-131. DOI: 10.1016/j.sjbs.2014.12.001

[14] Negrão S, Schmöckel S, Tester M. Evaluating physiological responses of plants to salinity stress. Annals of Botany. 2017;**119**(1):1-11. DOI: 10.1093/aob/mcw191

[15] Silveira JA, Carvalho FE. Proteomics, photosynthesis and salt resistance in crops: An integrative view. Journal of Proteomics. 2016;**143**:24-35. DOI: 10.1016/j.jprot.2016.03.013

[16] Latef AAHA, Alhmad MFA, Abdelfattah KE. The possible roles of priming with ZnO nanoparticles in mitigation of salinity stress in lupine (Lupinus termis) plants. Journal of Plant Growth Regulation. 2017;**36**(1):60-70. DOI: 10.1007/s00344-016-9618

[17] Rizwan M, Ali S, ur Rehman MZ, Adrees M, Arshad M, Qayyum MF, et al. Alleviation of cadmium accumulation in maize (Zea mays L.) by foliar spray of zinc oxide nanoparticles and biochar to contaminated soil. Environmental Pollution. 2019;**248**:358-367. DOI: 10.1016/j.envpol.2019.02.031

[18] Alabdallah NM, Alzahrani HS. The potential mitigation effect of ZnO nanoparticles on [Abelmoschus esculentus L. Moench] metabolism under salt stress conditions. Saudi Journal of Biological Sciences. 2020;**27**(11):3132-3137. DOI: 10.1016/j.sjbs.2020.08.005

[19] Farouk S, Al-Amri SM. Exogenous zinc forms counteract NaCl-induced damage by regulating the antioxidant system, osmotic adjustment substances, and ions in canola (Brassica napus L. cv. Pactol) plants. Journal of soil science and plant. Nutrition. 2019;**19**(4):887-899. DOI: 10.1007/s42729-019-00087

[20] Acosta-Motos JR, Ortuño MF, Bernal-Vicente A, Diaz-Vivancos P, Sanchez-Blanco MJ, Hernandez JA. Plant responses to salt stress: Adaptive mechanisms. Agronomy. 2017;**7**(1):18. DOI: 10.3390/agronomy7010018

[21] Venkatachalam P, Priyanka N, Manikandan K, Ganeshbabu I, Indiraarulselvi P, Geetha N, et al. Enhanced plant growth promoting role of phycomolecules coated zinc oxide nanoparticles with P supplementation in cotton (Gossypium hirsutum L.). Plant Physiology and Biochemistry. 2017;**110**:118-127. DOI: 10.1016/j.plaphy.2016.09.004

[22] Li J, Hu J, Xiao L, Gan Q, Wang Y. Physiological effects and fluorescence labeling of magnetic iron oxide nanoparticles on citrus (Citrus reticulata) seedlings. Water, Air, & Soil Pollution. 2017;**228**(1):52. DOI: 10.1007/s11270-016-3237-9

[23] Askary M, Talebi SM, Amini F, Bangan ADB. Effects of iron nanoparticles on Mentha piperita L. under salinity stress. Biologija. 2017;**63**(1):65-75. DOI: 10.6001/biologija.v63i1.3476

[24] Mozafari A-a, Ghaderi N. Grape response to salinity stress and role of iron nanoparticle and potassium silicate to mitigate salt induced damage under in vitro conditions. Physiology and Molecular Biology of Plants. 2018;**24**(1):25-35. DOI: 10.1007/s12298-017-0488

[25] Torabian S, Farhangi-Abriz S, Zahedi M. Efficacy of FeSO4 nano formulations on osmolytes and antioxidative enzymes of sunflower under salt stress. Indian Journal of

Plant Physiology. 2018;**23**(2):305-315. DOI: 10.1007/s40502-018-0366-8

[26] Pinedo-Guerrero ZH, Cadenas-Pliego G, Ortega-Ortiz H, González-MoralesS,Benavides-MendozaA, Valdés-Reyna J, et al. Form of silica improves yi-ld, fruit quality and antioxidant defense system of tomato plants under salt stress. Agriculture. 2020;**10**(9):367. DOI: 10.3390/agriculture10090367

[27] Siddiqui MH, Al-Whaibi MH, Firoz M, Al-Khaishany MY. Role of nanoparticles in plants. In: Siddiqui MH, Al-Whaibi MH, Firoz M, editors. Nanotechnology and Plant Sciences. 1st ed. Cham: Springer; 2015. p. 19-35. DOI: 10.1007/978-3-319-14502-0_2

[28] Kalteh M, Alipour ZT, Ashraf S, Marashi Aliabadi M, Falah NA. Effect of silica nanoparticles on basil (Ocimum basilicum) under salinity stress. Journal of Chemical Health Risks. 2014;**4**(3):49-55. DOI: 10.22034/JCHR.2018.544075

[29] Janmohammadi M, Sabaghnia N, Ahadnezhad A. Impact of silicon dioxide nanoparticles on seedling early growth of lentil (Lens culinaris Medik.) genotypes with various origins. Agriculture & Forestry. 2015;**61**(3):19-33. DOI: 10.17707/AgricultForest.61.3.02

[30] Gao X, Zou C, Wang L, Zhang F. Silicon decreases transpiration rate and conductance from stomata of maize plants. Journal of Plant Nutrition. 2006;**29**(9):1637-1647. DOI: 10.1080/01904160600851494

[31] Farhangi-Abriz S, Torabian S. Nano-silicon alters antioxidant activities of soybean seedlings under salt toxicity. Protoplasma. 2018;**255**(3):953-962. DOI: 10.1007/s00709-017-1202-0

[32] Mushtaq A, Rizwan S, Jamil N, Ishtiaq T, Irfan S, Ismail T, et al. Influence of silicon sources and controlled release fertilizer on the growth of wheat cultivars of Balochistan under salt stress. Pakistan Journal of Botany. 2019;**51**(5):1561-1567. DOI: 10.30848/PJB2019-5(44)

[33] Avestan S, Ghasemnezhad M, Esfahani M, Byrt CS. Application of nano-silicon dioxide improves salt stress tolerance in strawberry plants. Agronomy. 2019;**9**(5):246. DOI: 10.3390/agronomy9050246

[34] Cao Z, Rossi L, Stowers C, Zhang W, Lombardini L, Ma X. The impact of cerium oxide nanoparticles on the physiology of soybean (Glycine max (L.) Merr.) under different soil moisture conditions. Environmental Science and Pollution Research. 2018;**25**(1):930-939. DOI: 10.1007/s11356-017-0501-5

[35] Jurkow R, Sękara A, Pokluda R, Smoleń S, Kalisz A. Biochemical response of oakleaf lettuce seedlings to different concentrations of some metal (oid) oxide nanoparticles. Agronomy. 2020;**10**(7):997. DOI: 10.3390/agronomy10070997

[36] Mohammadi MHZ, Panahirad S, Navai A, Bahrami MK, Kulak M, Gohari G. Cerium oxide nanoparticles (CeO2-NPs) improve growth parameters and antioxidant defense system in Moldavian balm (Dracocephalum moldavica L.) under salinity stress. Plant. Stress. 2021;**1**:100006. DOI: 10.1016/j.stress.2021.100006

[37] Rossi L, Zhang W, Ma X. Cerium oxide nanoparticles alter the salt stress tolerance of Brassica napus L. by modifying the formation of root apoplastic barriers. Environmental Pollution. 2017;**229**:132-138. DOI: 10.1016/j.envpol.2017.05.083

[38] Rossi L, Zhang W, Lombardini L, Ma X. The impact of cerium oxide

nanoparticles on the salt stress responses of Brassica napus L. Environmental Pollution. 2016;**219**:28-36. DOI: 10.1016/j.envpol.2016.09.060

[39] Banan A, Kalbassi MR, Bahmani M, Sotoudeh E, Johari SA, Ali JM, et al. Salinity modulates biochemical and histopathological changes caused by silver nanoparticles in juvenile Persian sturgeon (Acipenser persicus). Environmental Science and Pollution Research. 2020;**27**:10658-10671. DOI: 10.1007/s11356-020-07687-7

[40] Almutairi ZM. Influence of silver nano-particles on the salt resistance of tomato (Solanum lycopersicum) during germination. International Journal of Agriculture and Biology. 2016;**18**(2):449-457. DOI: 10.17957/IJAB/15.0114

[41] Wahid I, Kumari S, Ahmad R, Hussain SJ, Alamri S, Siddiqui MH, et al. Silver nanoparticle regulates salt tolerance in wheat through changes in ABA concentration, ion homeostasis, and defense systems. Biomolecules. 2020;**10**(11):1506. DOI: 10.3390/biom10111506

[42] Khan I, Raza MA, Awan SA, Shah GA, Rizwan M, Ali B, et al. Amelioration of salt induced toxicity in pearl millet by seed priming with silver nanoparticles (AgNPs): The oxidative damage, antioxidant enzymes and ions uptake are major determinants of salt tolerant capacity. Plant Physiology and Biochemistry. 2020;**156**:221-232. DOI: 10.1016/j.plaphy.2020.09.018

[43] Buettner KM, Valentine AM. Bioinorganic chemistry of titanium. Chemical Reviews. 2012;**112**(3):1863-1881. DOI: 10.1021/cr1002886

[44] Carbajal-Vazquez VH, Gomez-Merino FC, Herrera-Corredor JA, Contreras-Oliva A, Alcantar-Gonzalez G, Trejo-TéLlez LI. Effect of titanium foliar applications on tomato fruits from plants grown under salt stress conditions. Notulae Botanicae Horti Agrobotanici Cluj-Napoca. 2020;**48**(2):924-937. DOI: 10.15835/nbha48211904

[45] Abdel Latef AAH, Srivastava AK, El-sadek MSA, Kordrostami M, Tran LSP. Titanium dioxide nanoparticles improve growth and enhance tolerance of broad bean plants under saline soil conditions. Land Degradation & Development. 2018;**29**(4):1065-1073. DOI: 10.1002/ldr.2780

[46] Gohari G, Mohammadi A, Akbari A, Panahirad S, Dadpour MR, Fotopoulos V, et al. Titanium dioxide nanoparticles (TiO 2 NPs) promote growth and ameliorate salinity stress effects on essential oil profile and biochemical attributes of Dracocephalum moldavica. Scientific Reports. 2020;**10**(1):1-14. DOI: 10.1038/s41598-020-57794-1

[47] Verma SK, Das AK, Gantait S, Kumar V, Gurel E. Applications of carbon nanomaterials in the plant system: A perspective view on the pros and cons. Science of the Total Environment. 2019;**667**:485-499. DOI: 10.1016/j.scitotenv.2019.02.409

[48] Baz H, Creech M, Chen J, Gong H, Bradford K, Huo H. Water-soluble carbon nanoparticles improve seed germination and post-germination growth of lettuce under salinity stress. Agronomy. 2020;**10**(8):1192. DOI: 10.3390/agronomy10081192

[49] Zhao G, Zhao Y, Lou W, Su J, Wei S, Yang X, et al. Nitrate reductase-dependent nitric oxide is crucial for multi-walled carbon nanotube-induced plant tolerance against salinity. Nanoscale. 2019;**11**(21):10511-10523. DOI: 10.1039/C8NR10514F

[50] Gohari G, Safai F, Panahirad S, Akbari A, Rasouli F, Dadpour MR, et al. Modified multiwall carbon nanotubes display either phytotoxic or growth promoting and stress protecting activity in Ocimum basilicum L. in a concentration-dependent manner. Chemosphere. 2020;**249**:126171. DOI: 10.1016/j.chemosphere.2020.126171

[51] Wan J, Wang R, Bai H, Wang Y, Xu J. Comparative physiological and metabolomics analysis reveals that single-walled carbon nanohorns and ZnO nanoparticles affect salt tolerance in Sophora alopecuroides. Environmental Science: Nano. 2020;**7**(10):2968-2981. DOI: 10.1039/D0EN00582G

[52] Shafiq F, Iqbal M, Ali M, Ashraf MA. Seed pre-treatment with polyhydroxy fullerene nanoparticles confer salt tolerance in wheat through upregulation of H2O2 neutralizing enzymes and phosphorus uptake. Journal of Soil Science and Plant Nutrition. 2019;**19**(4):734-742. DOI: 10.1007/s42729-019-00073-4

[53] Oliveira HC, Gomes BC, Pelegrino MT, Seabra AB. Nitric oxide-releasing chitosan nanoparticles alleviate the effects of salt stress in maize plants. Nitric Oxide. 2016;**61**:10-19. DOI: 10.1016/j.niox.2016.09.010

[54] Mujumdar N. Coping with water scarcity: An action framework for agriculture and food security/improving water use efficiency: New directions for water Management in India/Indian journal of agricultural economics. Indian Journal of Agricultural Economics. 2013;**68**(4):603

[55] Khot LR, Sankaran S, Maja JM, Ehsani R, Schuster EW. Applications of nanomaterials in agricultural production and crop protection: A review. Crop Protection. 2012;**35**:64-70. DOI: 10.1016/j.cropro.2012.01.007

[56] Gururaj SB, Krishna B. Water retention capacity of biochar blended soils. Journal of Chemical and Pharmaceutical Sciences. 2016;**9**(3):1438-1441

[57] Amjad S, Tiwari S, Serajuddin M. Applications of Nanobiotechnology in overcoming drought stress in crops. In: Nanobiotechnology. Switzerland: Springer; 2021. pp. 399-415. DOI: 10.1007/978-3-030-73606-4_17

[58] Sheng H, Chen S. Plant silicon-cell wall complexes: Identification, model of covalent bond formation and biofunction. Plant Physiology and Biochemistry. 2020;**155**:13-19. DOI: 10.1016/j.plaphy.2020.07.020

[59] Ashkavand P, Tabari M, Zarafshar M, Tomásková I, Struve D. Effect of SiO2 nanoparticles on drought resistance in hawthorn seedlings. Forest Research Papers/Leśne Prace Badawcze. 2015;**76**(4):350-359. DOI: 10.1515/frp-2015-0034

[60] Hattori T, Inanaga S, Araki H, An P, Morita S, Luxová M, et al. Application of silicon enhanced drought tolerance in Sorghum bicolor. Physiologia Plantarum. 2005;**123**(4):459-466. DOI: 10.1111/j.1399-3054.2005.00481.x

[61] Pei Z, Ming D, Liu D, Wan G, Geng X, Gong H, et al. Silicon improves the tolerance to water-deficit stress induced by polyethylene glycol in wheat (Triticum aestivum L.) seedlings. Journal of Plant Growth Regulation. 2010;**29**(1):106-115. DOI: 10.1007/s00344-009-9120-9

[62] Sedghi M, Hadi M, Toluie SG. Effect of nano zinc oxide on the germination parameters of soybean seeds under drought stress. Annales of West University of Timisoara Series of Biology. 2013;**16**(2):73

[63] Dimkpa C, Bindraban P, Fugice J, Agyin-Birikorang S, Singh U, Hellums D. Composite micronutrient nanoparticles and salts decrease drought stress in soybean. Agronomy for Sustainable Development. 2017;**37**:1-13. DOI: 10.1007/s13593-016-0412-8

[64] Taran N, Storozhenko V, Svietlova N, Batsmanova L, Shvartau V, Kovalenko M. Effect of zinc and copper nanoparticles on drought resistance of wheat seedlings. Nanoscale Research Letters. 2017;**12**(1):1-6. DOI: 10.1186/s11671-017-1839-9

[65] Yang K-Y, Doxey S, McLean JE, Britt D, Watson A, Al Qassy D, et al. Remodeling of root morphology by CuO and ZnO nanoparticles: Effects on drought tolerance for plants colonized by a beneficial pseudomonad. Botany. 2018;**96**(3):175-186. DOI: 10.1139/cjb-2017-0124

[66] Semida WM, Abdelkhalik A, Mohamed GF, Abd El-Mageed TA, Abd El-Mageed SA, Rady MM, et al. Foliar application of zinc oxide nanoparticles promotes drought stress tolerance in eggplant (Solanum melongena L.). Plants. 2021;**10**(2):421. DOI: 10.3390/plants10020421

[67] Sher A, Naveed K, Khan A. Grain zinc and iron enrichment through foliar application augments wheat yield under varying nitrogen regimes. Pakistan Journal of Botany. 2019;**52**(1):85-94. DOI: 10.30848/PJB2020-1(25)

[68] Saxena R, Tomar RS, Kumar M. Exploring nanobiotechnology to mitigate abiotic stress in crop plants. Journal of Pharmaceutical Sciences and Research. 2016;**8**(9):974

[69] Davar ZF, Roozbahani A, Hosnamidi A. Evaluation the effect of water stress and foliar application of Fe nanoparticles on yield, yield components and oil percentage of safflower (Carthamus tinctorious L). International Journal of Advanced Biological and Biomedical Research. 2014;**2**(4):1150-1159

[70] Martínez-Fernández D, Vítková M, Bernal MP, Komárek M. Effects of nano-maghemite on trace element accumulation and drought response of Helianthus annuus L. in a contaminated mine soil. Water, Air, & Soil Pollution. 2015;**226**(4):1-9. DOI: 10.1007/s11270-015-2365-y

[71] Yoon H, Kang Y-G, Chang Y-S, Kim J-H. Effects of zerovalent iron nanoparticles on photosynthesis and biochemical adaptation of soil-grown Arabidopsis thaliana. Nanomaterials. 2019;**9**(11):1543. DOI: 10.3390/nano9111543

[72] Kim J-H, Oh Y, Yoon H, Hwang I, Chang Y-S. Iron nanoparticle-induced activation of plasma membrane H+-ATPase promotes stomatal opening in Arabidopsis thaliana. Environmental Science & Technology. 2015;**49**(2):1113-1119. DOI: 10.1021/es504375t

[73] Jaberzadeh A, Moaveni P, Moghadam HRT, Zahedi H. Influence of bulk and nanoparticles titanium foliar application on some agronomic traits, seed gluten and starch contents of wheat subjected to water deficit stress. Notulae Botanicae Horti Agrobotanici Cluj-Napoca. 2013;**41**(1):201-207. DOI: 10.15835/nbha4119093

[74] Akbari G-A, Morteza E, Moaveni P, Alahdadi I, Bihamta M-R, Hasanloo T. Pigments apparatus and anthocyanins reactions of borage to irrigation, methylalchol and titanium dioxide. International Journal of Biosciences. 2014;**4**(7):192-208. DOI: 10.12692/ijb/4.7

[75] Kiapour H, Moaveni P, Habibi D. Evaluation of the application of gibbrellic

acid and titanium dioxide nanoparticles under drought stress on some traits of basil (Ocimum basilicum L). International Journal of Agronomy and Agricultural Research. 2015;**6**(4):138-150

[76] Aghdam MTB, Mohammadi H, Ghorbanpour M. Effects of nanoparticulate anatase titanium dioxide on physiological and biochemical performance of Linum usitatissimum (Linaceae) under well-watered and drought stress conditions. Brazilian Journal of Botany. 2016;**39**(1):139-146. DOI: 10.1007/s40415-015-0227-x

[77] Laware S, Raskar S. Effect of titanium dioxide nanoparticles on hydrolytic and antioxidant enzymes during seed germination in onion. International Journal of Current Microbiology and Applied Sciences. 2014;**3**(7):749-760

[78] Hassan SA, Hagrassi AME, Hammam O, Soliman AM, Ezzeldin E, Aziz WM. Brassica juncea L.(Mustard) extract silver nanoparticles and knocking off oxidative stress, proinflammatory cytokine and reverse DNA genotoxicity. Biomolecules. 2020;**10**(12):1650. DOI: 10.3390/biom10121650

[79] Hojjat SS, Ganjali A. The effect of silver nanoparticle on lentil seed germination under drought stress. International Journal of Farming and Allied Sciences. 2016;**5**(3):208-212

[80] Djanaguiraman M, Nair R, Giraldo JP, Prasad PVV. Cerium oxide nanoparticles decrease drought-induced oxidative damage in sorghum leading to higher photosynthesis and grain yield. ACS Omega. 2018;**3**(10):14406-14416. DOI: 10.1021/acsomega.8b01894

[81] García-Caparrós P, Hasanuzzaman M, Lao MT. Oxidative stress and antioxidant defense in plants under salinity. In: Hasanuzzaman M, Fotopoulos V, Nahar K, Fujita M, editors. Reactive Oxygen, Nitrogen and Sulfur Species in Plants: Production, Metabolism, Signaling and Defense Mechanisms. 1st ed. Chichester: Wiley; 2019. p. 291-309. DOI: 10.1002/9781119468677.ch12

[82] Szymańska R, Ślesak I, Orzechowska A, Kruk J. Physiological and biochemical responses to high light and temperature stress in plants. Environmental and Experimental Botany. 2017;**139**:165-177. DOI: 10.1016/j.envexpbot.2017.05.002

[83] Jalil SU, Ansari MI. Nanoparticles and abiotic stress tolerance in plants: Synthesis, action, and signaling mechanisms. In: Iqbal M, Khan R, Reddy PS, Ferrante A, Khan NA, editors. Plant Signaling Molecules. 1st ed. Woodhead Publishing; 2019. p. 549-561. DOI: 10.1016/B978-0-12-816451-8.00034-4

[84] Prażak R, Święciło A, Krzepiłko A, Michałek S, Arczewska M. Impact of Ag nanoparticles on seed germination and seedling growth of green beans in normal and chill temperatures. Agriculture. 2020;**10**(8):312. DOI: 10.3390/agriculture10080312

[85] Singh H, Kumar D, Soni V. Copper and mercury induced oxidative stresses and antioxidant responses of Spirodela polyrhiza (L.) Schleid. Biochemistry and Biophysics Reports. 2020;**23**:100781. DOI: 10.1016/j.bbrep.2020.100781

[86] Manaf A, Wang X, Tariq F, Jhanzab HM, Bibi Y, Sher A, et al. Antioxidant enzyme activities correlated with growth parameters of wheat sprayed with silver and gold nanoparticle suspensions. Agronomy. 2021;**11**(8):1494. DOI: 10.3390/agronomy11081494

[87] Giraldo JP, Landry MP, Faltermeier SM, McNicholas TP,

Iverson NM, Boghossian AA, et al. Plant nanobionics approach to augment photosynthesis and biochemical sensing. Nature Materials. 2014;**13**(4):400-408. DOI: 10.103/nmat3890

[88] Pavithra G, Reddy BR, Salimath M, Geetha K, Shankar A. Zinc oxide nano particles increases Zn uptake, translocation in rice with positive effect on growth, yield and moisture stress tolerance. Indian Journal of Plant Physiology. 2017;**22**(3):287-294. DOI: 10.1007/s40502-017-03032

[89] Kaveh R, Li Y-S, Ranjbar S, Tehrani R, Brueck CL, Van Aken B. Changes in Arabidopsis thaliana gene expression in response to silver nanoparticles and silver ions. Environmental Science & Technology. 2013;**47**(18):10637-10644. DOI: 10.1021/es402209w

[90] Mirzajani F, Askari H, Hamzelou S, Schober Y, Römpp A, Ghassempour A, et al. Proteomics study of silver nanoparticles toxicity on bacillus thuringiensis. Ecotoxicology and Environmental Safety. 2014;**100**:122-130. DOI: 10.1016/j.ecoenv.2013.10.009

[91] Sofy AR, Sofy MR, Hmed AA, Dawoud RA, Alnaggar AE-AM, Soliman AM, et al. Ameliorating the adverse effects of tomato mosaic tobamovirus infecting tomato plants in Egypt by boosting immunity in tomato plants using zinc oxide nanoparticles. Molecules. 2021;**26**(5):1337. DOI: 10.3390/molecules26051337

[92] Lei Z, Mingyu S, Xiao W, Chao L, Chunxiang Q, Liang C, et al. Antioxidant stress is promoted by nano-anatase in spinach chloroplasts under UV-B radiation. Biological Trace Element Research. 2008;**121**(1):69-79. DOI: 10.1007/s12011-007-8028-0

[93] Fatemi H, Pour BE, Rizwan M. Foliar application of silicon nanoparticles affected the growth, vitamin C, flavonoid, and antioxidant enzyme activities of coriander (Coriandrum sativum) plants grown in lead (Pb)-spiked soil. Environmental Science and Pollution Research. 2021;**28**(2):1417-1425. DOI: 10.1007/s11356-020-10549-x

[94] Banerjee J, Kole C. Plant nanotechnology: An overview on concepts, strategies, and tools. In: Kole C, Sakthi KD, Mariya VK, editors. Plant Nanotechnology: Principles and practices. 1st ed. Switzerland: Springer; 2016. p. 1-14. DOI: 10.1007/978-3-319-42154-4_1

[95] Singh A, Tiwari S, Pandey J, Lata C, Singh IK. Role of nanoparticles in crop improvement and abiotic stress management. Journal of Biotechnology. 2021;**337**:57-70. DOI: 10.1016/j.jbiotec.2021.06.022

[96] Kaphle A, Navya P, Umapathi A, Daima HK. Nanomaterials for agriculture, food and environment: Applications, toxicity and regulation. Environmental Chemistry Letters. 2018;**16**(1):43-58. DOI: 10.1007/s10311-017-0662-y

[97] Raliya R, Franke C, Chavalmane S, Nair R, Reed N, Biswas P. Quantitative understanding of nanoparticle uptake in watermelon plants. Frontiers in Plant Science. 2016;**7**:1288. DOI: 10.3389/fpls.2016.01288

[98] Zhao L, Lu L, Wang A, Zhang H, Huang M, Wu H, et al. Nano-biotechnology in agriculture: Use of nanomaterials to promote plant growth and stress tolerance. Journal of Agricultural and Food Chemistry. 2020;**68**(7):1935-1947. DOI: 10.1021/acs.jafc.pb06615

[99] Noori A, Donnelly T, Colbert J, Cai W, Newman LA, White JC. Exposure

of tomato (Lycopersicon esculentum) to silver nanoparticles and silver nitrate: Physiological and molecular response. International Journal of Phytoremediation. 2020;**22**(1):40-51. DOI: 10.1080/15226514.2019.1634000

[100] Zahedi SM, Karimi M, Teixeira da Silva JA. The use of nanotechnology to increase quality and yield of fruit crops. Journal of the Science of Food and Agriculture. 2020;**100**(1):25-31. DOI: 10.1002/jsfa.10004

[101] Arif N, Yadav V, Singh S, Tripathi DK, Dubey NK, Chauhan DK, et al. Interaction of copper oxide nanoparticles with plants: Uptake, accumulation, and toxicity. In: Nanomaterials in Plants, Algae, and Microorganisms. Academic Press, India: Elsevier; 2018. pp. 297-310. DOI: 10.106/B978-0-12-811487-2.00013-X

[102] Raliya R, Tarafdar JC, Biswas P. Enhancing the mobilization of native phosphorus in the mung bean rhizosphere using ZnO nanoparticles synthesized by soil fungi. Journal of Agricultural and Food Chemistry. 2016;**64**(16):3111-3118. DOI: 10.1021/acs.jafc.5b05224

[103] Larue C, Laurette J, Herlin-Boime N, Khodja H, Fayard B, Flank A-M, et al. Accumulation, translocation and impact of TiO2 nanoparticles in wheat (Triticum aestivum):Influence of diameter and crystal phase. Science of the Total Environment. 2012;**431**:197-208. DOI: 10.1016/j.scitotenv.2012.04.073

[104] Khan N, Bano A. Modulation of phytoremediation and plant growth by the treatment with PGPR, Ag nanoparticle and untreated municipal wastewater. International Journal of Phytoremediation. 2016;**18**(12):1258-1269. DOI: 10.1080/15226514.2014.1203287

[105] Ali S, Mehmood A, Khan N. Uptake, translocation, and consequences of nanomaterials on plant growth and stress adaptation. Journal of Nanomaterials. 2021:17 Article ID 6677616. DOI: 10.1155/2021/6677616

[106] Milewska-Hendel A, Sala K, Gepfert W, Kurczyńska E. Gold nanoparticles-induced modifications in cell wall composition in barley roots. Cell. 2021;**10**(8):1965. DOI: 10.3390/cells10081965

[107] Lv J, Zhang S, Luo L, Zhang J, Yang K, Christie P. Accumulation, speciation and uptake pathway of ZnO nanoparticles in maize. Environmental Science: Nano. 2015;**2**(1):68-77. DOI: 10.1039/C4EN00064A

[108] Hernandez-Viezcas JA, Castillo-Michel H, Andrews JC, Cotte M, Rico C, Peralta-Videa JR, et al. In situ synchrotron X-ray fluorescence mapping and speciation of CeO2 and ZnO nanoparticles in soil cultivated soybean (Glycine max). ACS Nano. 2013;7(2):1415-1423. DOI: 10.1021/nn305196q

[109] Da Costa M, Sharma P. Effect of copper oxide nanoparticles on growth, morphology, photosynthesis, and antioxidant response in Oryza sativa. Photosynthetica. 2016;**54**(1):110-119. DOI: 10.1007/s11099-015-0167-5

[110] Xiong T, Dumat C, Dappe V, Vezin H, Schreck E, Shahid M, et al. Copper oxide nanoparticle foliar uptake, phytotoxicity, and consequences for sustainable urban agriculture. Environmental Science & Technology. 2017;**51**(9):5242-5251. DOI: 10.1021/acs.est.6b05546

[111] Jahagirdar PS, Gupta PK, Kulkarni SP, Devarajan PV. Polymeric curcumin nanoparticles by a facile in situ method for macrophage targeted delivery. Bioengineering & Translational

Medicine. 2019;**4**(1):141-151. DOI: 10.1002/btm2.10112

[112] Rao S, Shekhawat GS. Phytotoxicity and oxidative stress perspective of two selected nanoparticles in Brassica juncea. 3 Biotech. 2016;**6**(2):1-12. DOI: 10.1007/s13205-016-0550-3

[113] Shende S, Rathod D, Gade A, Rai M. Biogenic copper nanoparticles promote the growth of pigeon pea (Cajanus cajan L.). IET Nanobiotechnology. 2017;**11**(7):773-781

[114] Apodaca SA, Tan W, Dominguez OE, Hernandez-Viezcas JA, Peralta-Videa JR, Gardea-Torresdey JL. Physiological and biochemical effects of nanoparticulate copper, bulk copper, copper chloride, and kinetin in kidney bean (Phaseolus vulgaris) plants. Science of the Total Environment. 2017;**599**:2085-2094. DOI: 10.1016/j.scitotenv.2017.05.095

[115] Thuesombat P, Hannongbua S, Akasit S, Chadchawan S. Effect of silver nanoparticles on rice (Oryza sativa L. cv. KDML 105) seed germination and seedling growth. Ecotoxicology and Environmental Safety. 2014;**104**:302-309. DOI: 10.1016/j.ecoenv.2014.03.022

[116] Adisa IO, Rawat S, Pullagurala VLR, Dimkpa CO, Elmer WH, White JC, et al. Nutritional status of tomato (Solanum lycopersicum) fruit grown in fusarium-infested soil: Impact of cerium oxide nanoparticles. Journal of Agricultural and Food Chemistry. 2020;**68**(7):1986-1997. DOI: 0.1021/acs.jafc.9b06840

[117] Raliya R, Nair R, Chavalmane S, Wang W-N, Biswas P. Mechanistic evaluation of translocation and physiological impact of titanium dioxide and zinc oxide nanoparticles on the tomato (Solanum lycopersicum L.) plant. Metallomics. 2015;**7**(12):1584-1594. DOI: 10.1039/c5mt00168d

[118] Shinde S, Paralikar P, Ingle AP, Rai M. Promotion of seed germination and seedling growth of Zea mays by magnesium hydroxide nanoparticles synthesized by the filtrate from aspergillus Niger. Arabian Journal of Chemistry. 2020;**13**(1):3172-3182. DOI: 10.1016/j.arabjc.2018.10.001

Chapter 10

Seed Nanopriming to Mitigate Abiotic Stresses in Plants

Afagh Yavari, Elham Ghasemifar and Mehdi Shahgolzari

Abstract

Abiotic stresses affect crop plants extensively during their life span, reducing productivity and threatening global food security. Stress conditions can result in failures of seed germination, uniformity, crop yield, cellular redox homeostasis, and the over-accumulation of reactive oxygen species. Seed nanopriming, pre-treating seeds with nanoparticles, is one way to overcome these limitations and successfully increase the tolerance of plants to future biotic and abiotic stress conditions. Nanopriming can play a significant role through the induction of several metabolic and physiological methods to better tolerate abiotic stresses. However, further research is needed to determine whether nanoparticles are stress promoters or stress inhibitors in plant systems. Here, we review how nanoparticle agents-based seed priming has the capacity to mitigate abiotic stresses.

Keywords: agriculture, nanoparticle, priming, seed, stress, tolerance

1. Introduction

The stressor factors negatively affect plant growth, development, and seed yield, which are commonly connected to biochemical, physiological, and molecular variations [1]. Priming is a technique to mitigate these stresses that allows plants to deploy a stronger and speedier defense response against of them [2]. Priming induces a collection of metabolic activities in seeds and seedlings which help them to tolerate various abiotic stresses. Therefore, under subsequent stimuli, plants can show better growth biomarkers and stress tolerance when repeatedly exposed to stress [3]. Pathogens, pests, useful microorganisms, natural and synthetic compounds, nanomaterials, and the existence of abiotic stresses at mild levels can trigger a priming event [4, 5].

There have been several advancements in use of nanoparticles (NPs) for improving sustainability in agriculture, such as nanopesticides, nanofertilizers, and nanosensors [6]. NPs can optimize depending on their unique physicochemical characteristics in order to increase the growth and development of plants, and resilience to stressful conditions [7, 8]. The utilization of NPs is also being investigated as a priming agent to ensure better germination and growth of the seedling, thereby increasing plant yields and nutritional value [9, 10]. The extremely small size of NPs, their surface area, and their slow release rate aid plants in increasing nutrient uptake [11]. There are many benefits associated with nanoparticle-based priming, including change in

IntechOpen

metabolism, physiology, enzyme activity, and their interplay with phytohormones, etc. [12, 13]. Here, we summarize the potential use of seed nanopriming to mitigate abiotic stresses.

2. Nanoparticles in agriculture

Nanoparticles on the nanoscale (less than 100 nm) have the capacity to contribute to a new technology-based agricultural revolution [14]. NPs exhibit unique physicochemical properties such as high surface area-to-volume ratios and high reactivity that make them suitable for several of agriculture applications. NPs can load and deliver agrochemicals (e.g., fertilizers and pesticides) with controlled releases, biomolecules (e.g., nucleotides, proteins, activators), and monitoring plant health (e.g., sensors) [5]. Agri-nanosustainability can use NPs to stimulate plant growth, increase crop productivity, protect plants, improve soil quality, and detect pathogens and pesticide residues [15]. During the last decade, NPs have been widely used as fertilizers or metal fertilizers [16].

NPs can enter into cell by direct diffusion, endocytosis, and channel process [4]. A key factor in NP delivery could be the electrical gradient across the cell membranes [17]. The efficacies of passage are related to several properties, namely particle size, hydrophobicity, structure, charge, and shape [18]. NPs also can transfer from the cell to the tissues *via* the apoplastic or symplastic route in foliar/shoots or roots [19]. The entry of NPs into plant cells can facilitate *via* the plasmodesmata [20], aquaporins [21], ion channels [22], cuticle membrane and stomata [23], vasculature [24]. For example, gold NPs (AuNPs) can be transported through plasmodesmata [25].

Positive impacts of NPs in plants can be achieved *via* foliar spray, root exposure, and seed priming to improve plant performance under biotic and abiotic stress conditions [17, 18]. For example, foliar use of ZnO nanoparticles and TiO2 in sunflower can induce physiological responses and increase resistance to drought and water depletion [18, 26]. Recent research has indicated that interfacing plant seeds with NPs has positive impacts on field crops under stress [5, 27].

3. Nanoparticles as seed priming agents (seed nanopriming)

Pre-treatment of seed and plant with chemical and biological agents can enter plants to the primed state (PS), which allows deploying faster and stronger responses compared with a non-primed [28]. Seed priming is pre-treating seeds before planting the seeds for a certain period of time in salt solutions (halo-priming), water (hydro-conditioning), osmotic agents (osmo-priming), plant hormone solutions (hormonal priming), valuable microbe solutions (bio-priming), under a magnetic field (magneto-priming), and solutions containing NPs (NPs) (nanopriming) [4]. In seed priming, various biochemical changes occur in the seed that increments the germination rate, consistency of development, abdicate, and resistance of seedlings against unfavorable natural conditions [3, 29]. Seed priming results in actuating chemicals dependable for fetus advancement and mining of the bland endosperm *via* increased water content. In addition, seed priming can initiate biochemical reactions of cell repair, increment RNA substance, and improve DNA replication. Seed priming can increase the activity of antioxidant enzymes such as superoxide dismutase, catalase, and glutathione reductase for improving the defense system [30]. Priming increases seed yield, crop uniformity, and germination, improves performance under various environmental conditions, and helps overcome dormancy. This causes variations in seed water content, regulation of cell cycle, and alteration of seed ultrastructure. Seed priming and foliar application reduce the stress response during seed development and seedling establishment [31].

Seed nanopriming, pre-treatment of seed with NPs, is an emerging method for seed priming [27]. Nanopriming involves soaking seeds in nanosuspensions or nanoformulations, and the seeds may or may not absorb the NPs [27, 32]. NPs are mostly absorbed, but mostly remain on the seed surface as coatings [27]. Seed nanopriming can promote seed germination *via* organization of nanopores in the seed coat for increasing uptake of water [3, 33]. The nanoparticle introduces reactive oxygen species (ROS) to the seed, makes aquaporin genes active, alters starch degradation enzymes, and alters the metabolism of seed tissue [3]. Seed coat uptake of NPs results in the aggregation of reactive oxygen species (ROS), resulting in a chain reaction [34]. ROS localization is important for the communication between cells in endosperm, as well as for the breakdown of glycosidic links between polysaccharides [35]. Superoxide dismutase (SOD) allows the interplay between H_2O_2 and the phytohormone gibberellic acid (GA). GA activates alpha-amylase to stabilize the hydrolysis of starch into highly soluble sugars to support embryo development and ultimately seed germination and thus seedling vigor [3, 36] (**Figure 1**). The absorption and transfer of nanoparticles into the seed compartment is affected by the anatomy of the seed coat. Fewer parenchyma cell layers and larger intracellular voids lead to faster uptake and translocation of NPs into seed compartments, i.e., seed coat, cotyledon, and radicle [27]. There have been several studies concerned with the application of NPs (NPs) as seed pre-treatment agents, including metal-based NPs (e.g., Ag NPs, Au NPs, Cu NPs, Fe NPs, FeS2 NPs, TiO2 NPs, Zn NPs, and ZnO NPs), carbon-based NPs (e.g., fullerene and carbon nanotubes), and polymeric NPs.

3.1 Polymeric nanoparticles

Natural and synthetic polymer nanoparticles are used for controlled release of fertilizers and pesticides in precision agriculture. For example, synthetic polysuccinimide polymeric NPs (PSI-NPs) were detailed to have awesome potential focusing on conveyance of anti-microbial in plants with negligible effect on soil quality. Recently, the Impacts of PSI-NPs on seed germination and seedlings of maize (Zea mays L.) demonstrate that PSI-NPs could mitigate the influence of the heavy metals stress (e.g. Cu) and phytotoxicity with the increase of antioxidant enzyme activities and storage of copper as Cu-PSI complexes [37].

Seed priming with natural chitosan nanoparticles (CSNP) increased salt tolerance in milk thistle seedlings by improving physiological mechanisms such as photosynthetic pigment synthesis, antioxidant enzyme activity, and free proline content [38]. Nanopriming of maize (*Z. mays* L.) seeds with chitosan NPs containing copper particles (NPCu) combined the properties of chitosan with the essentiality of Cu2+, advancing the enzymatic antioxidant reaction [39]. The result of distinctive concentrations of CSNPs appeared that adsorption of CSNPs on the surface of wheat seeds can initiate the auxin biosynthesis and increment seed germination and seedling development of wheat [40]. Seed priming of *Vicia faba* seeds cv. Sakha 1 study shows that the moderately low concentration of chitosan NPs improved the defense system of seeds by expanding total phenols and antioxidant enzyme activities [41].

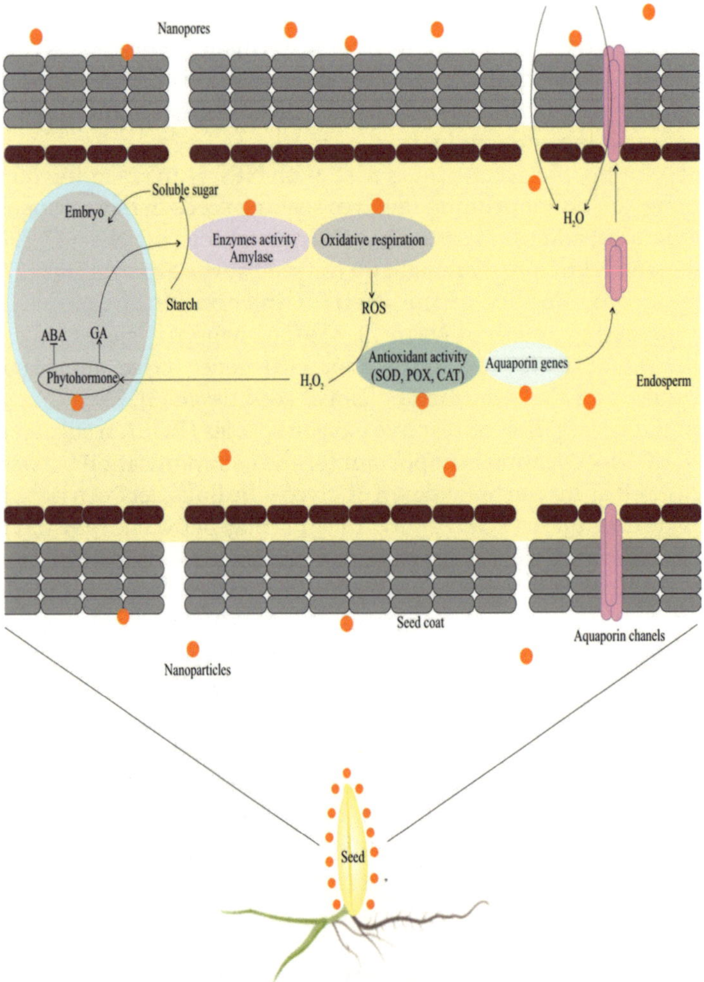

Figure 1.
Seed nanopriming can regulate abiotic stress tolerance via different mechanisms. Seed nanopriming can lead to the creation of nanopores and facilitate the uptake of nanoparticles (NPs) and water. NPs can enhance the expression of aquaporin genes and change metabolism. Nanopriming increases oxidative activity and produces reactive oxygen species (ROS, e.g., superoxide radical (O_2.–), hydrogen peroxide (H_2O_2)). NPs can activate antioxidant enzymes (e.g., SOD), and the conversion of ROS to H_2O_2. Diffusion of H_2O_2 to embryo and interaction with phytohormone gibberellic acid (GA) lead to GA activating, its impact on α-amylase to stabilize starch hydrolysis to highly soluble sugars to support embryo development and ultimately seed germination and thus seedling vigor.

3.2 Metallic nanoparticles

Metallic NPs (MNPs) consist of a metal core consisting of an inorganic metal or metal oxide [42]. Metallic nanomaterials in seeds can improve stress tolerance in plants. Seed nanopriming with Fe-NPs [43], TiO2-NPs [44], AgNPs [45], poly (acrylic acid)-coated cerium oxide NPs (PNC) [5], ZnO NPs [46] can induce defense responses to stressors. For example, it is demonstrated that priming seeds with PNC in cotton (*Gossypium hirsutum* L.) were associated with ROS, and Ca2+-related signaling pathways in salinity tolerance [5]. Similarly, nanopriming with TiO2 mitigates the salinity injury in maize *via* improving germination indexes, ion hemostasis, relative water content (RWC), non-enzymatic/enzymatic antioxidants as compared to non-primed [44]. Among the metal oxide NPs, zinc oxide NPs have attracted owing to unique photocatalytic and photo-oxidizing capacity against chemical and biological species the attention of many researchers [47]. For example, ZnO NPs alleviated the toxic impacts of cobalt (Co) by reducing its absorption and conferred stability to plant ultra-cellular structures and photosynthetic apparatus [46]. It has been demonstrated that nanoprimed seeds with solutions containing AgNPs can loosen seed coat cell walls and the endosperm, which in turn will induce seed germination [48].

3.3 Carbon nanoparticles

Carbon NPs (CNPs), such as single-walled carbon nanotubes (SWCNTs), multi-walled carbon nanotubes (MWCNTs), graphene (GR), and fullerenes, can stimulate a variety of positive responses owing to their unique chemical and physical properties, including accelerated growth and development, improved performance, and greater tolerance of stress [49]. It is known that CNTs can penetrate thick seed coats and support water absorption inside seeds, and may affect tomato seed germination and growth [50]. The penetration of MWCNTs into the seed coats of corn, barley, and soybean can induce the expression of genes encoding of water channel proteins [51]. MWCNTs functionalized with carboxylic acids can help to resolve seed dormancy in boreal forests by modulating lipid metabolism in cell membranes [52]. A significant increase in drought tolerance was achieved with SWCNTs at low concentrations by modifying water absorption and activating plant defense mechanisms, including up-regulating starch hydrolysis processes and reducing oxidative damage markers (e.g., H_2O_2, malondialdehyde concentrations) and electrolyte leakage [53]. A microscopic observation of seeds of *Sorbus luristanica* Bornm revealed seed endocarp abrasion and oxygen and moisture infiltration during MWCNT-priming [54].

The results indicate that MWCNTs enhanced tolerance of plant under Cd toxicity by active antioxidant enzymes (peroxidase (POD), superoxide dismutase (SOD), and catalase (CAT) activities) and reduced the malonaldehyde (MDA) [55]. The data presented demonstrate that sCNPs treatments can improve seed germination in plant species, including boreal forest native species, green alder, and lettuce [56, 57]. It was shown that fullerene nanopriming in wheat increments growth and productivity under salt stress [58]. Under drought stress, nanopriming of Caucasian alder seeds demonstrated that MWCNTs can be utilized to increment seed and seedling tolerance [59]. When tomato plants are subjected to salt stress, carbon nanomaterials can be added to seed by priming, modifying the bioactive compounds in the fruit and improving the antioxidant defenses. As a result, the plant may be protected from the negative effects of salinity stress [60].

Seeds treated with SWCNTs showed improved drought tolerance, and the combination of SOD, CAT, and POD activity can be responsible for improved antioxidant capacity under drought conditions [61].

4. Effects of seed nanopriming on plants

Nanopriming can improve the seed germination, stability, growth, and physiology of plant species by changing absorption, biochemical processes, antioxidants, photosynthesis [62, 63]. Various investigations demonstrate that nanoprimed seeds can better maintenance of cell balance and photosynthetic capacity [47, 62], increasing nutrient uptake and photosynthetic efficiency [59, 62], increased chlorophyll capacity and antioxidant activity, defense mechanisms (e.g., changes in osmotic pressure, stomatal movements) [64, 65]. It also removes the absorption of heavy metals (copper, cadmium, and zinc) and thus reduces toxicity [66].

Seed priming by NPs was discovered as a novel approach for regulating antioxidant enzymes in plants [67]. Plant antioxidant systems include non-enzymatic compounds and various enzymes, such as catalase (CAT), peroxidase (POX), ascorbate peroxidase (APX), superoxide dismutase (SOD), phenylalanine ammonia lyase (PAL), and glutathione. Priming of corn seeds with sodium metasilicate increased the activities of SOD, CAT, and POX under salt stress [68]. Similarly, maize seeds primed with TiO2 NPs increased SOD, CAT, and PAL activity [44]. Priming rice seeds with ZnO NPs enhances SOD and POD activity [67]. In addition, priming of Egyptian roselle (*Hibiscus sabdariffa* L.) seeds with Al2O3 NPs enhanced SOD, CAT, POD, and APX functions [69]. Priming of lavender (*Lavandula angustifolia* Mill.) by silver nanoparticles improved the performance of APX, POX, and SOD [70]. Wheat seeds primed with Si NPs increased SOD, POD, and CAT functions in cadmium stress [71]. Antioxidants produce ROS in response to environmental stressors. Antioxidant enzymes determine how ROS indirectly helps germination of nanoprimed seeds [72]. However, studies on the modulation of antioxidant enzymes by NPs priming are very few.

ROS as a by-product has a signaling role in germination and reducing seed dormancy [73]. This can occur by activating GA synthesis [74]. The accumulation of ROS such as superoxide (O2•-), hydrogen peroxide (H_2O_2), and hydroxyl radicals (OH) causes oxidative stress [73, 75]. Seeds generally receive NPs as extrinsic factors [72], and accumulation in the seed coat causes ROS production [34]. The increase of ROS in non-primed seeds is connected to the increase of abscisic acid (ABA), which caused disruption of seed dormancy and seed germination [76]. Furthermore, nanopriming increased ROS levels in plant cells, disrupted seed endosperm cell wall junctions, and promoted rapid and healthy seed germination [77]. Nanopriming can regulate ROS under normal and stress conditions, and seed nanopriming can regulate ROS production for faster seed germination. Under conditions, seeds accumulate ROS, and nanopriming can regulate ROS hemostasis *via* increased antioxidant activity for faster seed germination and improve plant stress tolerance [3]. Similarly, priming of maize seeds with copper NPs reduced ROS levels to drought stress [78]. Furthermore, *Lathyrus odoratus* seeds primed with Si-NPs reduced ROS and MDA levels under salt stress [79].

Stress and accumulation of ROS can affect membrane lipids, leading to lipid peroxidation, and loss of quality, germination, and seed viability [80, 81]. In stress conditions, an important lipid peroxidation reagent is malondialdehyde (MDA) [47, 82]. Studies have demonstrated that nanoparticle treatment by reducing lipid peroxidation stabilizes the cell membrane in various plants under abiotic stress [47], which is caused by the increased activity of antioxidant enzymes [83]. However, further research is necessary to elucidate the regulatory role of nanoparticle priming in ROS and membrane damage repair in different plants.

5. Seed nanopriming in abiotic stress mitigation

Due to the increase in pollution and climate changes, seeds are exposed to biotic and abiotic stress, which has a negative effect on their growth and development [83]. These stressors can cause physicochemical changes in various cellular levels.

Stress can sense *via* cellular compartments such as cell wall and membrane, cytoplasm, chloroplast, mitochondria, endoplasmic reticulum, and peroxisomes [84]. In response to stress, gene transcription, transcription, translation, and post-translational modifications (PTMs) of proteins are altered, leading to the production and modification of proteins that play various roles in stress response [84].

In non-stressed conditions, the seed activates the GA signaling pathway and the transcription factors of hydrolytic enzymes by absorbing water; both processes break down the endosperm and release soluble sugar for seed growth, whereas, in stress conditions, the seed is unable to absorb water. In contrast, it activates the ABA signaling pathway, overproduces ROS, and prevents endosperm breakdown, which directly either slows down or delays seed germination [27]. In stressful conditions, nanoparticles can reduce seed ROS levels and thus seed cell damage due to increased activity of enzymes such as superoxidase dismutase, catalase [14].

5.1 Seed nanopriming under salt stress

Salinity is abiotic stress that threatens to impede plant growth and thereby affect crop yield [85]. Salinity in seeds causes osmotic and oxidative stress, which is associated with slowing down and prolonging the germination period [27]. Seed priming with Mn nanoparticles increases root length, alters the redistribution of macro-/micronutrients including Mn, Na, and Ca, and increases salt tolerance of *Capsicum annuum* [10]. However, under the pressure of salinity, maize seeds and *Paeonia suffruticosa* were enhanced germination by TiO2, vigor index, shoot and root length, seedling biomass, RWC, total phenol, antioxidant enzyme function [44, 67]. In addition, priming rapeseed with cerium oxide increased germination, water absorption, SOD, POD, α-amylase activities, total soluble sugar content, and Na+/K+ ratio while reducing accumulation of ROS and improved salt tolerance [86–88]. Additionally, lentil seeds containing iron nanochelates [89], cucumber seeds containing NPs from water treatment residues [90], and milk thistle containing NPs of chitosan [38] reduced salt stress by enhancing physiological salt stress. Latef et al. (2017) showed that priming of lupine seed with ZnO NPs increased photosynthetic pigments, total phenols, Zn and APX, POD, SOD, and CAT activities, and decreased MDA content and Na + under salt stress conditions [91]. Similarly, ZnO nanopriming improved wheat salt tolerance by activating the antioxidants to reduce oxidative explosion and increased photosynthetic electron transport efficiency and sucrose production in plants under stress [9]. Priming lettuce seeds with water-soluble carbon NPs (CNPs) increased seed germination and Chl content under salt stress [57].

5.2 Seed nanopriming in drought stress

Drought stress inhibits plant growth and reduces crop yields [92]. NP-mediated priming had a great effect on the growth of different plants to reduce drought stress. Seed priming with multi-walled carbon nanotubes increased the germination rate, root index, and root-shoot growth of alnus subcordata (*Caucasian alder*) under dry conditions [93]. In addition, Cape (*Catharanthus roseus* L.) seeds under drought stress with chitosan nanoparticles improved aggregation, membrane integrity, and plant growth [94]. In addition, corn seeds prepared with Cu nanoparticles increased RWC, Chl, and carotenoid and anthocyanin content and decreased ROS accumulation during drought stress [78]. Priming marigold seeds (*Calendula officinalis* L.) with silicon NPs increased quercetin, total flavonoid content, and antioxidant activity under drought stress conditions [95]. Nanozymes, Fe_3O_4 magnetite, and γ-Fe_2O_3 magnetite are magnetic nanoparticles that are effective in the development of plant growth under drought stress. For example, γ-Fe_2O_3 nanoparticles in rapeseed reduced H_2O_2 and lipid peroxidation and increased growth under drought stress. It has been reported that nanozymes reduce plant stress by removing ROS and increasing enzyme capacity [96].

5.3 Seed nanopriming in heavy metal stress

Metal toxicity is one of the abiotic stressors that disrupts plant growth. NP priming reduces the accumulation of toxic metals and adverse effects in various agricultural productions. For example, in sunflower seeds primed with green synthetic sulfur under Mn stress (*Helianthus annuus* L.), it activates antioxidant enzymes and reduces ROS and lipid peroxidation [97]. Under cadmium (Cd) stress, priming of zinc nanoparticles increased amylase, POD and SOD activities, and seedling growth in rice [67]. Furthermore, wheat seeds pretreated with TiO_2 enhanced germination rate, seedling growth, and water holding capacity under Cd stress. In maize seed priming with zinc oxide nanoparticles under cobalt stress, ROS and MDA decreased and plant growth, biomass and photosynthesis were improved [46]. In general, NPs reduce the adverse effects of HM by modulating plant physiological and biochemical parameters [98].

6. Conclusion

Seed nanopriming is an effective perspective that enables us to provide seeds with nanoform micronutrients at the seed level that acts as initial fertilizers for seed and increases the seed germination process, plant growth, and yield. The nanopriming can modulate molecular mechanisms affecting plant morphology, and physiological and biochemical responses. More research is necessary to test the performance of nanopriming in plants under stress conditions. In this regard, the precise cascade of molecular changes and the specific genes induced to produce such an effect remain to be further elucidated. However, several issues such as nanotoxicity on cells, tissues, and organs, as well as long-term effects of NPs exposure still need to be studied.

Author details

Afagh Yavari[1], Elham Ghasemifar[1] and Mehdi Shahgolzari[2*]

1 Department of Biology, Payame Noor University (PNU), Tehran, Iran

2 Department of Medical Nanotechnology, Tabriz Medical University, Tabriz, Iran

*Address all correspondence to: mehdi.shahgolzari@gmail.com

References

[1] Hussain S et al. Aspirin priming circumvents the salinity-induced effects on wheat emergence and seedling growth by regulating starch metabolism and antioxidant enzyme activities. Acta Physiologiae Plantarum. 2018;**40**(4):1-12

[2] Filippou, P et al. Plant Acclimation to Environmental Stress Using Priming Agents. In: Tuteja N, Singh GS, editors. Plant Acclimation to Environmental Stress. New York, NY: Springer; 2013. pp. 1-27

[3] Kandhol N et al. Nano-priming: Impression on the beginner of plant life. Plant Stress. 2022;**5**:100091

[4] Nile SH et al. Nano-priming as emerging seed priming technology for sustainable agriculture—Recent developments and future perspectives. Journal of Nanobiotechnology. 2022;**20**(1):1-31

[5] An J et al. Emerging investigator series: Molecular mechanisms of plant salinity stress tolerance improvement by seed priming with cerium oxide nanoparticles. Environmental Science: Nano. 2020;**7**(8):2214-2228

[6] Lowry GV, Avellan A, Gilbertson LM. Opportunities and challenges for nanotechnology in the Agri-tech revolution. Nature Nanotechnology. 2019;**14**(6):517-522

[7] Sharma A et al. Synergistic action of silicon nanoparticles and indole acetic acid in alleviation of chromium (CrVI) toxicity in Oryza sativa seedlings. Journal of Biotechnology. 2022;**343**:71-82

[8] Ioannou A et al. Advanced nanomaterials in agriculture under a changing climate: The way to the future? Environmental and Experimental Botany. 2020;**176**:104048

[9] Wang Z et al. Nano-ZnO priming induces salt tolerance by promoting photosynthetic carbon assimilation in wheat. Archives of Agronomy and Soil Science. 2020;**66**(9):1259-1273

[10] Ye Y et al. Manganese nanoparticles control salinity-modulated molecular responses in Capsicum annuum L. through priming: A sustainable approach for agriculture. ACS Sustainable Chemistry & Engineering. 2020;**8**(3):1427-1436

[11] Rawat PS et al. Effect of nanoparticles on wheat seed germination and seedling growth. International Journal of Agricultural and Biosystems Engineering. 2018;**12**(1):13-16

[12] Masoomeh AK et al. Influence of nano-priming on seed germination and plant growth of forage and medicinal plants. Plant Growth Regulation. 2021;**93**(1):13-28

[13] Tian Y, et al. Responses of seed germination, seedling growth, and seed yield traits to seed pretreatment in maize (Zea mays L.). The Scientific World Journal. 2014;**2014**: pp. 1-8. Article ID 834630. DOI: 10.1155/2014/834630

[14] do Espirito Santo Pereira A et al. Nanotechnology potential in seed priming for sustainable agriculture. Nanomaterials. 2021;**11**(2):267

[15] Dziergowska K, Izabela M. The role of nanoparticles in sustainable agriculture. In Smart Agrochemicals for Sustainable Agriculture. Elsevier; Academic Press. 2022. pp. 225-278

[16] Younis M, Abdel-Aziz H, Heikal Y. Nanopriming technology enhances vigor and mitotic index of aged Vicia faba seeds using chemically synthesized silver

nanoparticles. South African Journal of Botany. 2019;**125**:393-401

[17] Wu H, Li Z. Nano-enabled agriculture: How nanoparticles cross barriers in plants? Plant. Communications. 2022;**2022**:100346

[18] Fiol DF et al. Nanotechnology in plants: Recent advances and challenges. Journal of Chemical Technology & Biotechnology. 2021;**96**(8):2095-2108

[19] Pérez-de-Luque A. Interaction of nanomaterials with plants: What do we need for real applications in agriculture? Frontiers in Environmental Science. 2017;**5**:12

[20] Roberts A. Plasmodesmata and the control of symplastic transport. Plant and Cell Environment. 2003;**26**:103-124

[21] Madeira A, Moura TF, Soveral G. Detecting aquaporin function and regulation. Frontiers in Chemistry. 2016;**4**:3

[22] Schwab F et al. Barriers, pathways and processes for uptake, translocation and accumulation of nanomaterials in plants–critical review. Nanotoxicology. 2016;**10**(3):257-278

[23] Chichiriccò G, Poma A. Penetration and toxicity of nanomaterials in higher plants. Nanomaterials. 2015;**5**(2):851-873

[24] Kolitsi LI, Yiantsios SG. Transport of nanoparticles in magnetic targeting: Comparison of magnetic, diffusive and convective forces and fluxes in the microvasculature, through vascular pores and across the interstitium. Microvascular Research. 2020;**130**:104007

[25] Zhai G et al. Transport of gold nanoparticles through plasmodesmata and precipitation of gold ions in woody poplar. Environmental Science & Technology Letters. 2014;**1**(2):146-151

[26] Kolenčík M et al. Foliar application of low concentrations of titanium dioxide and zinc oxide nanoparticles to the common sunflower under field conditions. Nanomaterials. 2020;**10**(8):1619

[27] Khan MN et al. Seed nanopriming: How do nanomaterials improve seed tolerance to salinity and drought? Chemosphere. 2022;**2022**:136911

[28] Balmer A et al. The 'prime-ome': Towards a holistic approach to priming. Trends in Plant Science. 2015;**20**(7):443-452

[29] Ashraf M, Foolad M. Pre-sowing seed treatment—A shotgun approach to improve germination, plant growth, and crop yield under saline and non-saline conditions. Advances in Agronomy. 2005;**88**:223-271

[30] Lemmens E et al. The impact of hydro-priming and osmo-priming on seedling characteristics, plant hormone concentrations, activity of selected hydrolytic enzymes, and cell wall and phytate hydrolysis in sprouted wheat (*Triticum aestivum* L.). ACS Omega. 2019;**4**(26):22089-22100

[31] Zafar S et al. Effect of zinc nanoparticles seed priming and foliar application on the growth and physio-biochemical indices of spinach (Spinacia oleracea L.) under salt stress. PLoS One. 2022;**17**(2):e0263194

[32] Acharya P et al. Green-synthesized nanoparticles enhanced seedling growth, yield, and quality of onion (*Allium cepa* L.). ACS Sustainable Chemistry & Engineering. 2019;**7**(17):14580-14590

[33] Mahakham W et al. Nanopriming technology for enhancing germination

and starch metabolism of aged rice seeds using phytosynthesized silver nanoparticles. Scientific Reports. 2017;7(1):1-21

[34] Guha T et al. Nanopriming with zero-valent iron synthesized using pomegranate peel waste: A "green" approach for yield enhancement in Oryza sativa L. cv. Gonindobhog. Plant Physiology and Biochemistry. 2021;**163**:261-275

[35] Oracz K, Karpiński S. Phytohormones signaling pathways and ROS involvement in seed germination. Frontiers in Plant Science. 2016;7:864

[36] Rai-Kalal P, Tomar RS, Jajoo A. H2O2 signaling regulates seed germination in ZnO nanoprimed wheat (*Triticum aestivum* L.) seeds for improving plant performance under drought stress. Environmental and Experimental Botany. 2021;**189**:104561

[37] Xin X et al. Use of polymeric nanoparticles to improve seed germination and plant growth under copper stress. Science of the Total Environment. 2020;**745**:141055

[38] Mosavikia AA et al. Chitosan nanoparticle and pyridoxine seed priming improves tolerance to salinity in milk thistle seedling. Notulae Botanicae Horti Agrobotanici Cluj-Napoca. 2020;**48**(1):221-233

[39] Gomes DG et al. Seed priming with copper-loaded chitosan nanoparticles promotes early growth and enzymatic antioxidant defense of maize (*Zea mays* L.) seedlings. Journal of Chemical Technology & Biotechnology. 2021;**96**(8):2176-2184

[40] Li R et al. Effects of chitosan nanoparticles on seed germination and seedling growth of wheat (Triticum aestivum L.). International Journal of Biological Macromolecules. 2019;**126**:91-100

[41] Abdel-Aziz H. Effect of priming with chitosan nanoparticles on germination, seedling growth and antioxidant enzymes of broad beans. Catrina: The International Journal of Environmental Sciences. 2019;**18**(1):81-86

[42] Khan SA. Metal nanoparticles toxicity: Role of physicochemical aspects. In: Metal Nanoparticles for Drug Delivery and Diagnostic Applications. Elsevier; 2020. pp. 1-11. DOI: 10.1016/B978-0-12-816960-5.00001-X

[43] Kasote DM et al. Seed priming with iron oxide nanoparticles modulate antioxidant potential and defense-linked hormones in watermelon seedlings. ACS Sustainable Chemistry & Engineering. 2019;7(5):5142-5151

[44] Shah T et al. Seed priming with titanium dioxide nanoparticles enhances seed vigor, leaf water status, and antioxidant enzyme activities in maize (Zea mays L.) under salinity stress. Journal of King Saud University-Science. 2021;**33**(1):101207

[45] Acharya P et al. Nanoparticle-mediated seed priming improves germination, growth, yield, and quality of watermelons (Citrullus lanatus) at multi-locations in Texas. Scientific Reports. 2020;**10**(1):1-16

[46] Salam A et al. Seed priming with zinc oxide nanoparticles downplayed ultrastructural damage and improved photosynthetic apparatus in maize under cobalt stress. Journal of Hazardous Materials. 2022;**423**:127021

[47] Rai-Kalal P, Jajoo A. Priming with zinc oxide nanoparticles improve germination and photosynthetic performance in wheat. Plant Physiology and Biochemistry. 2021;**160**:341-351

[48] Feizi H et al. Assessment of concentrations of nano and bulk iron oxide particles on early growth of wheat (*Triticum aestivum* L.). Annual Research & Review in Biology. 2013:752-761

[49] López-Vargas ER et al. Seed priming with carbon nanomaterials to modify the germination, growth, and antioxidant status of tomato seedlings. Agronomy. 2020;**10**(5):639

[50] Khodakovskaya M et al. Carbon nanotubes are able to penetrate plant seed coat and dramatically affect seed germination and plant growth. ACS Nano. 2009;**3**(10):3221-3227

[51] Lahiani MH et al. Impact of carbon nanotube exposure to seeds of valuable crops. ACS Applied Materials & Interfaces. 2013;**5**(16):7965-7973

[52] Ali MH et al. Carbon nanoparticles functionalized with carboxylic acid improved the germination and seedling vigor in upland boreal forest species. Nanomaterials. 2020;**10**(1):176

[53] Hatami M, Hadian J, Ghorbanpour M. Mechanisms underlying toxicity and stimulatory role of single-walled carbon nanotubes in Hyoscyamus niger during drought stress simulated by polyethylene glycol. Journal of Hazardous Materials. 2017;**324**:306-320

[54] Sayedena SV, et al. Effects of seed nano-priming with multiwall carbon nanotubes (MWCNT) on seed germination and seedlings growth parameters of mountain ash (Sorbus luristanica Bornm.). Iranian Journal of Forest and Poplar Research. 2018;**26**(2):202-214

[55] Chen J et al. Seed priming with multiwall carbon nanotubes (MWCNTs) modulates seed germination and early growth of maize under cadmium (Cd) toxicity. Journal of Soil Science and Plant Nutrition. 2021;**21**(3):1793-1805

[56] Sobze J-M et al. The potential of carbon nanoparticles as a stimulant to improve the propagation of native boreal forest species: A mini-review. Frontiers in Forests and Global Change. 2022;**2022**:102

[57] Baz H et al. Water-soluble carbon nanoparticles improve seed germination and post-germination growth of lettuce under salinity stress. Agronomy. 2020;**10**(8):1192

[58] Khan MN et al. Role of nanomaterials in plants under challenging environments. Plant Physiology and Biochemistry. 2017;**110**:194-209

[59] Rahimi D et al. Increasing drought resistance of Alnus subcordata CA Mey. Seeds using a nano priming technique with multi-walled carbon nanotubes. Journal of Forest Science. 2016;**62**(6):269-278

[60] González-García Y et al. Seed priming with carbon nanomaterials improves the bioactive compounds of tomato plants under saline stress. Plants. 2022;**11**(15):1984

[61] Sun W, Shahrajabian MH, Huang Q. Soybean seeds treated with single walled carbon nanotubes (SwCNTs) showed enhanced drought tolerance during germination. International Journal of Advanced Biological and Biomedical Research. 2020;**8**(1):9-16

[62] Salam A et al. Nano-priming against abiotic stress: A way forward towards sustainable agriculture. Sustainability. 2022;**14**(22):14880

[63] Tariq M et al. Role of nanoparticles in abiotic stress. Technology. Agriculture. 2021;**2021**:323

[64] Ivani R et al. Role of bulk and Nanosized SiO2 to overcome salt

stress during fenugreek germination (Trigonella foenum-graceum L.). Plant Signaling & Behavior. 2018;**13**(7):e1044190

[65] Hojjat SS, Kamyab M. The effect of silver nanoparticle on fenugreek seed germination under salinity levels. Russian Agricultural Sciences. 2017;**43**(1):61-65

[66] Konate A et al. Magnetic (Fe3O4) nanoparticles reduce heavy metals uptake and mitigate their toxicity in wheat seedling. Sustainability. 2017;**9**(5):790

[67] Liu L et al. Nanosized titanium dioxide seed priming enhances salt tolerance of an ornamental and medicinal plant Paeonia Suffruticosa. Pakistan Journal of Botany. 2021;**53**(4):1167-1175

[68] Abdel Latef AA, Tran L-SP. Impacts of priming with silicon on the growth and tolerance of maize plants to alkaline stress. Frontiers. Plant Science. 2016;**7**:243

[69] Abdel Latef AAH et al. The impact of priming with Al2O3 nanoparticles on growth, pigments, osmolytes, and antioxidant enzymes of Egyptian Roselle (*Hibiscus sabdariffa* L.) cultivar. Agronomy. 2020;**10**(5):681

[70] Jadczak P et al. Effect of AuNPs and AgNPs on the antioxidant system and antioxidant activity of lavender (Lavandula angustifolia Mill.) from in vitro cultures. Molecules. 2020;**25**(23):5511

[71] Hussain A et al. Seed priming with silicon nanoparticles improved the biomass and yield while reduced the oxidative stress and cadmium concentration in wheat grains. Environmental Science and Pollution Research. 2019;**26**(8):7579-7588

[72] Anand A et al. Hydrogen peroxide signaling integrates with phytohormones during the germination of magnetoprimed tomato seeds. Scientific Reports. 2019;**9**(1):1-11

[73] Kaur R, Chandra J, Keshavkant S. Nanotechnology: An efficient approach for rejuvenation of aged seeds. Physiology and Molecular Biology of Plants. 2021;**27**(2):399-415

[74] Dietz K-J, Mittler R, Noctor G. Recent progress in understanding the role of reactive oxygen species in plant cell signaling. Plant Physiology. 2016;**171**(3):1535-1539

[75] Kurek K, Plitta-Michalak B, Ratajczak E. Reactive oxygen species as potential drivers of the seed aging process. Plants. 2019;**8**(6):174

[76] Bailly C. The signalling role of ROS in the regulation of seed germination and dormancy. Biochemical Journal. 2019;**476**(20):3019-3032

[77] Chandrasekaran U et al. Are there unidentified factors involved in the germination of nanoprimed seeds? Frontiers in Plant Science. 2020;**11**:832

[78] Van Nguyen D et al. Copper nanoparticle application enhances plant growth and grain yield in maize under drought stress conditions. Journal of Plant Growth Regulation. 2022;**41**(1):364-375

[79] El-Maarouf-Bouteau H et al. Role of protein and mRNA oxidation in seed dormancy and germination. Frontiers in Plant Science. 2013;**4**:77

[80] Chandra J, Keshavkant S. Desiccation-induced ROS accumulation and lipid catabolism in recalcitrant Madhuca latifolia seeds. Physiology and Molecular Biology of Plants. 2018;**24**(1):75-87

[81] Agarwal S, Shaheen R. Stimulation of antioxidant system and lipid peroxidation by abiotic stresses in leaves of Momordica charantia. Brazilian Journal of Plant Physiology. 2007;**19**:149-161

[82] Arbona V et al. Antioxidant enzymatic activity is linked to waterlogging stress tolerance in citrus. Physiologia Plantarum. 2008;**132**(4):452-466

[83] Shelar A et al. Sustainable agriculture through multidisciplinary seed nanopriming: Prospects of opportunities and challenges. Cell. 2021;**10**(9):2428

[84] Zhang H et al. Abiotic stress responses in plants. Nature Reviews Genetics. 2022;**23**(2):104-119

[85] Isayenkov SV, Maathuis FJ. Plant salinity stress: Many unanswered questions remain. Frontiers in Plant Science. 2019;**10**:80

[86] Mohammadi MHZ et al. Cerium oxide nanoparticles (CeO2-NPs) improve growth parameters and antioxidant defense system in Moldavian Balm (*Dracocephalum moldavica* L.) under salinity stress. Plant. Stress. 2021;**1**:100006

[87] Liu J et al. Cerium oxide nanoparticles improve cotton salt tolerance by enabling better ability to maintain cytosolic K+/Na+ ratio. Journal of Nanobiotechnology. 2021;**19**(1):1-16

[88] Chen L et al. CeO2 nanoparticles improved cucumber salt tolerance is associated with its induced early stimulation on antioxidant system. Chemosphere. 2022;**299**:134474

[89] Nourafcan H, Shahmoradi M. The Effect of Seed Priming by Salicylic Acid and Nano-iron Chelate on Germination and Initial Growth of Lentil under Salinity Stress. Modern Science of Sustainable Agriculture Journal. 2014;**10**(2):65-75

[90] Mahdy AM et al. Seed priming in nanoparticles of water treatment residual can increase the germination and growth of cucumber seedling under salinity stress. Journal of Plant Nutrition. 2020;**43**(12):1862-1874

[91] Abdel Latef AAH, Abu Alhmad MF, Abdelfattah KE. The possible roles of priming with ZnO nanoparticles in mitigation of salinity stress in lupine (Lupinus termis) plants. Journal of Plant Growth Regulation. 2017;**36**(1):60-70

[92] Osakabe Y et al. Response of plants to water stress. Frontiers in Plant Science. 2014;**5**:86

[93] Rahimi S, Hatami M, Ghorbanpour M. Effect of seed priming with nanosilicon on morpho-physiological characterestics, quercetin content and antioxidant capacity in Calendula officinalis L. under drought stress conditions. Journal of Medicinal Plants. 2019;**18**(72):186-203

[94] Ali E et al. A vital role of chitosan nanoparticles in improvisation the drought stress tolerance in *Catharanthus roseus* (L.) through biochemical and gene expression modulation. Plant Physiology and Biochemistry. 2021;**161**:166-175

[95] Rahimi M et al. Identification of drought-responsive proteins of sensitive and tolerant tea (Camellia sinensis L) clones under Normal and drought stress conditions. Current Proteomics. 2020;**17**(3):227-240

[96] Palmqvist N et al. Maghemite nanoparticles acts as nanozymes, improving growth and abiotic stress tolerance in Brassica napus. Nanoscale Research Letters. 2017;**12**(1):1-9

[97] Ragab G, Saad-Allah K. Seed priming with greenly synthesized sulfur nanoparticles enhances antioxidative defense machinery and restricts oxidative injury under manganese stress in Helianthus annuus (L.) seedlings. Journal of Plant Growth Regulation. 2021;**40**(5):1894-1902

[98] Rajput VD et al. Coping with the challenges of abiotic stress in plants: New dimensions in the field application of nanoparticles. Plants. 2021;**10**(6):1221

Chapter 11

Toxic Aluminum and Water Deficit Interaction in Plants: Physiological Aspects and Chemical Soil Management to Improve Root Environment in the Context of Global Climate Change

Flávio José Rodrigues Cruz

Abstract

Human activity has contributed to the intensification of climate change. These climate changes cause a reduction in plant growth and agricultural production due to increasingly frequent periods of water restriction. This effect can be more severe in tropical regions where the acid pH of the soil and the toxic levels of aluminum have a natural origin due to the weathering of the soils. In this context, water deficiency and aluminum toxicity alone or together promote biochemical and physiological changes in plants. This suggests the need to adopt soil management strategies that minimize the joint impact of these two abiotic stresses. Thus, liming and gypsum contribute to improving the edaphic environment, because they reduce the availability of toxic aluminum but increase the soil pH. In this chapter, we propose a systematic review of the isolated and combined effects of water deficiency and aluminum toxicity in plants based on physiological, biochemical, and nutritional variables. Thus, the understanding of these responses will improve the understanding of the mechanisms of tolerance to the two abiotic stresses, indicating the need to use soil correctives to minimize the effects of water deficiency and toxic aluminum in the soil on plant growth.

Keywords: toxic metal, soil pH, global warming, plant growth, gas exchange

1. Introduction

Drought is a factor that leads to environmental degradation and has adverse effects on rural populations dependent on natural resources such as water and soil. Drought can eventually lead to the loss of livelihoods, promote migration in affected areas [1], and have a significant impact on the economy, society, and environment [2]. According to the IPCC, the global population exposed to extreme or exceptional total scarcity of stored water will be 3–8%. In this context, risks of drought are predicted throughout the 21st century in many regions, increasing the probability of an economic crisis [3]. This prediction is worrying because the world population reached 8 billion inhabitants in 2022 [4], a fact that increases the demand for food. However, global climate changes can negatively affect agricultural production by altering the rainfall regime, inducing water deficiency and, consequently, reducing agricultural production [5, 6].

In tropical regions, high precipitation over time promotes the leaching of exchangeable bases and increases the levels of toxic aluminum. These two edaphic aspects result from the intense weathering of the soil. However, in tropical regions the occurrence of periods of water deficiency is normal. Thus, the sum of the factors of water deficiency and toxic aluminum potentiate the reduction of plant growth and agricultural production, because they cause disorders in gas exchange, nitrogen metabolism, and antioxidant in plants [7–10].

The chemical management of weathered tropical soils with toxic levels of aluminum can contribute to minimizing the effect of water deficit because liming reduces the toxic aluminum content and increases the surface pH and the calcium and magnesium content of the soil. In addition, gypsum contributes to the reduction of subsurface aluminum content and the increase of sulfur content in the soil.

Together, liming and plastering can minimize the effects of water deficiency, because they create an edaphic environment that improves root growth in volume and depth and, consequently, increases the absorption of nutrients and water. Therefore, the use of these two agricultural inputs can mitigate the effects of water deficiency on plants in acid soils.

This chapter will address the problem of water deficiency and aluminum toxicity in plants in the context of global climate change, emphasizing plant responses to aluminum toxicity and water deficiency, and liming and gypsum management as mitigating agents of soil chemical stress.

2. Physiological mechanism of plants under water deficit

With the advancement of global climate change, the occurrence of longer periods of water deficiency becomes more frequent, causing climatic risks to plant growth and, consequently, to agricultural activity. In the environmental context, prolonged periods without precipitation cause a reduction in soil water content and a decrease in vegetation growth, except in plants adapted to conditions of water scarcity, a fact quite common in arid and semi-arid regions. In the semi-arid region of northeastern Brazil, the occurrence of prolonged periods of water scarcity is common, with a drastic reduction in the availability of water in the soil, with the permanence of only species adapted to the semi-arid climate (**Figure 1**).

Physiologically, plants under water stress manifest a set of responses that culminate in reduced plant growth. Thus, the decrease in water availability induces stomatal closure due to greater synthesis and physiological action of abscisic acid on stomata. These close as a physiological strategy to reduce water loss through perspiration. This physiological phenomenon induces stomatal limitation to photosynthesis due to the reduction in the intercellular concentration of carbon dioxide [11].

The drop in the intercellular concentration of carbon dioxide can decrease the consumption of ATP and the $NADPH_2$ reducing power by the Calvin cycle, allowing electrons from the electron transport chain to interact with free molecular oxygen forming superoxide radicals, since $NADPH_2$ is chemically reduced (**Figure 2**).

Figure 1.
Effect of rainfall seasonality on vegetation. (A) Atriplex nummularia in dry soil; (B) soils with cracks due to periods of intense water deficit; (C) plants tolerant to water deficit in the rainy season and (C) in the dry season. Images recorded in the semi-arid region. Northeast Brazil. Source: Author.

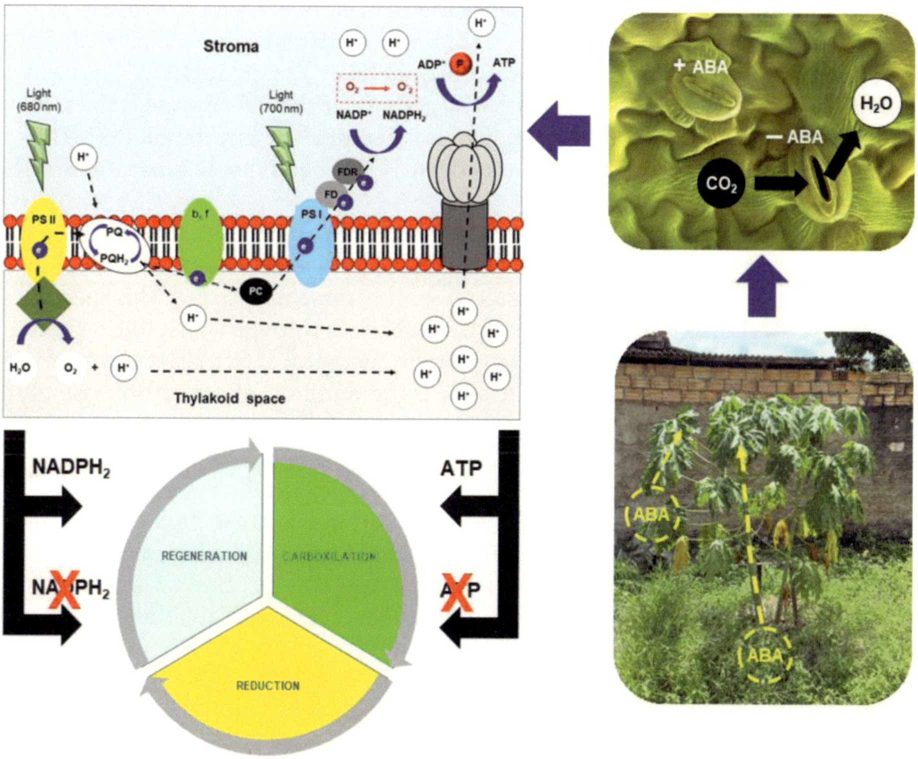

Figure 2.
Stomatal closure in response to water deficiency: Water deficiency induces the synthesis of abscisic acid in the root and leaves. ABA modulates stomatal closure, which reduces gas exchange. The decrease in CO_2 diffusion implies less consumption of $NADPH_2$ by the Calvin cycle. This allows O_2 to react with electrons from the electron transport chain and there is the formation of free radical O_2^- in the vicinity of photosystem I, triggering oxidative stress. Source: Author.

This set of events culminates in oxidative stress that causes lipid peroxidation and important cellular damage in plants under water deficit [12]. However, although water deficiency reduces the carboxylation activity of the enzyme RUBISCO (ribulose

1,5-bisphosphate carboxylase-oxygenase), its oxygenase function is increased, which allows temporary consumption of $NADPH_2$ and ATP, reducing the production of reactive oxygen species (ROSs). However, photorespiration allows the recycling of phosphoglycolate (a toxic compound) to phosphoglycerate during carbon fixation [13, 14].

Despite the cellular damage caused by ROS, they have a dual role in cells because they participate in cell signaling, but they are also toxic products of aerobic metabolism in plants [15, 16]. The main free radicals produced in plants under water deficit are the superoxide radical (O_2^-), peroxide radical (H_2O_2), and hydroxyl radical (OH^-). The O_2^- radical is synthesized in the apoplast, chloroplast, mitochondria, peroxisomes, and electron transport chain. The H_2O_2 radical, in turn, can be synthesized in peroxisomes, chloroplasts, mitochondria, cytosol, apoplast, and cytosol. The OH^- radical is synthesized from the H_2O_2 radical according to Fenton's reaction [16]. The oxidative stress resulting from water deficiency, in addition to increasing lipid peroxidation, reduces photosynthetic activity due to the harmful action of ROS on the photosynthetic machinery causing photoinhibition [17].

The mineral metabolism of plants is considerably affected by water deficiency, especially by nutrients that are absorbed through mass flow such as nitrogen [18]. The key enzyme present in plants that allows the entry of nitrogen into plants is nitrate reductase (RN, EC 1.6.6.1) which converts nitrate (NO_3^-) to nitrite (NO_2^-). One of the environmental factors that modulate the activity of the RN enzyme is the availability of nitrate, which is absorbed by the root system via transpiration [19]. Thus, water deficiency is a factor that indirectly decreases NR activity, because it limits the absorption of nitrate by the roots [20].

Water deficiency imposes limitations on plants regarding the acquisition of water in the environment in which they live since with the advancement of water restriction, the water potential of the soil tends to become more negative. In this sense, plants must maximize water use to avoid excessive loss through transpiration and maintain their water status favorable to their physiological activities. A biochemical strategy aimed at tolerating water deficiency is the synthesis of compatible osmolytes, which reduce the cellular osmotic potential for water influx into cells. Furthermore, compatible osmolytes preserve the conformational structure and maintain the biological activity of biomolecules [21, 22]. Amino acids (proline, glycine betaine, gamma-aminobutyric acid) and carbohydrates (sorbitol, sucrose, trehalose, mannitol, and raffinose) are compatible osmolytes used by plants during water deficit for osmotic adjustment and improvement of water status [21, 23]. Proline and glycine betaine are two important compatible osmolytes involved in modulating the response of plants to water stress [24, 25].

It should be emphasized that there is a negative correlation between the water content and the concentration of compatible osmolytes in plants under water deficit. However, increases in the concentration of compatible osmolytes do not necessarily imply an increase or stability in the plant growth rate in the face of water restriction experienced by the plants [7].

3. Physiological mechanisms of aluminum toxicity in plants

Metal toxicity is one of the world's biggest problems for agricultural production. Some metals are not essential to plants but are very toxic when present in certain forms in soil. Among metals, aluminum (Al) is one of the most toxic because it reduces the growth and production of many crops in acidic soils [26]. Around 50%

of arable land in the world is acidic [27] and around 60% of acidic soils are found in tropical and subtropical regions because in these regions the soil acidification process is natural [28].

The absorption of Al via symplast or apoplast can cause injuries to biomolecules in the cell wall, membrane, cytoplasm, and nucleus, affecting the structure of root cells and, consequently, the ability of root cells to absorb water and mineral salts from the soil [28]. Al bound to root cells appears to be localized to the apoplast cell wall and plasma membrane surface [29]. Therefore, the toxic effect of Al results from its external connection with root cells [27]. Thus, the initial site of Al toxicity occurs in the roots, which present physiological and biochemical changes that result in reduced root growth (**Figure 3**).

The physiological activity of the root system is affected by Al as it is the initial site of toxicity for this toxic metal. Thus, the ability to absorb water and mineral nutrients is compromised by toxic levels of Al. Therefore, toxic Al affects water relations in plants, reducing transpiration, water use efficiency, and intrinsic water use efficiency [30]. In addition, transpiration, root hydraulic conductivity, and leaf water potential are negatively affected by Al and these disorders in plant water relations coincide with increased levels of ABA in plants treated with toxic Al, indicating that this metal has a broad spectrum of action. Physiology in plants [31].

The mineral metabolism of plants is affected by the toxic action of Al, because this toxic metal inhibits the activity of the nitrate reductase enzyme and, consequently, the conversion of nitrate to nitrite in plants. In addition, Al reduces the levels of macronutrients (calcium, magnesium, phosphorus, and potassium) inducing nutritional disorders in plants that result in reduced plant growth [32, 33].

Although there is a greater accumulation of Al in the root system than in the aerial part of the plants, the physiological activity of the leaves is considerably affected by Al. This metal reduces the concentration of chlorophylls and, consequently, the

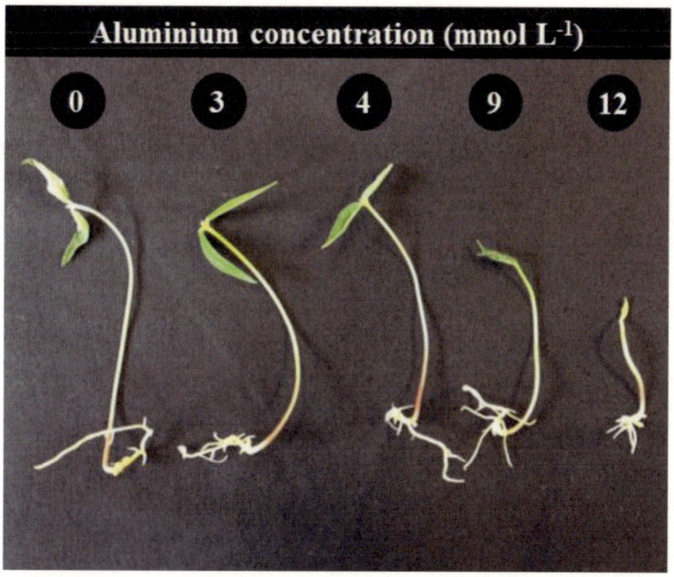

Figure 3.
Toxic aluminum primarily targets the root system. Aluminum concentrations above 4 mmol L^{-1} severely reduce root and shoot growth of Cajanus cajan seedlings. Source: Author.

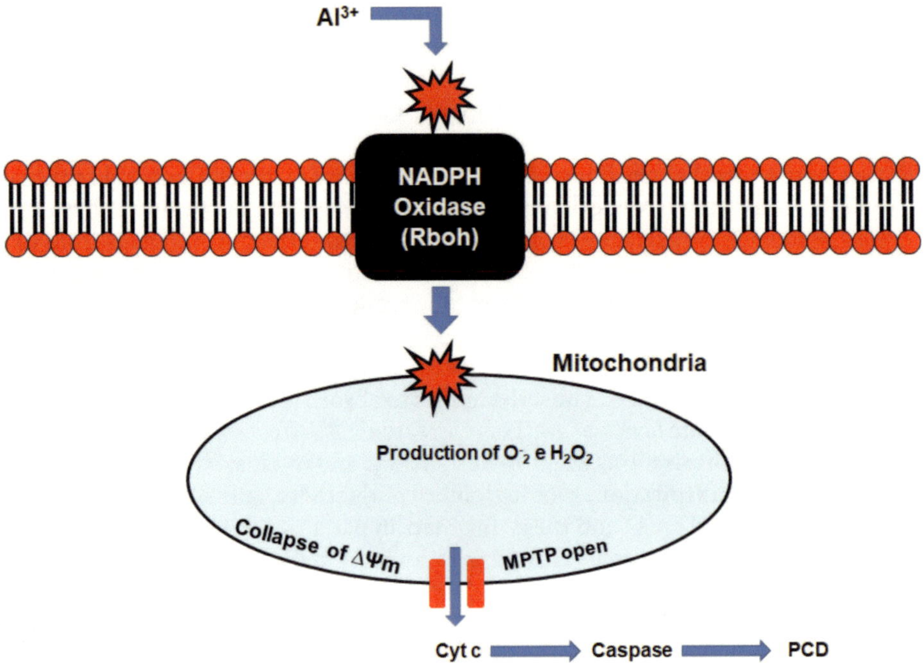

Figure 4.
Aluminum-modulated free radical production and programmed cell death in mitochondria. Aluminum activates the NADPH oxidase located in the mitochondrial membranes, inducing the overproduction of free radicals (O_2^- and H_2O_2) accelerating programmed cell death. Source: Figure adapted from hung et al. [39].

photosynthetic activity [34]. Al reduces the rate of photosynthesis because it is harmful to the functioning of the photosynthetic machinery by inhibiting electron transport mediated by photosystems I and II [34–36] and the carboxylate activity of the enzyme RUBISCO [37]. This negative effect of Al is reinforced by the fact that free radical production is located close to the reaction center of photosystems I and II of thylakoids in chloroplasts [15].

Al toxicity induces the production of O_2^-, H_2O_2, and OH- both in shoots and roots, causing lipid peroxidation and electrolyte leakage in plants [34, 38–40]. Although Al triggers oxidative stress in plants [34], the mechanism itself is indirect because Al activates NADPH oxidase (**Figure 4**) one of the main sources of ROS generation in plants under Al stress [39].

Respiratory burst oxidase homolog proteins (RBOHS) are integral plasma membrane proteins. They are formed by six transmembrane domains that support two heme groups, C-terminal FAD and NADPH hydrophilic domains, and two N-terminal calcium-binding domains (EF-hand). NADPH oxidase acts as a cytosolic electron donor to the extracellular O_2 electron acceptor, which is reduced to O_2^- via FAD and two independent hemes [41].

4. Interaction between stress: water deficiency and aluminum toxicity

In tropical countries, the occurrence of periods of water scarcity is common in regions where the soil is weathered and with high levels of toxic Al. However, with global climate changes, the occurrence of water deficiency has become more frequent.

Thus, the interaction between toxic Al and water deficiency potentiates the reduction of plant growth and production.

Under non-stressful conditions (absence of water deficiency and Al toxicity), cell growth can be explained by cell expansion resulting from the action of the enzyme xyloglucan endotransglycosylase—XET (EC 2.4.1.207). This enzyme promotes the cleavage and reformation of bonds between the xyloglucan chains (**Figure 5**) allowing cell expansion to occur due to the entry of water into the cell and an increase in cell pressure potential [41].

However, the interaction between water deficiency and toxic aluminum results in lower water influx into the cell and inhibition of XET activity. This set of events implies less cellular expansion of the root system with negative repercussions on plant growth and production (**Figure 5**). Therefore, toxic aluminum has as its primary target the root system whose elongation rate is considerably reduced when there is an interaction between toxic aluminum and water deficiency [8, 42].

The interaction between toxic Al levels and reduced water availability induces an increase in leaf contents of important compatible osmolytes such as glycine betaine, proline, and trehalose. However, a greater accumulation of compatible osmolytes does not prevent the decrease in plant growth [7, 43] represented by the dry mass of roots and shoots, and leaf area [43, 44].

The photosynthetic machinery is greatly affected by the interaction between aluminum toxicity and water deficiency because the concentration of photosynthetic pigments is considerably reduced [8] due to the production of free radicals that lead to lipid peroxidation, degradation of chlorophyll, and carotenoids [9, 45]. The degradation of chloroplast pigments may originate from lower root absorption and reduced

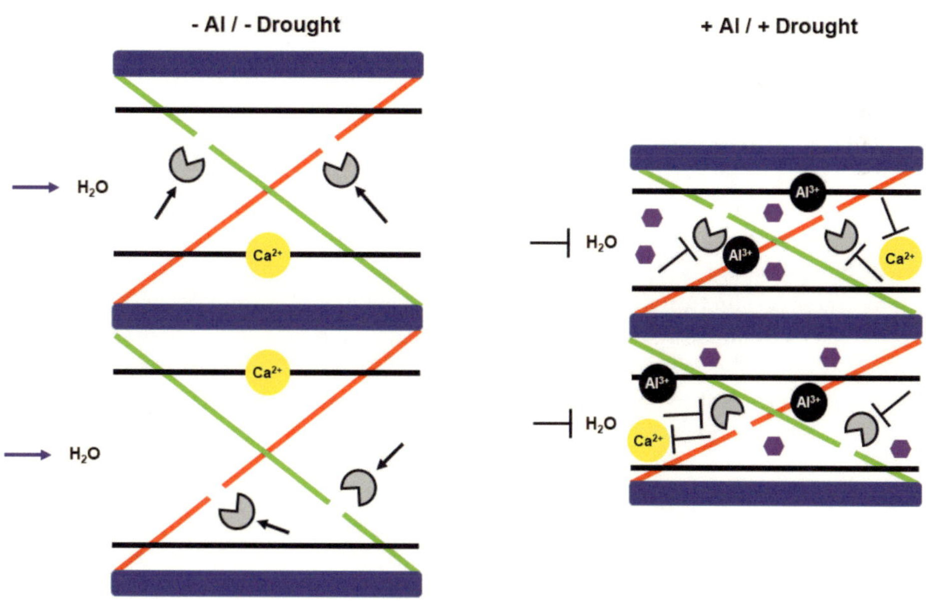

Figure 5.
Cell expansion in the presence (left) and absence (right) of water and aluminium. The symplastic and apoplastic influx of H_2O into cells increases the activity of the cell wall enzyme xyloglucan endotransglucosylase (XET) and the pressure potential on the cell wall. XET improves cell wall extensibility which is favored by intracellular water influx. These biochemical and physical phenomena imply cell expansion (left). Toxic aluminum reduces water influx into the cell and decreases XET activity. This implies less cell expansion (right). Source: Figure adapted from Yang et al. [42].

accumulation of magnesium in plant leaves due to toxic Al since magnesium is an integral part of the chlorophyll molecule [46].

Despite the large production of ROS such as O_2^-, H_2O_2, and OH^- due to the interaction between toxic Al and water deficiency, plants activate detoxification enzymatic mechanisms. In this context, the enzymes superoxide dismutase and guaiacol peroxidase have their activities increased to reduce the production of ROS and lipid peroxidation [9].

The mineral metabolism of plants is affected by the interaction between Al toxicity and water deficiency because the joint action of these two limiting factors reduces the calcium, magnesium, and phosphorus content in the leaves and roots of plants [47]. Essential macronutrients absorbed mainly by mass flow such as calcium and magnesium can be found in lower levels in plants due to the lower flow of water in the soil-plant-atmosphere system under conditions of water deficit.

This negative effect of water deficiency can be potentiated by Al toxicity. For example, in acid soil with an Al content of 12 mmol L^{-1} (soil depth of 0–20 cm) and with 50% of its pores filled with water, the contents of nitrogen, phosphorus, potassium, calcium, magnesium, zinc, and manganese were reduced in maize plants under the interaction of stresses [48]. Similarly, in soybean plants, the contents of nitrogen, phosphorus, potassium, calcium, magnesium, sulfur, copper, and manganese are reduced under the interaction of stresses. Furthermore, root density and biomass production of soybean and corn is strongly reduced by the interaction between toxic aluminum and water stress [48].

5. Liming and gypsum: Chemical soil management aimed at mitigating toxic aluminum and improving the water status of plants

Agricultural production can be limited by physical and chemical factors in the soil, which reduce root growth and restrict the uptake of mineral nutrients and water. In this context, toxic Al plays an important role in tropical regions where soil pH is not corrected because growth, development, and, consequently, agricultural production are negatively affected. This effect on plants is aggravated when there are dry spells or periods of water deficiency, an increasingly recurrent phenomenon in the context of global climate change.

The water conduction mechanisms in the soil-plant system represented by hydraulic conductivity and stomatal conductance are affected by the increase in the cellular concentration of abscisic acid induced by Al toxicity [31]. In addition, cellular and structural damage to the root system caused by toxic Al reduces cell turgor [49] and decreases plant root growth [31]. The negative effects of toxic Al on plants are potentiated by the action of water deficiency which, together, reduce plant growth [7].

A soil chemical management strategy aimed at neutralizing toxic Al consists of the adoption of liming based on soil chemical analysis. Liming raises the pH and reduces Al availability in the layers where limestone is applied [50], increases the availability of calcium and magnesium in the soil, and the assimilation of nutrients such as nitrogen, phosphorus, potassium, and sulfur by plants [51]. The increase in pH occurs through the exchange of H^+ for Ca^{2+} in the soil colloids and its neutralization (**Figure 6**). Similarly, the Al^{3+} is exchanged for Ca^{2+} in the soil colloids forming a little toxic compound ($AlOH_3$) that rushes into the soil (**Figure 7**).

Another strategy to neutralize toxic Al, but in subsurface soil layers, is plastering. Gypsum renders Al insoluble causing it to be leached into deep soil layers. In addition,

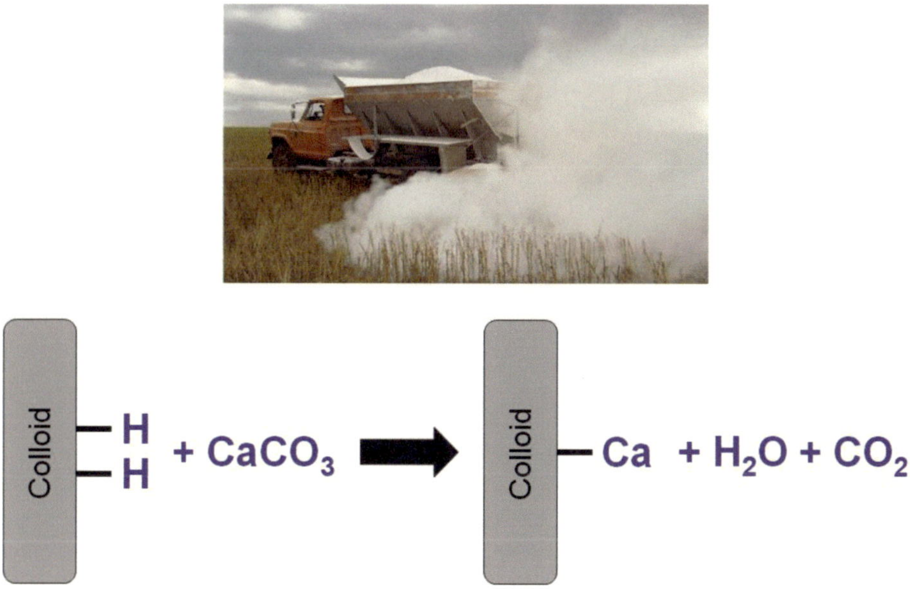

Figure 6.
Effect of limestone on the neutralization of soil acidity. Ca^{2+} displaces H^+ from soil colloids, neutralizing it and increasing soil pH. Source: Author.

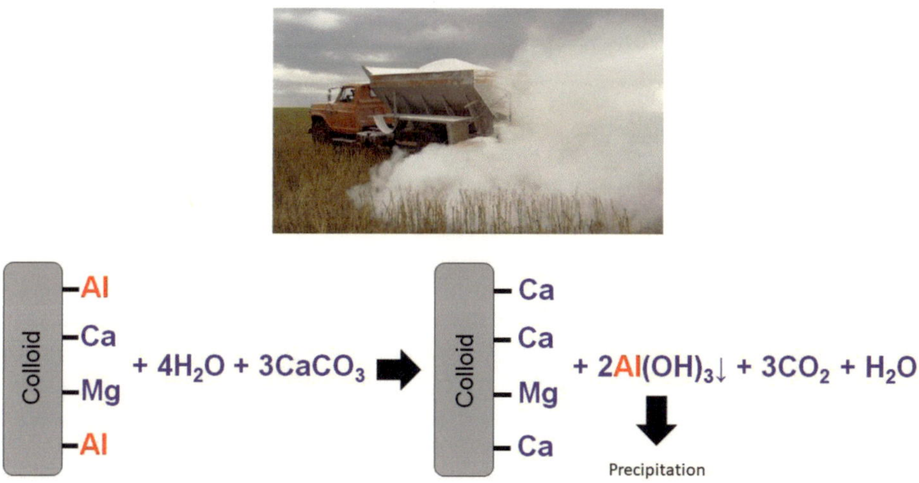

Figure 7.
Chemical mechanism of action of limestone in reducing soil aluminum saturation. Ca^{2+} displaces Al^{3+} from soil colloids, reducing its solubilization and causing precipitation. Source: Author.

gypsum improves the physical and chemical quality of the soil, creating better conditions for root development and growth. These changes promoted by gypsum nullify the effects of Al toxicity, increase root growth and attenuate the effects of water deficiency in plants [52]. These effects of gypsum indicate that the practice of gypsum can mitigate the effects of global climate change on water deficiency because there is greater root growth of plants in depth in the soil. Agricultural gypsum provides calcium and sulfur that improve soil fertility. In addition, the gypsum insolubilizes toxic aluminium, allowing its leaching to the subsurface layers of the soil (**Figure 8**).

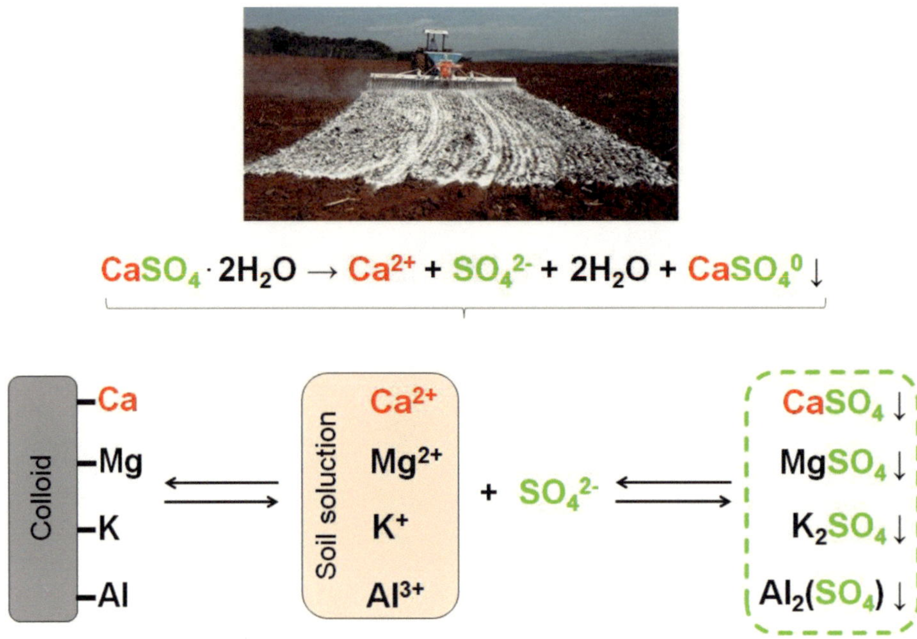

Figure 8.
Gypsum action on subsurface toxic aluminium. Gypsum supplies calcium and sulfur to the soil. Ca^{2+} from gypsum displaces Al^{3+}. This will form with sulfate ions a poorly soluble compound, $Al_2(SO_4)$, which is leached to deeper layers of the soil. Source: Author.

The benefit of gypsum is related to the greater root growth of plants. For example, the mean percentage distribution of the root system of sugarcane plants in the 0–20 cm, 20–40 cm, and 40–60 cm layers after gypsum soil application was 51.80, 29.72, and 18.64%, respectively. Furthermore, gypsum shows a positive effect on the root density of sugarcane plants [53]. Gypsum also increases the percentage of water absorbed by the root system at depths greater than 40 cm (**Figure 9**) because gypsum improves the physical and chemical attributes of the soil, allowing greater root growth in depth [53]. Gypsum improves the physical properties of the soil because it has a flocculating action on soil particles. This favors the aggregation of clay, reducing its dispersion. Thus, there is greater soil porosity, increased permeability, and water retention capacity in the soil. These changes promoted by gypsum create a favorable environment for the root growth of crops.

The interaction between limestone and gypsum in the soil shows a positive effect on root growth and agricultural production. For example, the application of gypsum and limestone in the soil shows positive effects on root growth because it improves the relative root distribution in depth [53], in addition to increasing the production of soybeans by 11.4% in conditions of water deficit [54]. Under conditions of water deficit, gypsum is shown to be efficient in increasing grain production in grasses such as maize and wheat [55].

In an oat cultivation area, liming and plastering increase grain production under water deficit [50]. These positive effects of gypsum, together or not with liming, result in a better edaphic environment that favors greater root growth with a positive impact on the absorption of water and mineral nutrients, two essential factors for plant production. Thus, the mechanism of action of gypsum and limestone in the soil suggests that these two agricultural inputs can be used to mitigate the effects of global

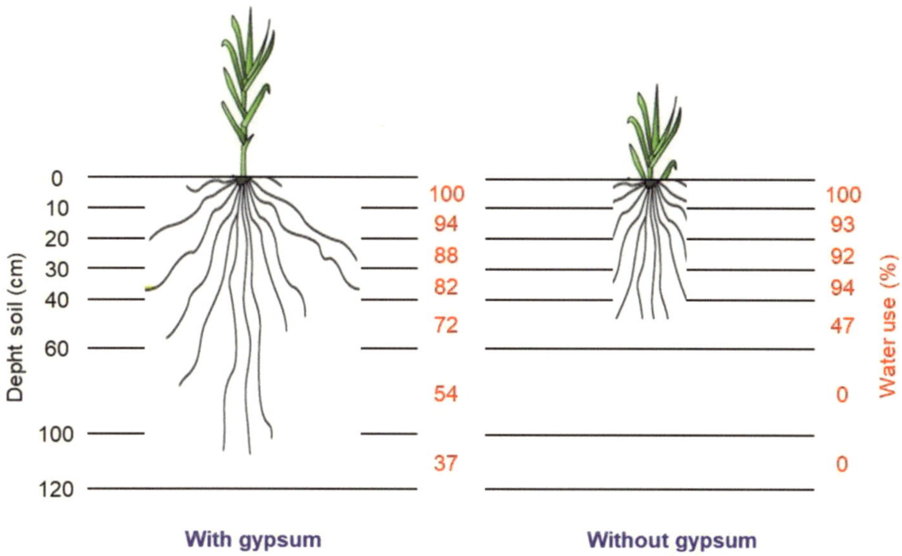

Figure 9.
Root growth and percentage of water use by plants in response to the presence or absence of gypsum. In the presence of gypsum, there is greater root growth in depth and, therefore, better use of water by the root system up to 100 cm deep in the soil. Source: Figure adapted from Sousa et al. [54].

climate change, particularly in tropical regions where the natural acidity of the soil is a limiting factor for plant growth and agricultural production [52].

6. Conclusion

In the context of global climate change, water deficiency is an increasingly frequent phenomenon in many regions of the world, directly impacting plant and animal production, which have water as a vital input. In tropical regions in particular, water deficiency is particularly serious because in these regions many soils are acidic and with high Aluminum saturation. This metal compromises plant growth and production because its target is the root system, the organ responsible for absorbing nutrients and water from the soil. Thus, water deficiency and Al toxicity together potentiate the reduction of plant growth and production, with considerable social and economic impacts. The increasingly frequent periods of water deficiency restrict soil water availability, inducing less water absorption and cellular water influx. These events culminate in the loss of cell turgor, lower cell expansion, and, consequently, lower plant growth with a negative impact on gas exchange, and mineral and antioxidant metabolism. These effects of water deficiency are potentiated by the action of toxic Al present in the soil because the primary site of Al action is the root system. Al reduces the fluidity of cell membranes, affecting their capacities related to the absorption of mineral nutrients and water, inducing nutritional and water deficiency in plants. Although the genetic improvement of plants aimed at tolerance to Al toxicity is an important tool to increase agricultural productivity in Al-affected soils, it occurs slowly and in few areas. A strategy that can be adopted to overcome the problem of acidity and toxic Al is the use of gypsum and agricultural limestone as soil amendments. Together, the use of gypsum and limestone reduces toxic Al in depth and surface, increases soil pH, and thus creates an edaphic environment favorable to greater root growth in volume and depth. Thus, crops will be more tolerant to periods of water deficit in soils treated with gypsum and limestone. However, the physiological and biochemical mechanisms of plant responses to liming and soil gypsum need further studies, since these two agricultural inputs, when applied to the soil, improve the absorption mechanisms of mineral nutrients and water, gas exchange, and the production of cultures.

Author details

Flávio José Rodrigues Cruz
Federal Institute of Education, Science and Technology of Amapá,
Laranjal do Jari, AP, Brazil

*Address all correspondence to: fjrc@bol.com.br

References

[1] Hermans K, McLeman R. Climate change, drought, land degradation, and migration: Exploring the linkages. Current Opinion in Environmental Sustainability. 2021;**50**:236-244

[2] Petersen-Perlman JD, Aguilar-Barajas I, Megdal SB. Drought and groundwater management: Interconnections, challenges, and policy responses. Current Opinion in Environmental Science & Health. 2022;**28**:100364

[3] IPCC. Climate Change 2023: Synthesis Report. A Report of the Intergovernmental Panel on Climate Change. Contribution of Working Groups I, II and III to the Sixth Assessment Report of the Intergovernmental Panel on Climate Change [Core Writing Team, Lee H, Romero J, editors]. Geneva, Switzerland: IPCC (in press) [Accessed: Apr 14, 2023]

[4] United Nations. World Populations Prospect 2022. Available from: http:// World Population Prospects 2022: Summary of results population division (un.org) [Accessed: December 10, 2022]

[5] Bahta YT. Social vulnerability to agricultural drought: Insights from northern cape, South Africa. Scientific African. 2022;**17**:e01324

[6] Eze E, Girma A, Zenebe A, Okolo CC, Kourouma JM, Negash E. Predictors of drought-induced crop yield/losses in two agroecologies of southern Tigray. Northern Ethiopia. Scientifc Reports. 2022;**12**:6284

[7] Carlin SD, Santos DMM. Indicadores fisiológicos da interação entre déficit hídrico e acidez do solo em cana-de-açúcar. Pesquisa Agropecuária Brasileira. 2009;**9**:1106-1113

[8] Pandey P, Srivastava RK, Dubey RS. Water deficit and aluminium tolerance are associated with a high antioxidative enzyme capacity in indica rice seedlings. Protoplasma. 2014;**251**:147-160

[9] Pandey P, Srivastava RK, Rajpoot R, Rani A, Pandey AK, Dubey RS. Water deficit and aluminium interactive effects on generation of reactive oxygen species and responses of antioxidative enzymes in the seedlings of two rice cultivars differing in stress tolerance. Environmental Science and Pollution Research. 2016;**23**:1516-1528

[10] Siecińska J, Wiącek D, Przysucha B, Nosalewicz A. Drought in acid soil increases aluminium toxicity especially of the Alsensitive wheat. Environmental and Experimental Botany. 2019;**165**:185-195

[11] Xingyang S, Guangsheng Z, Qijing H, Huailin Z. Stomatal limitations to photosynthesis and their critical water conditions in different growth stages of maize under water stress. Agricultural Water Management. 2020;**241**:106330

[12] Miller G, Suzuki N, Ciftci-Yilmaz S, Mittler R. Reactive oxygen species homeostasis and signalling during drought and salinity stresses. Plant, Cell and Environment. 2010;**33**:453-467

[13] Bauwe H, Hagemann M, Fernie AR. Photorespiration: Players, partners and origin. Trends in Plant Science. 2010;**6**:330-336

[14] Voss I, Sunil B, Scheibe R, Raghavendra AS. Emerging concept for the role of photorespiration as na important part of abiotic stress response. Plant Biology. 2013;**15**:713-722

[15] Asada K. Production and scavenging of reactive oxygen species in chloroplasts and their functions. Plant Physiology. 2006;**141**:391-396

[16] Mittler R. ROS are good. Trends in Plant Science. 2017;**1**:11-19

[17] Hajiboland R. Reactive oxygen species and photosynthesis. In: Ahmad P, editor. Oxidative Damage to Plants. Amsterdam, The Netherlands: Academic Press; 2014. pp. 1-63

[18] Marschner P, Rengel Z. Nutrient availability in soils. In: Marschner P, editor. Marschner's Mineral Nutrition of Higher Plants (Third Edition). San Diego: Academic Press; 2012. pp. 315-330

[19] Campbell WH. Nitrate reductase structure, function and regulation: Bridging the gap between biochemistry and physiology. Annual Review of Plant Physiology and Plant Molecular Biology. 1999;**50**:277-303

[20] Lobato AKS, Costa RCL, Neto CFO, Santos Filho BG, Gonçalves Vidigal MC, Vidigal Filho PS, et al. Consequences of the water deficit on water relations and symbiosis in Vigna unguiculata cultivars. Plant, Soil and Environment. 2009;**4**:139-145

[21] Yancey PH. Organic osmolytes as compatible, metabolic and counteracting cytoprotectants in high osmolarity and other stresses. The Journal of Experimental Biology. 2005;**208**:2819-2830

[22] Shukla N, Bembenek B, Taylor EA, Othon CM. Conformational consequences for compatible osmolytes on thermal denaturation. Life. 2021;**11**:1394

[23] Ghosh UK, Islam MN, Siddiqui MN, Khan MAR. Understanding the roles of osmolytes for acclimatizing plants to changing environment: A review of potential mechanism. Plant Signaling & Behavior. 2021;**16**:1913306

[24] Cruz FJR, Castro GLS, Silva Júnior DD, Festucci-Buselli RA, Pinheiro HA. Exogenous glycine betaine modulates ascorbate peroxidase and catalase activities and prevent lipid peroxidation in mild water-stressed Carapa guianensis plants. Photosynthetica. 2013;**1**:102-108

[25] Jales Filho RC, Melo YL, Viégas PRA, Oliveira APS, Neto VEA, Ferraz RLS, et al. Salicylic acid and proline modulate water stress tolerance in a traditional variety of cowpeas. Revista Brasileira de Engenharia Agrícola e Ambiental. 2023;**1**:18-25

[26] Rahman R, Upadhyaya H. Aluminium toxicity and its tolerance in plant: A review. Journal of Plant Biology. 2021;**64**:101-121

[27] Jaskowiak J, Kwasniewska J, Milewska-Hendel A, Kurczynska EU, Szurman-Zubrzycka M. Szarejko I (2019) aluminium alters the histology and pectin cell wall composition of barley roots. International Journal of Molecular Sciences. 2019;**20**:3039

[28] Kochian LV, Piñeros MV, Liu J, Magalhaes JV. Plant adaptation to acid soils: The molecular basis for crop aluminium resistance. Annual Review of Plant Physiology and Plant Molecular Biology. 2015;**66**:571-598

[29] Horst WJ, Wang Y, Eticha D. The role of the root apoplast in aluminium induced inhibition of root elongation and in aluminium resistance of plants: A review. Annals of Botany. 2010;**106**:185-197

[30] Konrad MLF, Silva JAB, Furlani PR, Machado EC. Trocas gasosas

e fluorescência da clorofila em seis cultivares de cafeeiro sob estresse de alumínio. Bragantia. 2005;**3**:339-347

[31] Gavassi MA, Dodd IC, Puértolas J, Silva GS, Carvalho RF, Habermann G. Aluminium-induced stomatal closure is related to low root hydraulic conductance and high ABA accumulation. Environmental and Experimental Botany. 2020;**179**:104233

[32] Cruz FJS, Lobato AKS, da Costa RCL, Lopes MJS, Neves HKB, Neto CFO, et al. Aluminium negative impact on nitrate reductase activity, nitrogen compounds and morphological parameters in sorghum plants. Australian Journal of Crop Science. 2011;**6**:641-645

[33] Cruz FJR, de Almeida HJ, dos Santos DMM. Growth, nutritional status and nitrogen metabolism in Vigna unguiculata (L.) Walp is affected by aluminium. Australian Journal of Crop Science. 2014;**7**:1132-1139

[34] Pereira LB, Mazzanti CMA, Gonçalves JF, Cargnelutti D, Tabaldi LA, Becker AG, et al. Aluminium-induced oxidative stress in cucumber. Plant Physiology and Biochemistry. 2010;**48**:683-689

[35] Roy AK, Sharma A, Talukder G. Some aspects of aluminium toxicity in plants. The Botanical Review. 1988;**54**:145-117

[36] Chen LS, Qi YP, Smith BR, Liu XH. Aluminium-induced decrease in CO2 assimilation in citrus seedlings is unaccompanied by decreased activities of key enzymes involved in CO2 assimilation. Tree Physiology. 2005;**25**:317-324

[37] Xiao X, Liu X, Yang Z, Chen L, Cai Y. Effect of aluminium stress on the photosynthesis of longan seedlings. Chinese Journal of Tropical Crops. 2004;**26**:63-69

[38] Achary VMM, Jena S, Panda KK, Panda BB. Aluminium induced oxidative stress and DNA damage in root cells of Allium cepa L. Ecotoxicology and Environmental Safety. 2008;**70**:300-310

[39] Huang W, Yang X, Yao S, LwinOo T, He H, Wang A, et al. Reactive oxygen species burst induced by aluminium stress triggers mitochondria-dependent programmed cell death in peanut root tip cells. Plant Physiology and Biochemistry. 2014;**82**:76-84

[40] Yamamoto Y, Kobayashi Y, Devi SR, Rikiishi S, Matsumoto H. Aluminum toxicity is associated with mitochondrial dysfunction and the production of reactive oxygen species in plant cells. Plant Physiology. 2002;**128**:63-72

[41] Sagi M, Fluhr R. (2006) production of reactive oxygen species by plant NADPH oxidases. Plant Physiology. 2006;**141**:336-340

[42] Siecińska J, Wiącek D, Przysucha B, Nosalewicz A. Drought in acid soil increases aluminum toxicity especially of the Al-sensitive wheat. 2019;**169**:185-195

[43] Marin A, dos Santos DMM. Interação da deficiência hídrica e da toxicidade do alumínio em guandu cultivado em hidroponia. Pesquisa Agropecuária Brasileira. 2008;**10**:1267-1275

[44] Krizek DT, Foy CD. Role of water stress indifferential aluminium tolerance of two barley cultivars grown in an acid soil. Journal of Plant Nutrition. 1988;**4**:351-367

[45] Zaifnejad M, Clark RB, Sullivan CY. Aluminium and water stress effects on growth and proline of sorghum. Journal of Plant Physiology. 1997;**150**:338-344

[46] Ravi RK, Krishna K, Naik GR. Effect of polyethylene glycolinduced water stress on physiological and biochemical responses in pigeonpea (Cajanus cajan L. Millsp.). Journal of Research in Science Teaching. 2011;**3**:148-152

[47] Vitorello VA, Haug A. Short-term aluminium uptake by tobacco cells: Growth dependence and evidence for internalization in a discrete peripheral region. Physiologia Plantarum. 1996;**97**:536-544

[48] Schier GA, Mcquattie CJ. Effect of water stress on aluminium toxicity in pitch pine seedlings. Journal of Plant Nutrition. 2000;**5**:637-647

[49] Joris HAW, Caires EF, Bini AR, Scharr DA, Haliski A. Effects of soil acidity and water stress on corn and soybean performance under a no-till system. Plant and Soil. 2013;**365**:409-424

[50] Barceló J, Poschenrieder C, Vázquez MD, Gunsé B. Aluminium phytotoxicity. Fertilizer Research. 1996;**43**:217-223

[51] Soratto RP, Crusciol CAC. Cátions hidrossolúveis na parte aérea de culturas anuais mediante aplicação de calcário e gesso em superfície. Revista Brasileira de Ciências do Solo. 2007;**31**:81-90

[52] Castro GSA, Crusciol CAC. Effects of superficial liming and silicate application on soil fertility and crop yield under rotation. Geoderma. 2013;**195-196**:234-242

[53] Clemente PRA, Bezerra BKL, da Silva VSG, dos Santos JCM, Endres L. Root growth and yield of sugarcane as a function of increasing gypsum doses. Pesquisa Agropecuária Tropical. 2017;**1**:110-117

[54] Sousa DMG, Lobato E, Rein TA. Uso de gesso agrícola nos solos do Cerrado. Planaltina: Distrito Federal; 2005. p. 19

[55] Caires EF, Feldhaus IC, Blum J. Crescimento radicular e nutrição da cevada em função da calagem e aplicação de gesso. Bragantia. 2001;**3**:213-223

Chapter 12

Microbial Mitigation of Drought Stress in Plants: Adaptations to Climate Change

Mohammad Sayyar Khan and Irba Zulfiqar

Abstract

The global temperature is constantly increasing due to the phenomenon of climate change. Plants have developed various mechanisms to defend themselves against environmental stresses including drought stress. Apart from indigenous biochemical, physiological, and molecular mechanisms of adaptation to stress, the plant-associated microbes may also play a crucial role in plant drought tolerance. The endophytic and rhizospheric microbes perform various functions and produce different enzymes and compounds that play an important role in plants' adaptation to various environmental stresses including drought stress. Some of the key mechanisms include production of growth hormones, siderophores, organic acids, induction of the ROS scavenging system, phosphate solubilization, and nitrogen fixation. However, the production of ACC deaminase in the plant-associated microbes has vital roles in reduction of ethylene levels under drought stress, resulting in improved plant growth and stress tolerance. Owing to the complex nature of drought tolerance, a multi-pronged approach would have to be adapted to further enhance the microbial-mediated drought tolerance in plants.

Keywords: microbial functions, rhizospheric microbes, ROS scavenging, ACC deaminase, osmolytes

1. Introduction

Climate change is the long-term shift in temperatures and weather patterns. These changes may be natural, such as through variations in the solar cycle or man-made activities. Different studies reveal that it mainly occurs due to anthropogenic activities taking place in a huge amount these days. Also, many other factors such as greenhouse effect, deforestation, urbanization, global warming, fossil fuels combustion, increased livestock farming, excessive use of fertilizers, and nitrous oxide emissions are included, which has resulted in increasing incidence of different abiotic and biotic stresses [1]. Currently, climate change is known to be the most serious contemporary challenge for humanity. The global climate conditions have been adversely affected due to different natural and unnatural activities, which have stimulated the rise in average temperature and carbon dioxide level. During the last century, the average global temperature soared up by 1.5 F, and it has been estimated that within a period of 100 years that it might surge higher by additional 0.5–8.6 F [2]. These changes in climatic conditions have led to erratic events such as increase in rainfall and snowfall worldwide, excessive release of greenhouse gases, increase in air temperature, which renders soil dry and decreases its moisture content forming drought-like conditions, water scarcity, severe fires, rising sea levels, flooding, melting polar ice, catastrophic storms, declining biodiversity and an increase in pollution due to excessive CO_2 emissions, air pollutants, ground-level ozone, aerosols, methane, and others. By altering wind patterns, it also induces seasonal variations. In other words, it involves the altering of climatic or weather patterns that occur by emission of greenhouse gases, which are likely to be more erratic and extreme in the forthcoming years [3]. Moreover, these climate-changing parameters are known to affect different terrestrial macroorganisms such as plants. However, according to recent studies, other organisms and ecosystems may be impacted as well [4]; that is, it affects human health, their ability to grow food, housing, safety, and work; for example, some people living in small islands in developing states are already more vulnerable to climate impacts.

As a result of ever-growing population, there is an increased demand of food, which has resulted in intensive agricultural practices including excessive use of agro-chemicals, livestock generation (for meat and other source of income), and exploitation of water resources, which have further worsen the situation by release of GHG (due to agricultural activities) and resulting in pollution of natural resources. The uncontrolled deforestation mainly for development and agriculture has created imbalance in the natural process of carbon cycle, which has led to a rise in the concentration of carbon footprint and brought uneven pattern of climate causing a variety of adverse effects that have huge effect on agricultural production. Due to climate change, a very high rate of land degradation has been observed causing enhanced desertification and nutrient-deficient soil, which is characterized as a major global threat. According to Global Assessment of Land Degradation and Improvement (GLADA), a quarter of land area around the globe can now be marked as degraded. Lives of 1.5 billion people have been adversely affected by land degradation caused by anthropogenic activities and climate change, and also 15 billion tons of fertile soil is lost every year, which results in mass migrations. According to a report published by United Nations Environment Programme in 2017, about 500 million hectares of farmland has been abandoned due to drought and desertification resulting in major social and environmental constraints [5].

The climate alterations highly influence the growth behavior of different crop species. For example, it has a huge impact on mineral accumulation and nutrient acquisition in the plants. Changes in morphology, physiology, and plant productivity are the direct effects observed on the plants body. The organs and membranes of plants are impaired due to different stress conditions. Due to increased oxidative stress, the production of carbohydrates, proteins, and secondary metabolite in the plants is adversely influenced. Furthermore, it also highly affects soil fertility, irrigation measures, occurrence of pests and diseases, and also, stress incidence such as heat and drought are detected. Other ill effects of climate change on production strategies observed are effects on food demand, trade opportunities, and unequal distribution of products [6]. These variations thus cause a very damaging impact on plant growth, and their developmental patterns also significantly impact the diversity and activities of different plant-associated microbial communities as they possess a huge variety of microbiomes in phyllosphere and endophytes from surrounding soil and air [7]. This beneficial microbiome present in the plants plays a vital role in regulating plant immunity, production of metabolites, nutrient uptake and acquisition, disease, and insect-pest management along with several other functions. The different microbes living in soil system comprise numerous PGPR, fungi, actinomycetes, algae, yeasts, cyanobacteria, and many more [8]. They help in transforming different organic substance into simpler form that can be assimilated by plants. Different microbiomes especially filamentous fungi and some bacterial species secrete a wide range of secondary metabolites, which are essential for plant development and many phenological responses [9]. Consequently, the microorganisms that are known for their positive effects on plant development and health might also be compromised due to climate change, in terms of exhibiting their desirable properties and their colonization capacity under certain conditions. Therefore, microorganisms present in the soil hold great importance in this regard, as they are vital constituent of nitrogen and carbon cycles and are also involved in the emission and removal of greenhouse gases such as CO_2 and CH_4, which in turn are mostly responsible for the climate change.

2. Drought stress effects on agricultural production

In nature, plants are exposed to various environmental stresses due to their sessile lifestyle. These different unfavorable factors negatively impact plant growth, productivity, and their geographic distribution. Plants may face many diverse stresses (e.g., drought, salinity, and pathogens) under field conditions individually or in combination, which might have a devastating effect on crop yield [10]. Water is the most essential resource for plants, and all the plant organs need to maintain 60–90% water content for sustainable activity. However, global climate change, caused by different anthropogenic activities and greenhouse gas emissions, has become more thoughtful worldwide, leading to drought conditions all over the world [11]. In agricultural technology, it is considered one of the main environmental stresses for plants. Due to the frequent changes in climate observed throughout the world, it has increased the severity of drought events for plants [12]. Important cereal crops are increasingly diminishing by over 10% yield due to drought stress, and it is still the main limiting factor of food production in numerous countries [13]. Decrease in plant metabolism and electrolyte disturbances in plant cells are major symptoms of drought stress, which automatically lead to their death. Because of inhibiting various morphological, physiological, and biochemical processes such as changes in leaf, root length, biomass photosynthesis, respiration, translocation, carbohydrate synthesis, nutrient metabolism, ion uptake, and growth promoters of plants are affected [14]. Also, it primarily prevents the photosynthesis system by causing an imbalance between light capture and utilization, due to which Rubisco activity is reduced and the amount of photosynthetic pigments, inhibiting leaf area and damaging the photosynthetic apparatus [15]. Similarly, it reduces the rate of carbon fixation by inhibiting metabolism or limiting carbon dioxide input into leaves. It also leads to various biochemical changes, such as an excessive accumulation of ROS including O^{2-}, and H^2O^2, inside the host, which can further damage various tissues and cellular constituents such as nucleic acids and other biomolecules, resulting in cell death [16]. Furthermore, drought also lowers seedling vigor and affects germination by reducing water intake. Wilting, yellowing, discoloration, and leaf burning are the phenotypic signs observed in plants under drought condition [17]. Also, leaf senescence, drooping, leaf rolling, brittleness, scorching, limp leaves, premature fall, etiolation, wilting, turgidity, flower sagging are the other symptoms observed [18]. Drought stress also alters carbon permeability and transport networks by lowering cation (Ca^{2+}, K^{+}, and Mg^{2+}) absorption by roots. They later can also limit development by preventing the activity of several critical enzymes that take part in nutrient digestion, uptake, translocation, and metabolism of plants [19]. It also has a negative impact on biogeochemical cycles, such as the nitrogen and carbon cycles, which further reduce the decomposition of organic matter that considerably lowers the uptake of water and minerals by the root system, thus increasing soil fertility. For instance, many drought-triggered plants decreasing in macronutrient absorption and translocation (K, N, and P) are found [20]. Many vital characteristics representing plant water relations in plants include relative water content (RWC), leaf water potential, stomatal conductance, transpiration rate, leaf, and canopy temperatures [21]. These traits have also been found to be affected considerably during drought stress in plants [22]. So the above information shows that water scarcity affects plants at all growth stages but causes maximum damage during critical growth phases, such as during the seed development stage or reproductive phase, thereby reducing seed size, number, and quality, which are primarily responsible for substantial yield losses [23].

3. Potential strategies to mitigate drought stress in plants

Plants incorporate a wide range of morphological, physiological, and molecular defense responses contrary to drought, which prevents water loss, maintaining cellular water content, and water supply to vital parts [24]. Drought stress can be reduced through breeding, mass screening, and exogenous phytohormone production. Different strategies are used by plants to minimize stress, for example, by producing phytohormones (e.g., abscisic acid (ABA) and gibberellins) and low-molecular-weight osmolytes (e.g., amino acids and polyols) and by modifying succulent leaves to reduce transpiration loss [25]. Also, a significant plant defense strategy in response to drought is the transcriptional and translational reprogramming of key genes and proteins that are involved in signal perception and transduction, transcription factors, and upregulation of drought tolerant genes, all of which drive drought resilience [26]. Plants protect themselves from drought-induced reactive oxygen species (ROS) and other radicals owing to their efficient antioxidant system. During extended drought stress, they also synthesize an array of osmoprotectants such as prolines, soluble sugars, betaine, and spermines, to maintain cell turgor pressure. Over the past two decades, researchers have focused on transgenic approaches and other molecular breeding tools to increase drought resilience in different crops [27]. For instance, various biotechnological tools, such as CRISPR/Cas, RNAi, and transgenics, have made significant contributions to improving drought-resilient traits in both model and crop plants. But due to their high costs, complexity, ethical considerations, and toxicity concerns, their accessibility to farmers has been limited [27]. Furthermore, adaptive responses in plants are driven by complex genetic features involving several pathways, which have proven to be major impediments to long-term drought-tolerant crop improvement. Furthermore, the development of climate-resilient crops is required by integrating modern technological methods. The use of next-generation breeding approaches (genomic selection and genomic editing) and high-throughput phenotyping is desirable to develop crops that are exposed to different stresses [28]. Also, different bioinformatics tools have also been reported to overcome stress responses [29].

The recent advancement in genomics and genome editing technologies has been coming across various approaches of genetic study to produce climate-resilient crops [30]. Many other strategies are incorporated to grow climate-resilient/smart crop including SNP genotype, trait mapping, and plant breeding methods. The CRISPR/Cas technology has also been efficiently used in enhancing productivity in rice crop in fluctuating climatic conditions [31].

4. Microbial mechanisms governing drought stress tolerance

Microbes have the potential to promote plant growth directly and indirectly through several mechanisms. The indirect activation of plant growth involves a series of events by which microbes prevent the inhibition of plant growth and development induced by pathogens [32]. During direct activation, microbes biosynthesize bacterial compounds that promote the uptake of nutrients from the soil and stimulate plant growth and development [33]. Microbes trigger local or systemic stress mitigation response mechanisms that enable plants to survive and overcome the negative effects of abiotic stress conditions. The mechanisms governing microbial-mediated stress tolerance may include such as drought stress and help plants sustain growth and development through the production, mobilization of nutrients, and induction of the levels of hormones and organic phytostimulants [34]. Below are the fundamental mechanisms governing drought stress tolerance in plants.

4.1 Microbial production of aminocyclopropane-1-carboxylate deaminase

Aminocyclopropane-1-carboxylic acid (ACC) is a precursor of ethylene, and its production increases in plants during stress conditions. Plants enhance their ethylene production under drought stress, which inhibits plant growth by affecting root enlargement and seed germination. The production of higher ACC levels in plants is a strategy to combat severe drought stress [35]. A group of beneficial microbes have the potential to produce ACC deaminase that regulates plant growth and development by sequestering the plant-produced ACC, responsible for ethylene production in plants. A large number of microorganisms have been reported to produce ACC deaminase that in turn reduces ACC, thereby lowering the increased ethylene levels in plants under stress conditions [36]. These microbes play a vital role in plants' adaptation to stress conditions. In particular, drought stress tolerance has been achieved in several plants through the production of ACC deaminase. Some of the prominent examples of the microbial production of ACC deaminase and its mitigation effects on drought stress in several plants have been summarized (**Table 1**). The ACC deaminase production by *Bacillus subtilis* Rhizo SF 48 strain conferred maximum seed and plant growth promotion in tomato plants under drought stress [37]. The underlying biochemical mechanisms for this improved drought stress tolerance included induction in the proline, SOD, and APX activities, whereas reduction in the MDA and H_2O_2 contents. The maize plant-associated rhizospheric microbial species; that is, *Pseudomonas aeruginosa*, *Enterobacter cloacae*, *Achromobacter xylosoxidans,* and *Leclercia adecarboxylata* were reported to produce ACC deaminase that resulted into enhanced drought stress tolerance. The plants showed improved grain yield plant^{-1}, photosynthetic rate, and stomatal conductance, enhanced chlorophyll a, total chlorophyll, and carotenoid contents under drought stress [38, 42]. Chandra et al. [40] reported ACC deaminase production in the wheat-associated microbes, that is, *Variovorax paradoxus* RAA3,

Plant type	Microbe	Effects on growth	Biochemical changes	Reference
Solanum lycopersicum	*Bacillus subtilis* Rhizo SF 48.	Maximum seed (laboratory) and plant growth promotion (greenhouse)	Increased Proline, SOD and APX activity. Decrease in MDA, H2O2 contents	Gowtham et al. [37]
Zea mays	*Leclercia adecarboxylata*	Enhanced shoot and root length, shoot fresh and dry weight and root fresh and dry weight in maize seedlings under axenic conditions	—	Danish et al. [38]
Capsicum anum	*Bulkhorderia cepacia*, *Citrobacter feurendii*	Plant growth promotion under stress conditions	—	Maxton et al. [39]
Triticum aestivum	*Variovorax paradoxus* RAA3; *Pseudomonas* spp. *Achromobacter* spp. *Ochrobactrum anthropi* DPC9	Improved wheat plant growth and foliar nutrient concentrations under glasshouse experiment	Positive changes in antioxidant properties	Chandra et al. [40]
grapevine (*Vitis vinifera* L.)	Strains of *Pseudomonas*, *Enterobacter*, and *Achromobacter*	Increased plant height, biomass of shoot and root organs, relative water contents, and net photosynthetic rate of leaves	Significant changes in IAA, abscisic acid, and malondialdehyde	Duan et al. [41]
Z. mays	*Pseudomonas aeruginosa*, *Enterobacter cloacae*, *Achromobacter xylosoxidans* and *L. adecarboxylata*	Grain yield plant-1, photosynthetic rate, stomatal conductance	Enhanced chlorophyll a, total chlorophyll and carotenoids contents under drought stress	Danish et al. [42]
T. aestivum	*Serratia marcescens* and *Pseudomonas* sp.	Improved harvest index	Improved water status, reactive oxygen species, osmolyte accumulation, chlorophyll and carotenoids content	Khan and Singh [43]
Cyamopsis tetragonoloba	Strains of *Pseudomonas*, *Enterobacter*, and *Stenotrophomonas*	—	—	Goyal et al. [44]
Vigna mungo L. and *Pisum sativum* L.	*Ochrobactrum pseudogrignonense*RJ12, *Pseudomonas* sp.RJ15 and *B. subtilis*RJ46	increase seed germination percentage, root length, shoot length, and dry weight of treated plants	An elevated production of ROS scavenging enzymes and cellular osmolytes; higher leaf chlorophyll content	Saikia et al. [45]

Table 1.
Microbial ACC deaminase production confers drought stress tolerance in plants.

Pseudomonas spp. *Achromobacter* spp. and *Ochrobactrum anthropi* DPC9 that improved plant growth and foliar nutrient concentrations in the wheat plants subjected to drought stress under glasshouse conditions. The plants also showed positive changes in the antioxidant properties under drought stress. Similarly, the rhizospheric *Serratia marcescens* and *Pseudomonas* sp. conferred drought stress tolerance to wheat plants, which showed improved harvest index, water status, reactive oxygen species scavenging, osmolyte accumulation, chlorophyll and carotenoid content [40]. Moreover, the rhizospheric *Ochrobactrum pseudogrignonense* RJ12, *Pseudomonas* sp. RJ15, and *B. subtilis*RJ46 exhibited increased seed germination percentage, root length, shoot length, and dry weight of treated plants, *Vigna mungo* L. and *Pisum sativum* L under drought stress. The ACC deaminase production triggered the induction of ROS scavenging enzymes and cellular osmolytes, and higher leaf chlorophyll content. The microbial production of ACC deaminase showed growth improvement and drought stress tolerance in other plants such as *Capsicum anum*, *Vitis vinifera,* and *Cyamopsis tetragonoloba* [39, 41, 43].

4.2 Microbial production of phytohormone

Phytohormones such as indole acetic acid (IAA), gibberellins, ABA, ethylene, and cytokinin are organic chemical messengers that coordinate cellular events in plants and, therefore, play a crucial role in plant development and drought stress tolerance [44]. These phytohormones are generally produced by plants; however, they also produced by some plant-associated microorganisms. Phytohormones such as auxins indirectly regulate drought stress through modification of root growth and root hairs in a manner that enable plants to absorb maximum water and nutrients from the soil. In plants, auxins are produced through the tryptophan-dependent pathways [46]. Downward in the tryptophan pathway, indole-3-acetamide is produced that is converted into indole-3-acetaldoxime and tryptamine that further give indole-3-pyruvic acid, the final product [47]. The role of IAA in drought stress mitigation through inducing the drought signaling pathways has been reported in several studies [48]. Auxin production has the potential to induce the elongation of stems and coleoptiles of plants under stress conditions; thus, its production in the microbe-treated plants may trigger such modifications [49]. In a similar passion, plant-associated microbes also induce the production of plant gibberellins. These are diterpeniods and are responsible for the hyperactive elongation of stems under stress conditions. Gibberellins are in association with carotenes and isoprene, bioactive compounds [50]. Carotenes protect the plant cells from harmful photodynamic reactions through triggering the photosynthesis, whereas isoprene regulates the turgor pressure that provides stability to the cell membranes [51]. In plants, the cytokinin biosynthesis increases in association with auxin and regulation of developmental responses under abiotic stress conditions. Cytokinin mediates in the phosphorylation of sugars leading to cellular accumulation in cells and also helps to prevent the reverse diffusion of sugars [52]. This cytokinin-mediated competitive phosphorylation enables plant cells to adapt to stress and provides protection against the damaging effects of oxidative stress [53]. Abscisic acid is generally known as a universal stress hormone because of its very important in plant adaptation to stress condition. In plants, the stomatal closure and inhibitor of stomatal opening are the underling mechanisms of the ABA-mediated stress tolerance [54]. The ABA production also confers desiccation tolerance through regulation of gene expression [55].

The ethylene production in plants plays a significant role in regulation of plant growth and senescence. In interaction with other hormones, ethylene serves as a messenger hormone that regulates developmental processes, ranging from seed germination to the plant vegetative and reproductive stages [56].

The plant-associated microbes have the potential to produce phytohormones, which in turn help plants to adapt to the stress condition. Phytohormones mitigate stress through triggering a complex signaling network of genes and production of metabolites, which protect the inner cellular machinery and function as a response to environmental stress condition. The underlying mechanisms may include hormone regulation, and production of osmolytes and antioxidant enzymes.

4.3 Osmotic adjustment

Plants experience a sudden osmotic shock upon exposure to abiotic stresses. However, the plant-associated microbial communities and their interactions assist in osmotic adjustments. In response to drought stress, the plant-microbe interactions are governed by a series of biochemical and molecular changes leading to secretion of metabolites such as glycine, proline, organic acids, sugars, betaine, trehalose, calcium, chloride, and potassium ions. Proline is one of the most important osmolytes that is accumulated in plants and provides maintenance and protection to vital cellular organs as an adaptation to osmotic stress. Proline production has been reported in several rhizospheric bacteria [57]. Trehalose, a non-reducing sugar, is produced in plants under stress condition. Trehalose contains two glucose molecules that store energy for utilization under stress conditions. Trehalose biosynthesis in microbes is accelerated through the TPS/trehalose-6-phosphate phosphatases (TPS/TPP) pathway. Trehalose stabilizes turgor pressure and maintains osmotic adjustment in plant cells [58]. Production of organic acids has been one of the key mechanisms that microbes utilize to benefit the associated plants. Microbial inoculation of plants triggers the secretion of organic acids such as oxalic acid, malic acid, citric acid, and minerals, for example, chlorine, potassium, and sodium. These are very important for metabolic reactions, maintenance of osmoregulation, and nutrient availability in plant cells [59].

4.4 Microbial production of exopolysaccharides for drought stress mitigation

Exopolysaccharides are long-chain polymers of repeating sugar units (e.g., glucose, galactose, and rhannose) [60]. Exopolysaccharides play a crucial role by forming hydrophilic biofilms, which provide protection against aridness during osmotic stress. Exopolysaccharides enhance the water-retaining potential and regulate the distribution of biological carbon sources in the soil. Microbes protect the roots from dehydration and maintain the moisture content by forming sheaths of exopolysaccharides [61]. Microbes release exopolysaccharides in the soil as slime ingredients comprising van der Waals linkages, anion adsorption interactions, and cation hydrogen bridges, which improve the biological properties of the soil [62]. The vital microbe-plant interaction regulates the production of biofilms, which facilitate microbial attachment to the plant roots, and imparts a strong root adhering capability. In conclusion, the microbial production of exopolysaccharide is one of the important strategies that protect plants against the damaging effects of abiotic stresses including drought stress [63].

4.5 Effects of microbial volatile organic compounds against osmotic stress

The plant growth-promoting microbes produce volatile compounds, which increase plant growth and development, iron uptake, photosynthesis, and overall crop productivity. Organic acid production helps plants against disease-causing pathogens. The stress-induced organic compounds such as 2-pentylfuran, 3-hydroxy-2-butanone, and 2,3-butanediol play an important role in plant growth and development [64]. These compounds regulate stomatal closure and impart systemic stress resistance, and thus ensure plant growth and development under abiotic stress conditions [65]. These volatile compounds help boost plant growth by acting as insect repellents owing to their strong odor. The microbial production of these compounds triggers stress tolerance in the associated plants through inducing the biosynthesis of ROS scavengers and gene expression.

4.6 Microbial induction of antioxidant machinery in plants

The production of reactive oxygen species (ROS) is induced in plants under osmotic stress. These include hydrogen peroxide (H_2O_2), superoxide, singlet oxygen, radicals, alkoxy radicals, and superoxide anion radicals. The ROS production triggers irreversible damages to lipids, proteins, and DNA, affecting the redox regulation [66]. Plants induce antioxidant defense systems involving enzymatic and non-enzymatic pathways to protect against the oxidative damage during osmotic stress. Several enzymes (e.g., glutathione reductase, superoxide dismutase (SOD), catalase, and ascorbate peroxidase) and non-enzymatic components (e.g., ABA, cysteine, and glutathione) catalyze the biosynthetic pathways of antioxidant pathways [67, 68]. The microbial inoculation of plants triggers the antioxidant biosynthetic pathways in the soil and, thus, confers drought stress tolerance. Kaushal and Wani [69] and Ilyas et al. [70] reported the secretion of phenolic components by various microbes (e.g., algae, zatinomycetes, and fungi) upon exposure of plants to stress condition. The microbe-plant interaction confers stress tolerance through regulating the ROS biosynthesis and maintaining a homeostatic balance between ROS and their removal. Overall, the ROS scavenging ability of PGPR regulates the antioxidant enzymes and may provide a solid barrier against abiotic stress.

4.7 Stimulation of stress-response genes by plant-microbe interactions

The microbial inoculation of plants confers stress tolerance by triggering the expression of genes involved in plant defense against abiotic stresses. Under drought stress in plants, the expression of stress-responsive genes is modulated, which is necessary for optimization of plant growth and development. There are numerous stress-responsive genes and proteins, which are involved in plant-microbe interactions and the resulting stress tolerance. These include sHSP, CaPR-10, dehydrin-like protein (Cadhn), 11-pyrroline-5-carboxylate reductase (P5CR), pyrroline-5-carboxylate dehydrogenase (P5CDH), and vacuolar ATPases [71]. Depending on the role of the encoded proteins, the expression of these stress-responsive genes can be categorized into functional or regulatory proteins as revealed by microarray studies. The stress-responsive functional proteins include water channel transporters, detoxification enzymes, osmolyte biosynthesis enzymes, macromolecule protection factors, and proteases [71, 72]. The stress-responsive genes encoding regulatory proteins include transcription factors, ABA biosynthetic factors, and phosphate kinases. The microbial-plant interaction-based expression and upregulation of stress-responsive genes can be harnessed as a powerful tool for enhancing plant drought stress tolerance.

5. Microbes and drought stress mitigation

5.1 Plant growth-promoting rhizospheric bacteria

The plant growth-promoting rhizobacteria (PGPR) have the potential to mitigate drought stress and alleviate the negative effects of climate change on plant growth and development in a sustainable way [73]. These microbes trigger the onset of biochemical changes, which enable the plant to set a response to alleviate drought stress [74]. The underlying mechanisms include optimization of exopolysaccharides and phytohormone production, antioxidant defense system, and cyclic metabolic pathways, involved in the deposition of sugars, polyamines, amino acids, and heat-shock protein synthesis [75]. Several studies have reported the positive effects of PGPR on plant growth and development under drought stress (**Table 2**). Particularly in wheat, the inoculation of PGPR mitigated drought stress. In one study, Li et al. [76] reported growth improvement in wheat plants upon inoculation with the Actinomycetes, *Streptomyces pactum* Act12. The bacterial inoculation significantly increased the overexpression of several genes including P5CS, EXPA6, SnRK2, and EXPA2. Overall, the root length, shoot length, and fresh biomass were significantly increased. Enhanced levels of sugars and antioxidant enzymes were detected in the exposed seedlings under stress condition. Inoculation of wheat plants with *Pseudomonas libanensis* EU-LWNA-33 increased the root length and biomass under drought stress [77]. The biochemical analysis revealed an increased production of osmolytes, that is, proline and glycine betaine. At the cellular level, proline and glycine betaine production regulates osmotic homeostasis, as well as the phosphorus solubilization and uptake. Phosphorus availability is a crucial factor in the overall growth and development of plants. In this study, the inoculated strains showed solubilization of phosphorus. In previous studies, Jochum et al. [79] reported drought stress tolerance in wheat and maize plants when inoculated with *Bacillus* sp. 12D6 and *Enterobacter* sp. 16i. The inoculation improved root length, surface area, and plant productivity. The study further revealed that *Bacillus* sp. 12D6 was comparatively more effective in countering drought stress. This enhanced drought stress mitigation was possible due to the production of phytohormones such as IAA and salicylic acid. In another study, Raheem et al. [80] isolated and investigated the impact of PGPR, namely *Bacillus*, *Moraxella*, *Enterobacter*, and *Pseudomonas,* on wheat plants under drought stress. Biochemical analysis of the inoculated stressed plants revealed production of increased levels of auxin that obviously helped plants to avoid the negative impact of drought stress. It was further concluded that the enhanced auxin production triggered by the *Bacillus* species improved the field capacity by 10% and crop yield by 34%. The drought stress mitigation effects of the plant growth-promoting bacteria (PGPB), *Azospirillum* were investigated in wheat plants [78]. The stressed plants showed drought tolerance, which was attributed to the microbial-mediated production of phytohormones, solutes, ACC deaminase, exopolysaccharides, chlorophyll synthesis, and increased mineral solubilization.

Drought stress imposes osmotic and oxidative stresses, which negatively affect the crops' growth and productivity. Microbial inoculation has been the most preferred strategy to reduce stress-associated losses in crop plants. In this connection, Kour et al. [81] investigated the effects of bacterial inoculation on foxtail millet crop subjected to drought stress. Inoculation of plants with *Acinetobacter calcoaceticus* EU-LRNA-72 and *Penicillium* sp. EU-FTF-6 showed drought stress tolerance. The drought stress mitigation was mainly due to the accumulation of osmolytes

Host plant	Microbe	Underlying mechanism	Reference
Triticum aestivum	*Streptomyces pactum* Act12	Increased overexpression of EXPA2, EXPA6, P5CS, and SnRK2, increased root length, shoot length, sugar content, MDA, and ABA	Li et al. [76]
T. aestivum	*Pseudomonas libanensis* EU-LWNA-33	Increased ACC deaminase, osmolytes, and P solubilization	Kour et al. [77]
T. aestivum	*Azospirillum*	Increased phytohormone, solute formation, and exopolysaccharide production	Priyanka et al. [78]
T. aestivum and *Zea mays*	*Bacillus* sp. (12D6) and *Enterobacter* sp. (16i)	Increased IAA and SA	Jochum et al. [79]
T. aestivum	PGPR (*Bacillus*, *Enterobacter*, *Moraxella*, and *Pseudomonas*)	Increased auxin	Raheem et al. [80]
Setaria italica	*Acinetobacter calcoaceticus* EU-LRNA-72 and EU-FTF-6	Increased glycine betaine, chlorophyll a and b, proline, and sugars, decreased LPO	Kour et al. [81]
Mentha piperita	*Pseudomonas fluorescens* WCS417"r and *Bacillus amyloliquefaciens* (GB03)	Increased phenolic compounds and antioxidant defense	Chiappero et al. [82]
Glycine max	*Bacillus strains* UFGS1, UFGS2, UFGRB2, and UFGRB3	Expression of Gmdreb1a, increased stomatal conductance, transpiration and Fv/Fm	Martins et al. [83]
Cicer arietinum	*Pseudomonas putida* and *B. amyloliquefaciens*	Increased chlorophyll, antioxidant enzymes, and protein content	Kumar et al. [84]

Table 2.
Rhizospheric plant-growth-promoting bacteria and drought stress mitigation in plants.

(proline and glycine betaine) and increased levels of chlorophyll a and b contents. In this study, the increased proline and glycine betaine levels improved osmotic adjustment and membrane integrity, while the increased chlorophyll content resulted in plant growth and development.

In one study, Chiappero et al. [82] investigated the positive impact of PGPR inoculation on peppermint subjected to drought stress. Two rhizospheric bacteria, *Pseudomonas fluorescens* WCS417r and *Bacillus amyloliquefaciens* GB03, were used in the inoculation and drought experiment. The results revealed a significant improvement in drought stress tolerance, which was attributed mainly to the upregulation of the antioxidant defense system and phenolic components.

The effects of bacterial inoculation on drought stress were further tested in soybean plants [83]. The plants were inoculated with *Bacillus thuringiensis, Bacillus cereus*, and *B. subtilis* strains. The inoculation of plants with these strains resulted into an improved efficiency of the photosystem II (PS-II) and maintained the overall photosynthetic rates of the plants, transpiration rate, and stomatal conductance, which in turn improved the overall growth of inoculated plants compared with that of control plants. In addition, the genomic analysis revealed that the overexpression of *Gmdreb1a* might partly be responsible for drought stress mitigation.

The PGPBs have proven their potential as ecofriendly biofertilizers that can alleviate the negative effects of drought stress on plants. In a previous study, the PGPR strains, *Pseudomonas putida* and *B. amyloliquefaciens,* were isolated from alkaline soils and then were used in the inoculation of chickpea plants under drought stress in the greenhouse and *in vitro* experiments [84]. The inoculation of plants with the strains in combination showed increased chlorophyll content, osmolyte production, and improved photosynthesis and biomass compared with plants inoculated with a single strain. In conclusion, the PGPR enhances the overall growth and development, as well as the biotic and abiotic stress tolerance of plants through a wide range of mechanisms.

5.2 Endophytic bacteria and fungi

Endophytes bacteria and fungi reside in different organs and tissues of plants and establish symbiotic relationship. The endophytes get their prepared food, while plants are benefited in different ways such as access to limited nutrients in the soil and biotic and abiotic stress tolerance. Endophytes have been specifically focused due to their crucial role in abiotic stress tolerance of plants [85]. It was previously reported that endophytes assist their host plants to increase their biomass under stress conditions [86]. However, different plant species showed variable levels of endophytic-mediated biomass accumulation under stress condition. For example, eudicots and C4 plants exhibited increased biomass accumulation compared with C3 and monocots [87].

Endophytic microbes play a very important role in reducing the damaging effects of abiotic stresses on plants. Several studies have demonstrated the drought stress mitigation in plants with endophytic bacterial inoculation (**Table 3**). Previously, Singh et al. [88] investigated the inoculation effects of endophytic bacterial strains, *Trichoderma* T42 and *Pseudomonas* on the growth and metabolic alterations in rice plants subjected to drought stress. The inoculated plants showed significantly improved metabolic activity such as induction of the antioxidant enzymes, and increases in the total polyphenolic content, which in turn, conferred oxidative stress tolerance. In another study, rice seedlings were treated with *Gluconacetobacter diazotrophicus* strain Pal5, and the drought stress tolerance was tested in the inoculated and uninoculated plants [89]. The plants were subjected to various drought stress levels for 15 days. The inoculated plants showed increased levels of proline and glycine betaine, which conferred the plants drought stress tolerance. Molecular analysis revealed relative expression of several genes such as cat, gor, sod, BADH, and P5CR. In conclusion, inoculation with *G. diazotrophicus* mitigated the effects of drought stress on rice plants.

The effects of bacterial inoculation on the growth of *Glycine max* were evaluated under drought stress [90]. Inoculation with bacterial strains, LHL10 and LHL06, produced positive improvement in plant growth under stress condition. The inoculated plants showed increased roots, shoot length, leaf area, and dry biomass. The underlying mechanisms included an increase in *HSP90* expression levels: lipid peroxidation, increased calcium levels, and phosphate solubilization. In a study, Kour et al. [91] investigated the effects of *Streptomyces laurentii* EU-LWT3–69 and *Penicillium* sp. EU-DSF-10 on sorghum plants subjected to drought stress. Bioavailability of phosphorus to plants is reduced under drought stress. However, the plant-associated microbes contain the active form of phosphate that is provided to plants to counter the stress effects. Both the strains used in the study might solubilize phosphate and ensure its availability to the plants. The biochemical alterations mediated by bacterial inoculation included an increase in the proline and glycine betaine levels and chlorophyll content, while a decrease in the lipid peroxidation. Overall, the study suggested that bacterial inoculation enabled the plants to grow better under the drought stress.

Microbial inoculation confers drought stress tolerance to the plants partly by an induction in the growth hormone levels and acquisition to soil mineral content. This was demonstrated in a study conducted by Kang et al. [100], in which the alfalfa plants were inoculated with two *Enterobacter ludwigii* strains, namely AFFR02 and Mj1212. The inoculated plants were assessed under drought stress for hormones and mineral concentrations. The results showed that the inoculated plants were more drought-tolerant than the uninoculated plants. Growth attributes such as fresh and dry biomass, root/shoot elongation, and stalk diameter were significantly higher in the inoculated plants than in the uninoculated plants. It was also observed that the treated plants accumulated higher levels of flavonoids, minerals, and ABA than those of untreated plants. In a previous study, Silambarasan et al. [92] used *Rhodotorula mucilaginosa* strain CAM4 in the inoculation of *Lactuca sativa* subjected to drought stress. The inoculated plants showed drought tolerance at various developmental stages. The treated plants showed a clear increase in the growth, dry biomass, root proliferation, and stem elongation as compared with the untreated plants. The inoculation caused a significant increase in the content of chlorophyll, carotenoids, and proline, while a decrease in the malondialdehyde (MDA) levels, indicating lipid peroxidation.

Fungal endophytic species have been extensively studied for their positive effects on plant growth, stress tolerance, and disease resistance. They exert their positive effects through production of growth hormones, siderophores, secondary metabolites, and phosphate solubilization. Several studies have demonstrated the drought stress-mitigating effects of fungal inoculation on plants (**Table 3**). Fungal endophytes, specifically isolated from desert plants, have shown promising results when used for stress mitigation in crop plants. Desert plants are usually exposed to high magnitudes of drought conditions and thus may harbor fungal endophytes that may confer drought and salt stress tolerance under arid environments. Jain et al. [101] used halotolerant fungal endophytic strains, namely *Neocamarosporium chichastianum*, *Neocamarosporium goegapense*, and *Periconia macrospinosa* in the inoculation of tomato and cucumber seedlings. The treated plants showed stress tolerance, which was evident from the increased plant growth, chlorophyll content, proline, and antioxidant enzyme levels.

Osmotic adjustment is one of the key physiological mechanisms, which have been observed in the drought-tolerant plants. Endophytic fungal strains confer drought tolerance in plants by maintaining osmotic balance and water uptake efficiency. In one study, Dastogeer et al. [94] investigated the role of the fungal endophyte *Neotyphodium coenophialum* in the drought tolerance of *Lolium arundinaceum*. The results revealed that the treated plants had high drought tolerance than the untreated plants. The underlying mechanism of this enhanced tolerance was dependent on the osmotic balance and improved water uptake efficiency, which in turn enhanced the gene expression and photosynthesis rate. In addition, drought tolerance was achieved in *Nicotiana benthamiana*, when inoculated with fungal endophytes isolated from a *Nicotiana* plant. Overall, the fungal endophytes contribute to the drought tolerance trait mainly by increasing the water-use efficiency, nutrient uptake and maintaining the ion homeostasis to induce stress tolerance in the associated plants.

5.3 Mycorrhizae

Mycorrhizae are fungal species that establish a symbiotic relationship with higher plants and play a significant role in plant growth, nutrient acquisition, soil fertility,

Microbe type	Host		Underlying mechanism	Reference
Endophytic bacteria	*Oryza sativa L.*	Trichoderma T42 and Pseudomonas	PAL expression, polyphenols	Singh et al. [88]
	O. sativa L.	Gluconacetobacter diazotrophicus Pal5	Expression of gor, cat, P5CR, sod, and BADH Increased proline and glycine betaine production	Filgueiras et al. [89]
	Glycine max	Endophytes (LHL10 and LHL06)	Expression of GmHsp90A2 and GmHsp90A1 increased SOD and decreased LPO	Bilal et al. [90]
	Foxtail millet	EU- LRNA-72 and EU-FTF-6	Increased glycine betaine, chlorophyll a and b, proline, and sugars	Kour et al. [81]
	Sorghum bicolor L.	Streptomyces laurentii EU-LWT3–69 and Penicillium sp. strain EU-DSF-10	Increased osmolytes, ACC and deaminase, solubilize P and reduced LPO	Kour et al. [91]
	Lactuca sativa	CAM4	Reduced MDA, and increased proline, chlorophyll a and b and carotenoids	Silambarasan et al. [92]
Endophytic fungi	Broccoli	YNA59	Increased sugar, protein, chlorophyll content, ABA, JA, and SA	Kim et al. [93]
	Lolium Arundinaceum, *Nicotiana benthamiana*	Neotyphodium coenophialum	Increased gaseous exchange and antioxidant enzyme production, altered gene expression and osmotic balance	Dastogeer et al. [94]
Mycorrhizae	*G. max*	Arbuscular mycorrhizal fungus	Increased proline, glycine, and soluble sugars, reduced MDA content	Grümberg et al. [95]
	Fragaria ananassa Duch.	Arbuscular mycorrhizal fungus	Increased Zn, AA enzyme and water uptake	Moradtalab et al. [96]
	Trifoliate orange	Funneliformis mosseae	Increased phenolic contents, terpenes, root exudates and coumarins, reduced alkanes, ester, and amides	Cheng et al. [97]
	Triticum spp.	Glomus mosseae	Decreased 6-SFT, SOD and sulfur metabolism; increased genetic diversity	Bernardo et al. [98]
	Ephedra foliata Boiss	AMF (Glomus etunicatum, Glomus intraradices, and Glomus mosseae)	Increased gene expression, mineral solubilization, hormone production, expression, osmolytes and antioxidant enzymes	Al-Arjani et al. [99]

Table 3.
Microbial mitigation of drought stress in plants.

and biotic and abiotic stress tolerance. They have interspecific functionality and are generally host-specific [102]. The endophytic microbes induce stress tolerance in the associated plants mainly through producing phytohormones and induction of the synthesis of secondary metabolites. On the contrary, arbuscular mycorrhizae confer stress tolerance to host plants by maintaining a steady flow of water and nutrient absorption from the soil [103].

The role of arbuscular mycorrhizae in drought stress tolerance of plants has been documented in several previous studies (**Table 3**). In one study, the impact of inoculation of arbuscular mycorrhizal strains, namely *Septoglomus constrictum*, *Glomus* sp., and *Glomus aggregatum* was studied in soybean, which is highly sensitive to abiotic stress [95]. The plants were subjected to drought stress after inoculation, and the biochemical, physiological, and molecular attributes were investigated. The treated plants showed increased levels of soluble sugars, proline, and glycine betaine and reduced MDA levels. The increased osmolyte levels in the treated plants conferred increased protection against the drought stress, while the lowered MDA content reduced the osmotic stress. The induction of phenolic compounds is one of the key mechanisms through which plants generate a response not only to infectious diseases but also to drought stress. In one study, Cheng et al. [97] used *Funneliformis mosseae* in the inoculation of trifoliate orange. The inoculated plants exhibited a marked increase in several growth attributes such as stem elongation, leaf number, leaf area, and root architecture. Biochemical analysis revealed induction in the contents of coumarin, terpene, and phenolic contents in the root exudates of the treated plants as compared with those of untreated plants. The drought stress tolerance in the inoculated plants was attributed to the induction of phenolic components, as they reduce oxidative stress in plants. In a similar study, a mycorrhizal fungal strain, *Glomus mosseae* was used to inoculate bread and durum wheat cultivars, and the plants were then exposed to drought stress [98]. The drought stress tolerance mechanism was evaluated through measurement of growth parameters and proteomics analysis. The inoculated plants showed increased dry weight, and the two genotypes responded differently to the fungal inoculation in terms of stress tolerance. A significant upregulation in the osmolytes concentrations was observed. Moreover, the inoculated plants accumulated lower ethylene levels an indication of stress tolerance.

Oxidative stress triggers the production of reactive oxygen species (ROS), which further causes irreversible damages to the macromolecules and key enzymes. Microbes have an ameliorating impact on oxidative stress in plants. Drought stress imposes oxidative stress on plants with associated growth and yield reduction. Plants respond to ROS generation by triggering the induction of ROS scavengers, which protect the cellular machinery. In one study, Zou et al. [104] used *Gigaspora margarita* and *Glomus intraradices* strains in the inoculation of host plants subjected to drought stress. Molecular analysis revealed upregulation of the expression of *GintSOD*, *GmarCuZnSOD*, *GintPDX1*, and *GintMT1* in the inoculated plants. Moreover, it was concluded that the drought stress tolerance mechanism also involved reduction in the cytoplasmic protein levels and regulation of redox status through synthesis of pyridoxamine. The drought-associated secondary stresses negatively impact both quality and quantity of crop plants. However, these negative effects can be efficiently mitigated through inoculation of various AMF strains. In one study, Al-Arjani et al. [99] isolated three AMF strains, namely *Glomus mosseae*, *Glomus etunicatum*, and *Glomus intraradices* from the rhizosphere of *Acacia gerrardii*. These strains were used to inoculate the *Ephedra foliata* Boiss plants, subjected to drought stress. Compared with the untreated plants, the treated plants showed a significant increase in the chlorophyll and carotenoid contents. In addition, the treated plants showed increased levels of sucrose-phosphate synthase and osmolyte levels, which might be responsible for the enhanced drought stress tolerance. In another study, Moradtalab et al. [96] inoculated strawberry seedlings with AMF and silicon to evaluate their combined effects against drought stress. It was observed that the AMF and silicone inoculation caused a marked increase in the water uptake, mineral content, and overall biomass. The antioxidant defense system was also triggered, which reduced the drought-associated damages and conferred stress tolerance.

6. Conclusion and prospects

Climate change tends to increase the global temperatures, which have devastated impacts on agricultural production. Furthermore, the rapid industrialization and increase in the world population have put a lot of burden on agriculture to produce more food and feed from the existing crop varieties, and land and water resources. Drought tolerance in plants is a multigenic trait and can be enhanced in a meaningful way by adoption of multi-pronged strategy. Endophytic and rhizospheric microbes have well-established mechanisms to support plants in nutrient acquisition, stress tolerance, and disease resistance. In this connection, microbes with high potential of osmolytes and siderophore production, phosphate solubilization, and nitrogen assimilation should be selected for plant inoculation. Also, the microbial induction of the genes involved in ROS scavenging may help plants to overcome the negative effects of drought stress. Inoculation experiments should involve mixtures or consortium of microbes rather than individual microbial strains. Some recent experiments have used mixtures of diverse microbial strains, and it was suggested that a microbial consortium would have broad impacts on the plant growth and productivity under drought stress. Furthermore, the modern and state-of-the-art gene sequencing and editing tools could be used in genomic studies, which would involve identification, cloning, and functional characterization of target genes in the selected microbes with high potential of conferring drought stress tolerance. These efforts could be combined with approaches of system biology studies, which would further explore the microbial-mediated alterations in metabolic profiles under drought stress. An important consideration would be to combine the complex genetic networks with those of metabolic events, which are lying at the core of plant-microbe interaction under environmental stresses including drought stress.

Acknowledgements

The authors are thankful to Fatma, Amelia, Zaran, and Zartasha for their precious time and patience in the accomplishment of this project.

Conflict of interest

The authors declare no conflict of interest.

Author details

Mohammad Sayyar Khan* and Irba Zulfiqar
Division of Microbiology and Microbial Biotechnology, Institute of Biotechnology and Genetic Engineering (IBGE), The University of Agriculture Peshawar, Khyber Pakhtunkhwa, Pakistan

*Address all correspondence to: sayyarkhankazi@aup.edu.pk

References

[1] Bastiaansen R, Doelman A, Eppinga MB, Rietkerk M. The effect of climate change on the resilience of ecosystems with adaptive spatial pattern formation. Ecology Letters. 2020;**23**(3):414-429

[2] Makridakis S, Bakas N. FORTHCOMING: Risk and decision analysis journal forecasting and uncertainty: A survey. Risk and Decision Analysis. 2016;**6**(1):37-64

[3] Intergovernmental Panel on Climate Change (IPCC). Climate change and land: An IPCC special report on climate change, desertification, land degradation, sustainable land management, food security, and greenhouse gas fluxes in terrestrial ecosystems. 2019. Available from: https://www.ipcc.ch/srccl/download/

[4] Le Houérou HN. Climate change, drought and desertification. Journal of Arid Environments. 1996;**34**(2):133-185

[5] Arora NK. Impact of climate change on agriculture production and its sustainable solutions. Environmental Sustainability. 2019;**2**:95-96. DOI: 10.1007/s42398-019-00078-w

[6] Raza A, Razzaq A, Mehmood SS, Zou X, Zhang X, Lv Y, et al. Impact of climate change on crops adaptation and strategies to tackle its outcome: A review. Plants. 2019;**8**(2):34

[7] Finkel OM, Castrillo G, Paredes SH, González IS, Dangl JL. Understanding and exploiting plant beneficial microbes. Current Opinion in Plant Biology. 2017;**38**:155-163

[8] Manzar N, Singh Y, Kashyap AS, Sahu PK, Rajawat MV, Singh A, et al. Biocontrol potential of native Trichoderma spp. against anthracnose of great millet (sorghum bicolour L.) from Tarai and hill regions of India. Biological Control. 2021;**152**:104474

[9] Egamberdieva D, Wirth SJ, Alqarawi AA, Abd_Allah, Elsayed F, & Hashem, Abeer. Phytohormones and beneficial microbes: Essential components for plants to balance stress and fitness. Frontiers in Microbiology. 2017;**8**:2104

[10] Teshome DT, Zharare GE, Naidoo S. The threat of the combined effect of biotic and abiotic stress factors in forestry under a changing climate. Frontiers in Plant Science. 2020;**11**:601009. DOI: 10.3389/fpls.2020.601009

[11] Shen M, Huang W, Chen M, Song B, Zeng G, Zhang Y. (micro) plastic crisis: Un-ignorable contribution to global greenhouse gas emissions and climate change. Journal of Cleaner Production. 2020;**254**:120138

[12] Canarini A, Schmidt H, Fuchslueger L, Martin V, Herbold CW, Zezula D, et al. Ecological memory of recurrent drought modifies soil processes via changes in soil microbial community. Nature Communications. 2021;**12**(1):1-14

[13] Lau JA, Lennon JT. Rapid responses of soil microorganisms improve plant fitness in novel environments. Proceedings of the National Academy of Sciences. 2012;**109**(35):14058-14062

[14] Kapoor D, Bhardwaj S, Landi M, Sharma A, Ramakrishnan M, Sharma A. The impact of drought in plant metabolism: How to exploit tolerance mechanisms to increase crop production. Applied Sciences. 2020;**10**(16):5692

[15] Farooq M, Wahid A, Kobayashi N, Fujita D, Basra SMA. Plant Drought Stress: Effects, mechanisms and management. In: Lichtfouse E, Navarrete M, Debaeke P, Veronique S, Alberola S, editors. Sustainable Agriculture. Springer, Dordrecht; 2009. pp. 153-188. DOI: 10.1007/978-90-481-2666-8_12

[16] Gill SS, Tuteja N. Reactive oxygen species and antioxidant machinery in abiotic stress tolerance in crop plants. Plant Physiology and Biochemistry. 2010;**48**(12):909-930

[17] Ullah A, Sun H, Yang X, Zhang X. Drought coping strategies in cotton: Increased crop per drop. Plant Biotechnology Journal. 2017;**15**(3):271-284

[18] Khan A, Pan X, Najeeb U, Tan DK, Yuen F, Shah Z, et al. Coping with drought: Stress and adaptive mechanisms, and management through cultural and molecular alternatives in cotton as vital constituents for plant stress resilience and fitness. Biological Research. 2018;**51**:47. DOI: 10.1186/s40659-018-0198-z

[19] Hussain SS, Mehnaz S, Siddique KHM. Harnessing the plant microbiome for improved abiotic stress tolerance. In: Egamberdieva D, Ahmad P, editors. Plant Microbiome: Stress Response. Microorganisms for Sustainability. Vol 5. Singapore: Springer; 2018. pp. 21-43. DOI: 10.1007/978-981-10-5514-0_2

[20] Suriyagoda L, De Costa WAJM, Lambers H. Growth and phosphorus nutrition of rice when inorganic fertiliser application is partly replaced by straw under varying moisture availability in sandy and clay soils. Plant and Soil. 2014;**384**(1):53-68

[21] Farooq M, Aziz T, Wahid A, Lee D-J, Siddique KHM. Chilling tolerance in maize: Agronomic and physiological approaches. Crop and Pasture Science. 2009;**60**(6):501-516

[22] Naylor D, Coleman-Derr D. Drought stress and root-associated bacterial communities. Frontiers in Plant Science. 2018;**8**:2223

[23] Queiroz MS, Oliveira CES, Steiner F, Zuffo AM, Zoz T, Vendruscolo EP, et al. Drought stresses on seed germination and early growth of maize and sorghum. Journal of Agricultural Science. 2019;**11**(2):310-318

[24] Gupta A, Rico-Medina A, Caño-Delgado AI. The physiology of plant responses to drought. Science. 2020;**368**(6488):266-269

[25] Siddique MRB, Hamid AIMS, Islam MS. Drought stress effects on water relations of wheat. Botanical Bulletin of Academia Sinica. 2000;**41**:35-39

[26] Rahdari P, Hoseini SM. Drought stress: A review. International Journal of Agronomy and Plant Production. 2012;**3**(10):443-446

[27] Fang Y, Xiong L. General mechanisms of drought response and their application in drought resistance improvement in plants. Cellular and Molecular Life Sciences. 2015;**72**(4):673-689

[28] Qaim M. Role of new plant breeding technologies for food security and sustainable agricultural development. Applied Economic Perspectives and Policy. 2020;**42**(2):129-150

[29] Varshney RK, Singh VK, Kumar A, Powell W, Sorrells ME. Can genomics deliver climate-change ready crops? Current Opinion in Plant Biology. 2018;**45**:205-211

[30] Laha A, Chakraborty P, Banerjee C, Panja AS, Bandopadhyay R. Application of Bioinformatics for Crop Stress Response and Mitigation.

In: Roychowdhury R, Choudhury S, Hasanuzzaman M, Srivastava S, editors. Sustainable Agriculture in the Era of Climate Change. Springer, Cham; 2020. pp. 589-614. DOI: 10.1007/978-3-030-45669-6_25

[31] Scheben A, Yuan Y, Edwards D. Advances in genomics for adapting crops to climate change. Current Plant Biology. 2016;**6**:2-10

[32] Biswal AK, Mangrauthia SK, Raghurami Reddy M, Yugandhar P. CRISPR mediated genome engineering to develop climate smart rice: Challenges and opportunities. Seminars in Cell Developmental Biology. 2019;**96**:100-106. DOI: 10.1016/j.semcdb.2019.04.005

[33] Ahmad A, Aslam Z, Naz M, Hussain S, Javed T, Aslam S, et al. Exogenous salicylic acid-induced drought stress tolerance in wheat (Triticum aestivum L.) grown under hydroponic culture. PLoS One. 2021;**16**:e0260556. DOI: 10.1371/journal.pone.0260556

[34] Choudhary DK, Varma A, Tuteja N. Plant-Microbe Interaction: An Approach to Sustainable Agriculture. Berlin: Springer; 2016

[35] Kour D, Yadav AN. Microbe mediated mitigation of drought stress in crops. Agricultural Letters. 2020;**1**:79-82

[36] Houda T, Issam BS, Naïma KB. Effectiveness of the plant growth-promoting rhizobacterium Pantoea sp. Soil Science. 2013;**158**:442-453

[37] Gowtham HG, Brijesh SS, Murali M, Shilpa N, Prasad M, Aiyaz M, et al. Induction of drought tolerance in tomato upon the application of ACC deaminase producing plant growth promoting rhizobacterium Bacillus subtilis Rhizo SF 48. Microbiological Research. 2020;**234**(2020):126422

[38] Danish S, Zafar-Ul-Hye M, Hussain S, Riaz M, Qayyum MF. Mitigation of drought stress In maize through inoculation with drought tolerant ACC deaminase containing PGPR under axenic conditions. Pakistan Journal of Botany. 2020;**52**(1):49-60

[39] Maxton A, Singh P, Masih SA. ACC deaminase-producing bacteria mediated drought and salt tolerance in *Capsicum annuum*. 2017. Journal of Plant Nutrition. 2018;**5**:574-583. DOI: 10.1080/01904167.2017.1392574

[40] Chandra D, Srivastava R, Gupta VVSR, Franco CMM, Sharma AK. Evaluation of ACC-deaminase-producing rhizobacteria to alleviate water-stress impacts in wheat (Triticum aestivum L.) plants. Canadian Journal of Microbiology. 2019;**65**(5):387-403. DOI: 10.1139/cjm-2018-0636

[41] Duan B, Li L, Chen G, Su-Zhou C, Li Y, Merkeryan H, et al. 1-Aminocyclopropane-1-carboxylate deaminase-producing plant growth-promoting Rhizobacteria improve drought stress tolerance in grapevine (Vitis vinifera L.). Frontiers in Plant Science. 2021;**12**:706990. DOI: 10.3389/fpls.2021.706990

[42] Danish S, Zafar-Ul-Hye M, Mohsin F, Hussain M. ACC-deaminase producing plant growth promoting rhizobacteria and biochar mitigate adverse effects of drought stress on maize growth. PLoS One. 2020;**15**:e0230615. DOI: 10.1371/journal.pone.0230615

[43] Khan A, Singh AV. Multifarious effect of ACC deaminase and EPS producing pseudomonas sp. and Serratia marcescens to augment drought stress tolerance and nutrient status of wheat. World Journal of Microbiology and Biotechnology. 2021;**37**:198

[44] Goyal D, Kumar S, Meena D, Swaroop SS, Pandey J. Selection of ACC

deaminase positive, thermohalotolerant and drought tolerance enhancing plant growth-promoting bacteria from rhizospheres of Cyamopsis tetragonoloba grown in arid regions. Letters in Applied Microbiology. 2021;**74**(4):519-535. DOI: 10.1111/lam.13633

[45] Saikia J, Sarma RK, Dhandia R, Yadav A, Bharali R, Gupta VK, et al. Alleviation of drought stress in pulse crops with ACC deaminase producing rhizobacteria isolated from acidic soil of Northeast India. Scientific Reports. 2018;**8**:3560. DOI: 10.1038/s41598-018-21921-w

[46] Jogawat A, Yadav B, Lakra N, Singh AK, Narayan OP. Crosstalk between phytohormones and secondary metabolites in the drought stress tolerance of crop plants: A review. Physiologia Plantarum. 2021;**172**:1106-1132. DOI: 10.1111/ppl.13328

[47] Zhao Y. Auxin biosynthesis and its role in plant development. Annual Review of Plant Biology. 2010;**61**:49-64. DOI: 10.1146/annurev-arplant-042809-112308

[48] Skubacz A, Daszkowska-Golec A, Szarejko I. The role and regulation of ABI5 (ABA-insensitive 5) in plant development, abiotic stress responses and phytohormone crosstalk. Frontiers in Plant Science. 2016;7:1884. DOI: 10.3389/fpls.2016.01884

[49] Fahad S, Nie L, Chen Y, Wu C, Xiong D, Saud S, et al. Crop plant hormones and environmental stress. In: Lichtfouse E, editor. Sustainable Agriculture Reviews. Cham: Springer; 2015. pp. 371-400. DOI: 10.1007/978-3-319-09132-7_10

[50] Ullah A, Nisar M, Ali H, Hazrat A, Hayat K, Keerio AA, et al. Drought tolerance improvement in plants: An endophytic bacterial approach. Applied Microbiology and Biotechnology. 2019;**103**:7385-7397. DOI: 10.1007/s00253-019-10045-4

[51] Alagoz Y, Nayak P, Dhami N, Cazzonelli CI. Cis-carotene biosynthesis, evolution and regulation in plants: The emergence of novel signaling metabolites. Archives of Biochemistry and Biophysics. 2018;**654**:172-184. DOI: 10.1016/j.abb.2018.07.014

[52] Mollah A, Pratiwi E. Application of various types of local microorganisms to the growth and production of two rice varieties. IOP Conference Series: Earth Environmental Science. 2020;**575**:012148. DOI: 10.1088/1755-1315/575/1/012148

[53] Mok DW, Mok MC. Cytokinin metabolism and action. Annual Review of Plant Biology. 2001;**52**:89-118. DOI: 10.1146/annurev.arplant.52.1.89

[54] Wybouw B, De Rybel B. Cytokinin–a developing story. Trends in Plant Science. 2019;**24**:177-185. DOI: 10.1016/j.tplants.2018.10.012

[55] Bright J, Desikan R, Hancock JT, Weir IS, Neill SJ. ABA induced NO generation and stomatal closure in Arabidopsis are dependent on H2O2 synthesis. The Plant Journal. 2006;**45**:113-122. DOI: 10.1111/j.1365-313X.2005.02615.x

[56] Sakata Y, Komatsu K, Takezawa D. ABA as a universal plant hormone. In: Lüttge U, Beyschlag W, Cushman J, editors. Progress in Botany. Berlin: Springer; 2014. pp. 57-96. DOI: 10.1007/978-3-642-38797-5_2

[57] Wen C-K. Ethylene in Plants. Berlin: Springer; 2014

[58] Mohammadkhani N, Heidari R. Drought-induced accumulation of soluble sugars and proline in two maize

varieties. World Applied Sciences Journal. 2008;**3**:448-453

[59] Kahraman M, Sevim G, Bor M. The role of proline, glycinebetaine, and trehalose in stress-responsive gene expression. In: Hossain M, Kumar V, Burritt D, Fujita M, Mäkelä P, editors. Osmoprotectant- Mediated Abiotic Stress Tolerance in Plants. Cham: Springer; 2019. pp. 241-256. DOI: 10.1007/978-3-030-27423-8_11

[60] Chen H, Jiang J-G. Osmotic adjustment and plant adaptation to environmental changes related to drought and salinity. Environmental Reviews. 2010;**18**:309-319. DOI: 10.1139/a10-014

[61] Bhagat N, Raghav M, Dubey S, Bedi N. Bacterial exopolysaccharides: Insight into their role in plant abiotic stress tolerance. Journal of Microbiology and Biotechnology. 2021;**31**:1045-1059. DOI: 10.4014/jmb.2105.05009

[62] Nadeem SM, Ahmad M, Tufail MA, Asghar HN, Nazli F, Zahir ZA. Appraising the potential of EPS-producing rhizobacteria with ACCdeaminase activity to improve growth and physiology of maize under drought stress. Physiologia Plantarum. 2021;**172**:463-476. DOI: 10.1111/ppl.13212

[63] Ilyas N, Mumtaz K, Akhtar N, Yasmin H, Sayyed R, Khan W, et al. Exopolysaccharides producing bacteria for the amelioration of drought stress in wheat. Sustainability. 2020;**12**:8876. DOI: 10.3390/su12218876

[64] Naseem H, Ahsan M, Shahid MA, Khan N. Exopolysaccharides producing rhizobacteria and their role in plant growth and drought tolerance. Journal of Basic Microbiology. 2018;**58**:1009-1022. DOI: 10.1002/jobm.201800309

[65] Ali F, Bano A, Fazal A. Recent methods of drought stress tolerance in plants. Plant Growth Regulation. 2017;**82**:363-375. DOI: 10.1007/s10725-017-0267-2

[66] Mohammadipanah F, Zamanzadeh M. Bacterial mechanisms promoting the tolerance to drought stress in plants. In: Singh HB, Keswani C, Reddy MS, Sansinenea E, García-Estrada C, editors. Secondary Metabolites of Plant Growth Promoting Rhizomicroorganisms. Singapore: Springer; 2019. pp. 185-224. DOI: 10.1007/978-981-13-5862-3_10

[67] Ahmad P, Jamsheed S, Hameed A, Rasool S, Sharma I, Azooz M, et al. Drought stress induced oxidative damage and antioxidants in plants. In: Ahmad P, editor. Oxidative Damage to Plants. New York, NY: Elsevier; 2014. pp. 345-367

[68] Li B, Feng Y, Zong Y, Zhang D, Hao X, Li P. Elevated CO2-induced changes in photosynthesis, antioxidant enzymes and signal transduction enzyme of soybean under drought stress. Plant Physiology and Biochemistry. 2020;**154**:105-114. DOI: 10.1016/j.plaphy.2020.05.039

[69] Kaushal M, Wani SP. Plant-growth-promoting rhizobacteria: Drought stress alleviators to ameliorate crop production in drylands. Annales de Microbiologie. 2016;**66**:35-42. DOI: 10.1007/s13213-015-1112-3

[70] Ilyas M, Nisar M, Khan N, Hazrat A, Khan AH, Hayat K, et al. Drought tolerance strategies in plants: A mechanistic approach. Journal of Plant Growth Regulation. 2021;**40**:926-944

[71] Shinozaki K, Yamaguchi-Shinozaki K. Gene networks involved in drought stress response and tolerance. Journal of Experimental Botany. 2007;**58**:221-227. DOI: 10.1093/jxb/erl164

[72] Shinozaki K, Yamaguchi-Shinozaki K. Molecular Responses to Drought Stress. Amsterdam: Elsevier Science; 1998

[73] Aslam MM, Idris AL, Zhang Q, Weifeng X, Karanja JK, Wei Y. Rhizosphere microbiomes can regulate plant drought tolerance. Pedosphere. 2022;**32**:61-74. DOI: 10.1016/s1002-0160(21)60061-9

[74] Ansari FA, Ahmad I. Alleviating drought stress of crops through PGPR: Mechanism and application. In: Singh DP, Gupta VK, Prabha R, editors. Microbial Interventions in Agriculture and Environment. Singapore: Springer; 2019. pp. 341-358. DOI: 10.1007/978-981-13-8383-0_11

[75] Sinha D, Mukherjee S, Mahapatra D. Multifaceted potential of plant growth promoting Rhizobacteria (PGPR): An overview. In: Ahmad Malik J, editor. Handbook of Research on Microbial Remediation and Microbial Biotechnology for Sustainable Soil. Pennsylvania: IGI Global; 2021. pp. 205-268. DOI: 10.4018/978-1-7998-7062-3.ch008

[76] Li H, Guo Q, Jing Y, Liu Z, Zheng Z, Sun Y, et al. Application of Streptomyces pactum Act12 enhances drought resistance in wheat. Journal of Plant Growth Regulation. 2020;**39**:122-132. DOI: 10.1007/s00344-019-09968-z

[77] Kour D, Rana KL, Sheikh I, Kumar V, Yadav AN, Dhaliwal HS, et al. Alleviation of drought stress and plant growth promotion by Pseudomonas libanensis EU-LWNA-33, a drought-adaptive phosphorussolubilizing bacterium. Proceedings of the National Academy of Sciences India Section B Biology Sciences. 2020b;**90**:785-795. DOI: 10.1007/s40011-019-01151-4

[78] Priyanka JP, Goral RT, Rupal KS, Saraf M. Rhizospheric microflora: A natural alleviator of drought stress in agricultural crops. In: Sayyed RZ, editor. Plant Growth Promoting Rhizobacteria for Sustainable Stress Management. Singapore: Springer; 2019. pp. 103-115. DOI: 10.1007/978-981-13-6536-2_6

[79] Jochum MD, Mcwilliams KL, Borrego EJ, Kolomiets MV, Niu G, Pierson EA, et al. Bioprospecting plant growth-promoting rhizobacteria that mitigate drought stress in grasses. Frontiers in Microbiology. 2019;**10**:2106. DOI: 10.3389/fmicb. 2019.02106

[80] Raheem A, Shaposhnikov A, Belimov AA, Dodd IC, Ali B. Auxin production by rhizobacteria was associated with improved yield of wheat (Triticum aestivum L.) under drought stress. Archives of Agronomy and Soil Science. 2018;**64**:574-587. DOI: 10.1080/03650340.2017.1362105

[81] Kour D, Rana KL, Yadav AN, Sheikh I, Kumar V, Dhaliwal HS, et al. Amelioration of drought stress in foxtail millet (Setaria italica L.) by P-solubilizing drought-tolerant microbes with multifarious plant growth promoting attributes. Environmental Sustainability. 2020c;**3**:23-34. DOI: 10.1007/s42398-020-00094-1

[82] Chiappero J, Del Rosario Cappellari L, Alderete LGS, Palermo TB, Banchio E. Plant growth promoting rhizobacteria improve the antioxidant status in Mentha piperita grown under drought stress leading to an enhancement of plant growth and total phenolic content. Industrial Crops and Products. 2019;**139**:111553. DOI: 10.1016/j.indcrop.2019.111553

[83] Martins SJ, Rocha GA, De Melo HC, De Castro Georg R, Ulhôa CJ, De Campos Dianese É, et al. Plant-associated bacteria mitigate drought stress in soybean. Environmental Science and Pollution Research. 2018;**25**:13676-13686. DOI: 10.1007/s11356-018-1610-5

[84] Kumar M, Mishra S, Dixit V, Kumar M, Agarwal L, Chauhan PS,

et al. Synergistic effect of pseudomonas putida and bacillus amyloliquefaciens ameliorates drought stress in chickpea (Cicer arietinum L.). Plant Signaling & Behavior. 2016;**11**:e1071004. DOI: 10.1080/15592324.2015.1071004

[85] Verma H, Kumar D, Kumar V, Kumari M, Singh SK, Sharma VK, et al. The potential application of endophytes in management of stress from drought and salinity in crop plants. Microorganisms. 2021;**9**:1729. DOI: 10.3390/microorganisms9081729

[86] Barman J, Samanta A, Saha B, Datta S. Mycorrhiza. Resonance. 2016;**21**:1093-1104

[87] Strobel GA. Endophytes as sources of bioactive products. Microbes and Infection. 2003;**5**:535-544. DOI: 10.1016/s1286-4579(03)00073-x

[88] Singh DP, Singh V, Gupta VK, Shukla R, Prabha R, Sarma BK, et al. Microbial inoculation in rice regulates antioxidative reactions and defense related genes to mitigate drought stress. Scientific Reports. 2020;**10**:4818. DOI: 10.1038/s41598-020-61140-w

[89] Filgueiras L, Silva R, Almeida I, Vidal M, Baldani JI, Meneses CHSG. Gluconacetobacter diazotrophicus mitigates drought stress in Oryza sativa L. Plant and Soil. 2020;**451**:57-73. DOI: 10.1007/s11104-019-04163-1

[90] Bilal S, Shahzad R, Imran M, Jan R, Kim KM, Lee I-J. Synergistic association of endophytic fungi enhances Glycine max L. resilience to combined abiotic stresses: Heavy metals, high temperature and drought stress. Industrial Crops and Products. 2020;**143**:111931. DOI: 10.1016/j.indcrop.2019.111931

[91] Kour D, Rana KL, Kaur T, Sheikh I, Yadav AN, Kumar V, et al. Microbe-mediated alleviation of drought stress and acquisition of phosphorus in great millet (sorghum bicolour L.) by drought-adaptive and phosphorussolubilizing microbes. Biocatalysis and Agricultural Biotechnology. 2020a;**23**:101501

[92] Silambarasan S, Logeswari P, Cornejo P, Abraham J, Valentine A. Simultaneous mitigation of aluminum, salinity and drought stress in Lactuca sativa growth via formulated plant growth promoting Rhodotorula mucilaginosa CAM4. Ecotoxicology and Environmental Safety. 2019;**180**:63-72. DOI: 10.1016/j.ecoenv.2019.05.006

[93] Kim Y-N, Khan MA, Kang S-M, Hamayun M, Lee I-J. Enhancement of drought-stress tolerance of Brassica oleracea var. italica L. by newly isolated Variovorax sp. YNA59. Journal of Microbiology and Biotechnology. 2020;**30**:1500-1509. DOI: 10.4014/jmb.2006.06010

[94] Dastogeer KMG, Chakraborty A, Sarker MSA, Akter MA. Roles of fungal endophytes and viruses in mediating drought stress tolerance in plants. International Journal of Agricultural Biology. 2020;**24**:1497-1512

[95] Grümberg BC, Urcelay C, Shroeder MA, Vargas-Gil S, Luna CM. The role of inoculum identity in drought stress mitigation by arbuscular mycorrhizal fungi in soybean. Biology and Fertility of Soils. 2015;**51**:1-10. DOI: 10.1007/s00374-014-0942-7

[96] Moradtalab N, Hajiboland R, Aliasgharzad N, Hartmann TE, Neumann G. Silicon and the association with an arbuscular-mycorrhizal fungus (Rhizophagus clarus) mitigate the adverse effects of drought stress on strawberry. Agronomy. 2019;**9**:41. DOI: 10.3390/agronomy9010041

[97] Cheng H-Q, Zou Y-N, Wu Q-S, Kuča K. Arbuscular mycorrhizal fungi alleviate drought stress in trifoliate Orange by regulating H^{+}-ATPase activity and gene expression. Frontiers in Plant Science. 2021;**12**:659694. DOI: 10.3389/fpls.2021.659694

[98] Bernardo L, Morcia C, Carletti P, Ghizzoni R, Badeck FW, Rizza F, et al. Proteomic insight into the mitigation of wheat root drought stress by arbuscular mycorrhizae. Journal of Proteomics. 2017;**169**:21-32. DOI: 10.1016/j.jprot.2017.03.024

[99] Al-Arjani A-BF, Hashem A, Abd Allah EF. Arbuscular mycorrhizal fungi modulates dynamics tolerance expression to mitigate drought stress in Ephedra foliata Boiss. Saudi Journal of Biological Science. 2020;**27**:380-394. DOI: 10.1016/j.sjbs.2019.10.008

[100] Kang S-M, Khan M-A, Hamayun M, Kim L-R, Kwon E-H, Kang Y-S, et al. Phosphate-solubilizing Enterobacter ludwigii AFFR02 and bacillus megaterium Mj1212 rescues Alfalfa's growth under post-drought stress. Agriculture. 2021;**11**:485. DOI: 10.3390/agriculture11060485

[101] Jain D, Phurailatpam L, Mishra S. Microbes-mediated mitigation of drought stress in plants: Recent trends and future challenges. In: Yadav A, Rastegari A, Yadav N, Kour D, editors. Advances in Plant Microbiome and Sustainable Agriculture. Singapore: Springer; 2020. pp. 199-218. DOI: 10.1007/978-981-15-3204-7_9

[102] Sheteiwy MS, Ali DFI, Xiong Y-C, Brestic M, Skalicky M, Hamoud YA, et al. Physiological and biochemical responses of soybean plants inoculated with arbuscular mycorrhizal fungi and Bradyrhizobium under drought stress. BMC Plant Biology. 2021;**21**:195. DOI: 10.1186/s12870-021-02949-z

[103] Agapitos A, Togelius J, Lucas SM, Schmidhuber J, Konstantinidis A. Generating diverse opponents with multiobjective evolution. In: Proceedings of the IEEE Symposium on Computational Intelligence and Games, 2008. CIG'08. Piscataway, NJ: IEEE; 2008. pp. 135-142

[104] Zou YN, Wu QS, Kuèa K. Unravelling the role of arbuscular mycorrhizal fungi in mitigating the oxidative burst of plants under drought stress. Plant Biology. 2021;**23**:50-57. DOI: 10.1111/plb.13161

Chapter 13

Plant-Growth Promoting Endophytic Bacteria and Their Role for Maize Acclimatation to Abiotic Stress

Víctor Manuel González-Mendoza, Mayra de la Torre and Jorge Rocha

Abstract

In order to grow, reproduce, and defend themselves, maize plants use various strategies to obtain adaptive advantages in varying conditions, for example, to tolerate abiotic stress (e.g., drought or heat due to climate change). One of these strategies is the establishment of interactions with plant-growth-promoting bacteria. Bacteria can be associated with plants in the rhizosphere, rhizoplane, or as endophytes. Recent evidence suggest that modern agricultural practices are detrimental to these beneficial plant-microbe interactions, and reservoirs like traditional agroecosystems called *milpas*, emerge as sources of microbiota associated with maize crops, with increased diversity and beneficial functions. Particularly, bacterial endophytes associated with native maize from *milpas* show promising features for their use as plant-growth-promoting inoculates, however, it is necessary to first understand the mechanisms known for beneficial functions of endophytes associated with maize and other plants. Here, we review the mechanisms of beneficial interactions between plants and endophytic bacteria, with emphasis on maize and with mentions of recent findings on maize landraces from *milpa* systems.

Keywords: plant associated-microbial communities, *milpa*, stress responses, amelioration, climate change

1. Introduction

Plants are constantly challenged with a plethora of stressful conditions and require several response mechanisms, including the interactions between the roots and soil microbes, which allow for nutrient availability, growth promotion, and disease suppression. Microbiota can be found associated to plants in different degrees and locations, which are divided into (1) endosphere (inter or intracellular tissues), (2) rhizoplane microbial (on the root surface and possibly attached to root hairs), and rhizosphere microbial (soil close to the root surface). The composition of each of these communities is influenced by the host genotype, soil source, cultivation practice, and so on [1–4] (**Figure 1**). Bacteria present in the endosphere have recently been considered as potential agents for acclimatation to abiotic stress response. One example is endophytic bacteria isolated from native maize from *milpa* traditional systems, which has been proposed as a bacteriome fraction that could be useful for obtaining products toward the bio-fertilization of maize crops [5]. *Milpas* are polyculture systems that include domesticated, semi-domesticated, and tolerated plant species that combine native maize landraces with almost any other crop, tree,

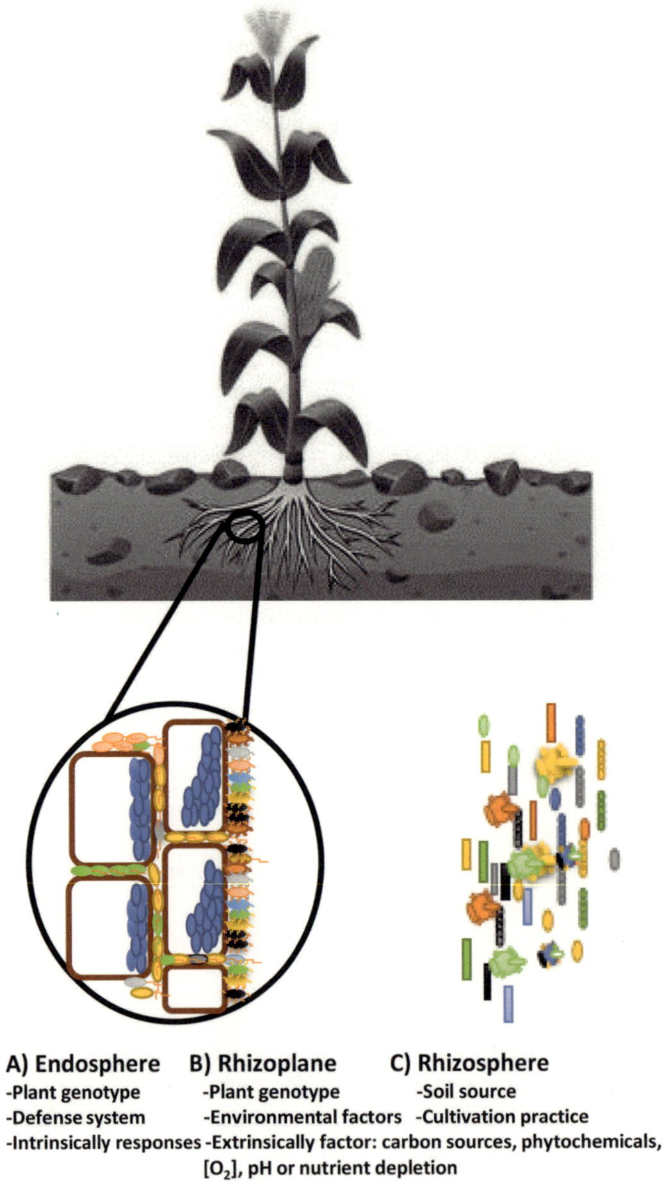

Figure 1.
Localization of maize associated-microbial communities. (A) Endosphere; here, communities are affected by plant genotype, its defenses system, and intrinsic responses; (B) in the rhizoplane, extrinsic factors like carbon sources or phytochemicals, [O_2], pH, or nutrient depletion, affect microbial community composition; (C) in the rhizosphere, communities are strongly affected by soil source, cultivation practice, and others.

or shrub species. *Milpas* constitute a dynamic system with diverse genetic resources, used as the main crop of native maize (*Zea mays* L.) and as associated crops to beans (*Phaseolus* spp.), squashes (*Cucurbita* sp.), chili peppers (*Capsicum* sp.), tomatoes (*Solanum lycopersicum*), among others. Modern practices of agriculture affect bacterial diversity and functions, therefore, *milpas* have retained unique beneficial microorganisms that interact with native maize, which could have been lost in modern hybrid maize with monocultures, high-tillage, and large agrochemical input [6].

With the current worldwide agricultural practices trend, it is necessary to promote a reduction of agrochemical use and a reintroduction of ancestral agricultural practices and/or the use of microorganism-based bio-formulations that generate a beneficial interaction between plants and microbes and thereby influence plant growth. One option is the use of plant-growth-promoting endophytes, which could be regarded as a new approach to mitigate the detrimental effect of pests and/or diseases, low rainfall, and current climate change [5].

Endophytic bacterial species have been identified in a vast number of plants; notably, they are symbionts that usually do not cause negative effects on plant growth [1, 7, 8]. The endophyte community within the plant is very as dynamic but usually is enriched with specialized types of bacteria with features such as flagella, plant-polymer-degrading enzymes, type V and VI protein secretion systems, iron acquisition and storage, quorum sensing, detoxification of reactive oxygen species (ROS), degradation of aromatic compounds, among others [9]. Colonization of endophytic bacteria can be also classified into three main categories in accordance with lifestyle strategies as (1) obligate endophytes, unable to proliferate outside of plants and are likely transmitted via seed; (2) facultative endophytes, as free living in soil but will colonize plants once conditions are appropriate, and (3) passive endophytes, as do not actively seek to colonize the plant [7, 9, 10].

Recent work suggests that microorganisms from the bulk soil in *milpas* are selected by native maize roots. Native maize selectively recruits, including strains from phyla like Acidobacteria, Actinobacteria, and Bacteroidetes, with a higher relative abundance in comparison to soil adjacent to the roots [6]. Recently, Gastélum *et al.* reported a greater presence of endophytic microbial load in native maize vs. hybrid landraces, which include more bacterial strains with antagonistic activity against soil-borne bacteria, and overall harbor more diverse bacterial communities [5]. There are many factors influencing endophytes to colonize and penetrate the root tissue, as well as the resulting interaction. After colonization, the role of microbes could modulate plant growth and development by diverse mechanisms [11]. In order to grasp maize endophytic microbes from *milpas* for plant growth promotion, we first need to understand the mechanisms for growth promotion. Here, we will discuss how maize-related endophytes can potentially have used in the alleviation of abiotic stress and/or climate change.

2. Endophytic bacteria

Bacterial endophytes inhabit the tissues of plants for at least a part of their life cycle without harming or inhibiting the growth of the plant and establishing symbiotic associations than can result in great benefits for plant health. Endophytes are microorganisms that survive within healthy plant tissues and promote plant growth under stress. A large proportion of endophytic bacteria groups are shared between leaves and roots, suggesting that they are inoculated from the soil. These communities are also dynamic, as they shift when the plants age [12]. Reinhold-Hurek et al. defined a community of least complexity where specific bacterial traits are required for internalization and establishment inside of this compartment, and host genotype likely has the strongest influence on community structure here, in comparison with other compartments such as the roots [9].

2.1 Plant-growth-promoting endophytes (PGPEs)

The PGPEs as well as PGPRs (plant-growth-promoting rhizobacteria) promote plant growth by regulating plant hormones, improving nutrition acquisition, siderophore production, and enhancing the antioxidant system [13]. Bacterial endophytes associated with plants can be classified into three groups, based on the ecological interaction: beneficial, deleterious, and neutral. Various genera of *Pseudomonas*, *Enterobacter*, *Bacillus*, *Klebsiella*, and *Burkholderia*, (which are normally considered pathogenic), are also present as PGPEs, promoting plant growth and development under both normal and stress conditions [14]. In most cases, these are indirect mechanisms, such as preventing the deleterious effects of other phytopathogenic microorganisms, and this function is achieved by antibiosis, induction of systemic resistance (ISR), and competitive exclusion [13–15]. There are several mechanisms for plant growth stimulation by PGPEs, such as nitrogen fixation; synthesis of auxin, 1-aminocyclopropane-1-carboxylate (ACC)-deaminase, siderophores production, and phosphate solubilization, and most of them are well documented [16]. In the case of endophytic bacteria isolated in native or hybrid genotypes of maize, genera, such as *Pantoea*, *Bacillus*, *Burkholderia*, *Klebsiella*, and others were found [6, 8]. There are mechanisms by PGPE to mitigate stress responses, and these are described in the following sections [17].

2.1.1 Nutrient fixation

PGPEs act as a direct enhancer to increase the accessibility and concentration of chemical elements by fixing or locking their nutritional supplies for plant growth and development [14]. To date, there are few nutrient-fixing bacteria reported as associated with maize.

2.1.2 Nitrogen fixation

Biological nitrogen fixation is a biological process that is carried out either by symbiotic or non-symbiotic interactions between microbes and plants. Symbiotic PGPRs, which are most frequently reported to fix atmospheric N_2 in soil, include strains of *Rhizobium* sp., *Azoarcus* sp., *Beijerinckia* sp., *Pantoea agglomerans*, and *Klebsiella pneumoniae* [10, 13]. Bio-fixation of atmospheric nitrogen is carried out by genera like *Azospirillum*, *Klebsiella*, *Burkholderia*, *Bacillus*, and *Pseudomonas* [18]. Someone of these species has been reported as endophytes in native maize.

2.1.3 Phosphate solubilization

Solubilization and mineralization of phosphorus by phosphate-solubilizing bacteria is an important trait that can be achieved by PGPR. Phosphate solubilizing PGPR includes genera, such as *Arthrobacter*, *Bacillus*, *Beijerinckia*, *Burkholderia*, *Enterobacter*, *Microbacterium*, *Pseudomonas*, *Erwinia*, *Rhizobium*, *Mesorhizobium*,

Flavobacterium, *Rhodococcus*, and *Serratia,* and some of these have been found associated to maize [19].

2.1.4 Potassium solubilization

Potassium solubilizing PGPR, such as *Acidothiobacillus* sp., *Bacillus edaphicus*, *Ferrooxidans* sp., *Bacillus mucilaginosus*, *Pseudomonas* sp., *Burkholderia* sp., and *Paenibacillus* sp., have been reported to release potassium in accessible form from potassium-bearing minerals in soils; some of them have been reported in interaction with maize [20].

2.1.5 Phytohormone production

One process employed by PGPRs is phytohormones production. Bacterial species, such as *Pantoea agglomerans*, *Rhodospirillum rubrum*, *Pseudomonas fluorescens*, *Bacillus subtilis*, *Paenibacillus polymyxa*, *Pseudomonas* sp., and *Azotobacter* sp., were found to carry out this mechanism and they have been tested in maize development [13].

2.1.6 Siderophore production

The production of siderophores by microbes is crucial for plant growth since these compounds chelate iron in the soil. This process is performed by a bacterium, such as *Pseudomonas* sp. and *Streptomyces* sp., and it is useful for generating soluble complexes that can be absorbed by plants, and some of these bacteria have been tested in maize [10].

2.1.7 Exopolysaccharide production (EPS)

EPS-producing PGPR, such as *Azotobacter vinelandii*, *Bacillus drentensis*, *Enterobacter cloacae*, *Agrobacterium* sp., *Xanthomonas* sp., and *Rhizobium* sp., play a vital role in maintaining water potential, aggregating soil particles, and ensuring an obligate contact between plant roots and rhizobacteria [21].

2.2 Acclimatation to abiotic stresses

It has been proposed that stress conditions cause the recruitment of particular microbial taxa from the soil. In this sense, environmental factors, such as drought, pH, and temperature have a significant impact on the microbiota associated to roots [15]. Advantages of bacterial endophytes on plant growth include protection from competing bacteria and fungi, a constant and reliable source of nutrition, and protection from exposure to a wide range of potentially deleterious environmental conditions, such as extreme temperature and the presence of inhibitory chemicals in the soil. Some of these responses are described below.

2.2.1 Chemical responses

Plant exudate compounds through their roots and these are key factors for the assembly of microbial communities in the rhizosphere. Some of these compounds are sugars, amino acids, organic acids, fatty acids, and secondary metabolites such as triterpenes [22]. On one hand, the composition of the root exudate profiles changes in different plant species, genotypes, and developmental stages. Thus, it is suggested that variations in the composition of root-associated endophytic microbial are caused by changes in root exudation [15, 22]. On the other hand, endophytic bacteria also synthesize a varied array of secondary metabolites with unique chemical structures that have been exploited as biocontrol agents. Additionally, these bioactive compounds can be beneficial as they can stimulate plant growth and development. The composition of secondary metabolites produced by endophytic bacteria depends on the physiological status and species of plants and microorganisms. The bacterial genera include *Azotobacter*, *Serratia*, *Azospirillum*, *Bacillus, Caulobacter*, *Chromobacterium*, *Agrobacterium*, *Erwinia*, *Flavobacterium*, *Arthrobacter*, *Micrococcous*, *Pseudomonas*, and *Burkholderia*. These carry out mechanisms like nutrient fixation, neutralizing biotic and abiotic stress, and producing volatile organic compounds (VOCs) and enzymes to prevent diseases. However, the mode of action is different depending on PGPR-types and varies according to the type of host plant [23]. Our unpublished data report that some native maizes have been shown to induce the synthesis of anthocyanin and phenolic compounds in response to drought or waterlogging (unpublished data), cold, high salinity, or nutrient deficiency stresses. This response is a protective strategy to alleviate these adverse impacts [24]; (**Figure 2**). Studies have shown some specific changes in root exudation of primary and secondary metabolites as follows: (1) high sugar levels exuded in early plant developmental stages may attract a wide range of microbes that can consume sugar substrates, and (2) high levels of phenolics exuded in later plant developmental stages induce specialized pathways, where these compounds are used as specific substrates or signaling molecules in ways that vary across taxa [25].

2.2.2 Nutritional starvation responses

Plants exposed to nutrient limitations exhibit a wide range of responses that include changes to the quantity and composition of the compounds released by roots. In maize, N-deficit causes a reduced exudation of amino acids; P-deficiency

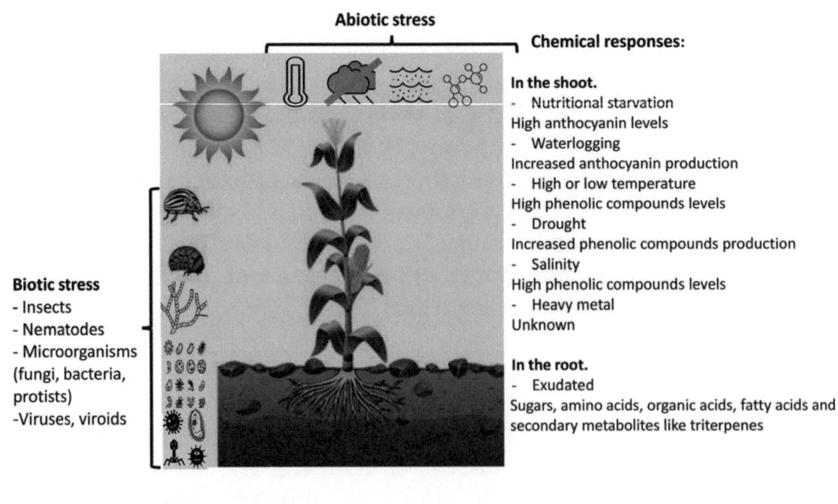

Figure 2.
Putative chemical response to abiotic stress.

stimulates the release of gamma-aminobutyric acid (GABA) and carbohydrates; whilst K-deficient plants release less sugars. Moreover, Fe deficiency causes increased release of glutamate, glucose, ribitol, and citrate [26]. Accordingly, plants exposed to different nutrient limitations show differences in the microbial structure composition. For example, P-deficient plants release compounds involved in bacterial chemotaxis and motility, whilst exudates released by Fe and K-deficient plants did not cause dramatic changes in bacterial composition [4, 26]. Interestingly, native maize landraces from *Los Tuxtlas*, Mexico show varying mycorrhizal dependency for P uptake, but there is still no data on bacterial composition [27].

2.2.3 High or low-temperature responses

Plants growing under high or low-temperature stress exhibit responses such as a decline in photochemical efficiency, stomatal conductance, and net CO_2 fixation. High temperatures cause changes in the plasma membrane, water content (transpiration), impaired photosynthesis activity, enzyme functioning, cell division, and plant growth. Some strategies to overcome this stress include the production and accumulation of enzymes and osmolytes. Temperature plays a significant role in the regulation of physiological and metabolic responses. Bacterial endophytes also possess effective mechanisms to protect the structure of proteins, membranes, and nucleic acid molecules, and in this way, they can survive under high temperatures or low temperatures. These phenomena have been studied in genera like *Pseudomonas cedrina*, *Brevundimonas terrae*, and *Arthrobacter nicotianae*, among others [14].

2.2.4 Waterlogging and water deficit responses

Waterlogging stress adversely impacts the physiology and photosynthetic capacity of the plant, and prolonged exposure generates severe damage to plant growth or productivity. Some strategies that are adopted by plants under water deficit response are reduction in transpiration loss through altering stomatal conductance and distribution, leaf rolling, root-to-shoot ratio dynamics, root length increment, accumulation of compatible solutes, enhancement in transpiration efficiency, osmotic and hormonal regulation, and delayed senescence [28]. In addition, bacterial endophytes can enhance plant tolerance through the maintenance of cell homeostasis and diminishing the adverse effects of oxidative stress [29].

2.2.5 Drought responses

Drought stress cause as responses a decline in turgor and water potentials, a suppression in photosynthesis, a decrease in the contents of the chlorophyll, and increased accumulation of proline in most plants [30]. Several studies suggest that the use of distinct endophytic bacteria could produce beneficial effects on their host plants if their co-inoculation does not generate antagonistic responses. They show mechanisms that involve the maintenance of the cell water homeostasis under drought conditions, allowing diminished water loss and increasing water inlet, carbon sequestration, nutrient cycling, resulting in health of crops, and rhizosphere ecosystem functioning [31, 32]. Many bacterial groups have been related to these mechanisms, for example, *Acinetobacter*, *Azospirillum*, *Azotobacter*, *Arthrobacter*, *Bacillus*, *Beijerinckia*, *Brevundimonas*, *Burkholderia*, *Clostridium*, *Delftia*, *Duganella*, *Erwinia*, *Enterobacter*, *Flavobacterium*, *Hydrogenophaga*, *Methylobacterium*, *Paenibacillus*,

Pantoea, *Proteus*, *Providencia*, *Pseudomonas*, *Psychrobacter*, *Rhizobium*, *Serratia*, *Stenotrophomonas*, *Streptococcus*, and *Streptomyces* [33, 34].

2.2.6 Salinity responses

Salinity stress induce a condition that prevents water uptake by the plant and relate to a decline in photosynthesis, growth, and uptake of other nutrients. Salinity adversely affects plant growth and development. Halophytic bacteria have several adaptations to mitigate salinity stress that include a reduced stomatal conductance of the host, lower water potential, uptake of inorganic ions, a salt discharge from roots, and accumulation of organic acids, among others [16]. Salinity can disrupt water uptake and ion equilibrium and lead to oxidative damage due to the production of ROS. Halophytic bacteria can keep these ROS at minimal levels due to the presence of an antioxidant system that consists of enzymes like catalase (CAT), peroxidase (POD), and superoxide dismutase (SOD). The salt-tolerant endophytes include genera like *Pseudomonas*, *Kocurias*, *Cronobacter*, *Gracilibacillus*, *Staphylococcus*, *Virgibacillus*, *Salinicoccus*, *Bacillus*, *Zhihengliuella*, *Brevibacterium*, *Oceanobacillus*, *Exiguobacterium*, *Arthrobacter*, and *Halomonas*. These bacterial groups possess an ACC deaminase activity with the potential to ameliorate plant salinity stress [16, 35, 36].

2.2.7 Heavy metal responses

Heavy metal stress is characterized by the inhibition of processes like photosynthesis, respiration, nitrogen and protein metabolism, and nutrient uptake [37]. There is a wide range of heavy metal-tolerant microorganisms and plant-associated microbes that involve various mechanisms such as efflux, impermeability to metals, volatilization, EPS sequestration, metal complexation, and enzymatic detoxification [14, 38]. The microorganisms with tolerance to heavy metals include genera such as *Rhizobacteria* and the phylum *Firmicutes* that promote plant growth and development during metal stress conditions. They carry out mechanisms to reduce ethylene concentration, production of plant growth regulators such as auxin indole-3-acetic acid (IAA), ACC deaminase, and disease suppression [14].

2.2.8 Pathogen responses

Biotic stress can be caused by different pathogens or plagues, such as bacteria, viruses, fungi, nematodes, protists, insects, and viroids. These result in a significant reduction in plant growth and development. Endophytic bacteria have been used as antagonists against plant pathogens and species like *Bacillus* spp. and *Pseudomonas* sp., produce a wide variety of compounds such as antibiotics, antifungal compounds, antivirals, and so on [39]. In plants, some defense mechanisms are activated by pathogenic or non-pathogenic invasion, that results in the activation of enzymes, such as chitinase, β-1, 3-glucanase, phenylalanine ammonia-lyase, polyphenol oxidase, peroxidase, lipoxygenase, SOD, CAT, and ascorbate peroxidase (APX). After these encounters, plants remain primed, which means that they are better prepared for future attacks by pathogens. PGPEs promote plant growth by producing metabolites that control phytopathogenic agents. These metabolites include β-1,3-glucanase, ACC-deaminase, and chitinase, which are generally involved in lysing cell walls and neutralizing pathogens [40]. Finally, species from diverse genera, including *Pseudomonas*, *Bacillus*, *Arthrobacter*, *Stenotrophomonas*, and *Serratia* can produce VOCs that impact plant growth and development [13].

2.3 Abiotic stress amelioration by plant growth-promoting endophytes

The use of bacterial strains from rhizosphere, phyllosphere, or endosphere has been suggested to promote an amelioration of abiotic stress. Endophytes promote plant growth through nitrogen fixation, phytohormone production, nutrient acquisition, and by conferring tolerance to abiotic and biotic stresses. These mechanisms have been reported across many genera such as *Bacillus*, *Pantoea*, *Klebsiella*, *Burkholderia*, *Gluconobacter*, and *Pseudomonas*, among others [13, 41, 42]. Specifically, for maize endophytes, these functions have been associated with genera, such as *Massilia*, *Burkholderia*, *Ralstonia*, *Dyella*, *Chitinophaga*, and *Sphingobium*. However, the bacterial community structure significantly changes through different growth or development process. For example, *Massilia*, *Flavobacterium*, *Arenimonas,* and *Ohtaekwangia* were enrichment at early growth stages, whilst genera like *Burkholderia*, *Ralstonia*, *Dyella*, *Chitinophaga*, *Sphingobium*, *Bradyrhizobium,* and *Variovorax* were dominant at later stages [43]. In *milpa*, studies have reported the presence of genera such as *Flavitalea*, *Sphingomonas*, *Blastococcus*, *Luteitalea,* and *Vicinamibacter*, among others groups that are uncommon in hybrid maize [5]. Moreover, endophytes bacteria like *Bacillus*, *Enterobacter*, *Pseudomonas*, *Azotobacter*, *Arthrobacter*, *Streptomyces*, and *Isoptericola* were related to the alleviation of drought, heat, and salt stress in different crop plants, **Figure 3** [11, 44].

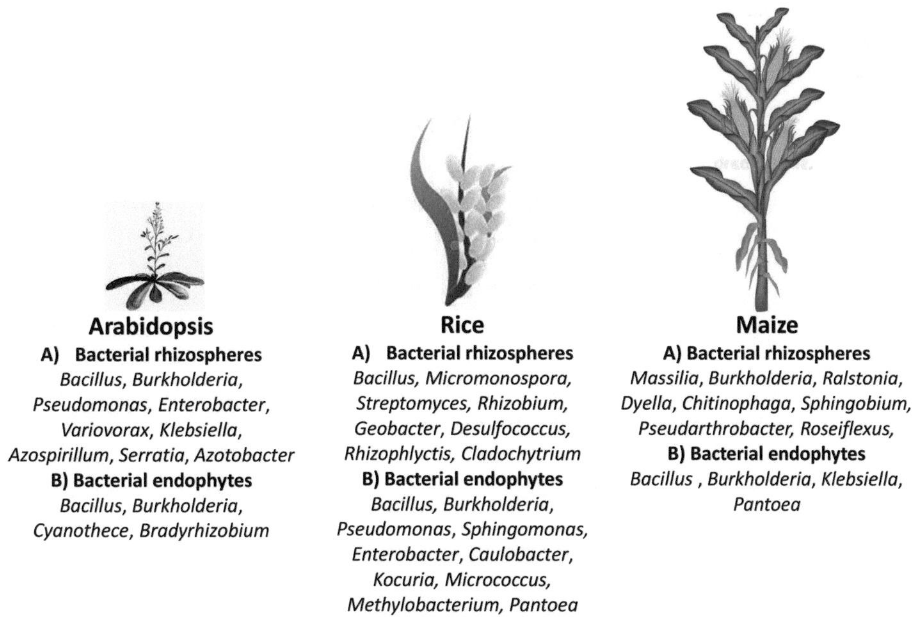

Figure 3.
Potential bacterial endophytes from Arabidopsis, Rice, and maize.

Several studies have proposed that different rhizosphere bacterial types may serve as initial inoculum populations. It was shown that bacterial communities, such as epiphytic and endophytic, are highly similar in both leaves and roots, respectively, supporting the hypothesis that the communities are recruited from the soil [12]; (**Figure 3**). Firstly, rhizosphere microbial could be defined by exudates released from the host; this is because concentration gradients of carbon sources and phytochemicals function as attractants, while the modulation of oxygen and pH in the soil acts as limiter strategy, and finally, nutrient depletion works as selective mode [9] (**Figure 1**). Secondly, rhizoplane microbes could be recruited by favoring specific functions like attachment or adhesion. Swimming and other types of motility and chemotaxis are the first step to colonization. These depend on cell structures, such as flagella or pili, while colonization requires biofilm formation and adhesins. These are all important features for gaining access to the plant surface and to colonize in susceptible areas caused by wounds or mechanical injuries [42]; (**Figure 1**), Endophytic microbes are found in inter or intracellular spaces in the plant, and it is hypothesized that they require properties such as flagella and twitching motility that contribute to access and colonize at the host. On the other hand, lipopolysaccharide production (LPS), ROS detoxification, plant polymer degradation, quorum sensing, and type VI secretion system are important for the establishment inside the plant host [9, 42]. Finally, it has been reported that some mechanisms are central features in abiotic stress alleviation by plant growth-promoting endophytes. One important example is ACC deaminase activity to keep the stress ethylene concentration below growth inhibitory effects [16, 45–47].

3. Conclusions

The plant associated-microbial communities play important roles in nutrient competition, and in general, in plant health and crop yields. It is therefore important to clearly differentiate between the structure and diversity of bacterial communities found as endophytes vs. on the roots (rhizosphere). Important progress has been achieved in the study of beneficial microbes in maize, but recent studies show that there are unexpected differences between hybrid and native landraces, probably due to opposing farming systems between conventional modern crops and traditional *milpas* system with ancestral practices. It has been hypothesized that *milpa* practices favor the maintenance of specific microorganisms that could have been lost in modern hybrid maize with monocultures, high-tillage, and large agrochemical input techniques. Particularly, endophytic microbes appear to be important communities with unique diversity and functions in native maize. In order to understand *milpa* microbial diversity for developing sustainable solutions, it is first necessary to review our general knowledge on the role of endophytes for the health of plants in general, but specifically for maize. Today, it is increasingly recognized that these endophytic bacterial communities could be significant restore to plant health and they could have offered an opportunity to maximize crop productivity and mitigate the effects of climate change without huge impacts on the ecosystems.

Acknowledgements

We thank to Montserrat Rios Romero for the anthocyanin and phenolic compounds analyses under abiotic stress.

Author contributions

VMGM did the writing and drew the graphs. JR and MT revised this manuscript.

Conflict of interest

The authors declare that the research was conducted in the absence of any commercial or financial relationships that could be construed as a potential conflict of interest.

Author details

Víctor Manuel González-Mendoza[1], Mayra de la Torre[2] and Jorge Rocha[3]*

1 Unidad Regional Hidalgo, CONACyT-Centro de Investigación en Alimentación y Desarrollo, Mexico

2 Department of Food Science, Centro de Investigación en Alimentación y Desarrollo A.C, Mexico

3 Unidad Regional Hidalgo, CONACyT-Centro de Investigación en Alimentación y Desarrollo, Mexico

*Address all correspondence to: jorge.rocha@ciad.mx

References

[1] Bulgarelli D, Rott M, Schlaeppi K, Loren V, van Themaat E, Ahmadinejad N, et al. Revealing structure and assembly cues for *Arabidopsis* root-inhabiting bacterial microbiota. Nature. 2012;**488**:91-95. DOI: 10.1038/nature11336

[2] Peiffer JA, Spor A, Koren O, Jin Z, Tringe SG, Dangl JL, et al. Diversity and heritability of the maize rhizosphere microbiome under field conditions. Proceedings of the National Academy of Sciences. 2013;**110**(16):6548-6553. DOI: 10.1073/pnas.1302837110

[3] Edwards J, Johnson C, Santos-MedellínC, LurieE, PodishettyNK, Bhatnagar S, et al. Structure, variation, and assembly of the root-associated microbiomes of rice. Proceedings of the National Academy of Sciences of the United States of America. 2015;**112**(8):E911-E920. DOI: 10.1073/pnas.1414592112

[4] Brisson VL, Schmidt JE, Northen TR, Vogel JP, Gaudin ACM. Impacts of maize domestication and breeding on rhizosphere microbial community recruitment from a nutrient depleted agricultural soil. Scientific Reports. 2019;**9**(1):15611. DOI: 10.1038/s41598-019-52148-y

[5] Gastélum G, Aguirre-von-Wobeser E, de la Torre M, Rocha J. Interaction networks reveal highly antagonistic endophytic bacteria in native maize seeds from traditional milpa agroecosystems. Environmental Microbiology. 2022;**24**(11):5583-5595. DOI: 10.1111/1462-2920.16189

[6] Aguirre-von-Wobeser E, Rocha-Estrada J, Shapiro LR, de la Torre M. Enrichment of Verrucomicrobia, Actinobacteria and Burkholderiales drives selection of bacterial community from soil by maize roots in a traditional milpa agroecosystem. PLoS One. 2018;**13**(12):e0208852. DOI: 10.1371/journal.pone.0208852

[7] Gaiero JR, McCall CA, Thompson KA, Day NJ, Best AS, Dunfield KE. Inside the root microbiome: Bacterial root endophytes and plant growth promotion. American Journal of Botany. 2013;**100**(9):1738-1750. DOI: 10.3732/ajb.1200572

[8] Ikeda AC, Bassani LL, Adamoski D, Stringari D, Cordeiro VK, Glienke C, et al. Morphological and genetic characterization of endophytic bacteria isolated from roots of different maize genotypes. Microbial Ecology. 2013;**65**(1):154-160. DOI: 10.1007/s00248-012-0104-0

[9] Reinhold-Hurek B, Bünger W, Burbano CS, Sabale M, Hurek T. Roots shaping their microbiome: Global hotspots for microbial activity. Annual Review of Phytopathology. 2015;**53**:403-424. DOI: 10.1146/annurev-phyto-082712-102342

[10] Liu H, Carvalhais LC, Crawford M, Singh E, Dennis PG, Pieterse CMJ, et al. Inner plant values: Diversity, colonization and benefits from Endophytic Bacteria. Frontiers in Microbiology. 2017;**8**:2552. DOI: 10.3389/fmicb.2017.02552

[11] Liu X, Liu L, Gong J, Zhang L, Jiang Q, Huang K, et al. Soil conditions on bacterial wilt disease affect bacterial and fungal assemblage in the rhizosphere. AMB Experimental. 2022;**12**:110. DOI: 10.1186/s13568-022-01455-1

[12] Wagner MR, Lundberg DS, Del Rio TG, Tringe SG, Dangl JL, Mitchell-Olds T. Host genotype and age shape the leaf and root microbiomes of a wild perennial plant. Nature Communications. 2016;7:12151. DOI: 10.1038/ncomms12151

[13] Gouda S, Kerry RG, Das G, Paramithiotis S, Shin HS, Patra JK. Revitalization of plant growth promoting rhizobacteria for sustainable development in agriculture. Microbiological Research. 2018;**206**:131-140. DOI: 10.1016/j.micres.2017.08.016

[14] Kumar A, Verma JP. Does plant-microbe interaction confer stress tolerance in plants: A review? Microbiological Research. 2018;**207**:41-52. DOI: 10.1016/j.micres.2017.11.004

[15] Chen C, Zhong C, Gao X, Tan C, Bai H, Ning K. Root-associated microbiota: The multifaceted hubs associated with environmental factors, growth status and accumulation of secondary metabolites. Environmental Microbiome. 2022;**17**(1):23. DOI: 10.1186/s40793-022-00418-0

[16] Aslam F, Ali B. Halotolerant bacterial diversity associated with Suaeda fruticosa (L.) Forssk. Improved growth of maize under salinity stress. Agronomy. 2018;**8**(8):131. DOI: 10.3390/agronomy8080131

[17] Egamberdieva D, Wirth SJ, Alqarawi AA, Abd Allah EF, Hashem A. Phytohormones and beneficial microbes: Essential components for plants to balance stress and fitness. Frontiers in Microbiology. 2017;**8**:2104. DOI: 10.3389/fmicb.2017.02104

[18] Islam S, Akanda AM, Prova A, Islam MT, Hossain MM. Isolation and identification of plant growth promoting Rhizobacteria from cucumber rhizosphere and their effect on plant growth promotion and disease suppression. Frontiers in Microbiology. 2016;**6**:1360. DOI: 10.3389/fmicb.2015.01360

[19] Oteino N, Lally RD, Kiwanuka S, Lloyd A, Ryan D, Germaine KJ, et al. Plant growth promotion induced by phosphate solubilizing endophytic Pseudomonas isolates. Frontiers in Microbiology. 2015;**6**:745. DOI: 10.3389/fmicb.2015.00745

[20] Abdelaal K, AlKahtani M, Attia K, Hafez Y, Király L, Künstler A. The role of plant growth-promoting Bacteria in alleviating the adverse effects of drought on plants. Biology. 2021;**10**(6):520. DOI: 10.3390/biology10060520

[21] Mahmood S, Daur I, Al-Solaimani SG, Ahmad S, Madkour MH, Yasir M, et al. Plant growth promoting Rhizobacteria and silicon synergistically enhance salinity tolerance of mung bean. Frontiers in Plant Science. 2016;7:876. DOI: 10.3389/fpls.2016.00876

[22] Gholizadeh S, Mohammadi SA, Salekdeh GH. Changes in root microbiome during wheat evolution. BMC Microbiology. 2022;**22**:64. DOI: 10.1186/s12866-022-02467-4

[23] Narayanan Z, Glick BR. Secondary metabolites produced by plant growth-promoting bacterial endophytes. Microorganisms. 2022;**10**(10):2008. DOI: 10.3390/microorganisms10102008

[24] Kovinich N, Kayanja G, Chanoca A, Riedl K, Otegui MS, Grotewold E. Not all anthocyanins are born equal: Distinct patterns induced by stress in Arabidopsis. Planta. 2014;**240**(5):931-940. DOI: 10.1007/s00425-014-2079-1

[25] Chaparro JM, Badri DV, Bakker MG, Sugiyama A, Manter DK, Vivanco JM. Root exudation of phytochemicals in Arabidopsis follows specific patterns that are developmentally programmed and correlate with soil microbial functions. PLoS One. 2013;**8**(2):e55731. DOI: 10.1371/journal.pone.0055731

[26] Wu Y, Sun J, Yu P, Zhang W, Lin Y, Ma D. The rhizosphere bacterial community contributes to the nutritional competitive advantage of weedy rice over cultivated rice in paddy soil. BMC Microbiology. 2022;**22**:232. DOI: 10.1186/s12866-022-02648-1

[27] Sangabriel-Conde W, Negrete-Yankelevich S, Maldonado-Mendoza IE, Trejo-Aguilar D. Native maize landraces from Los Tuxtlas, Mexico show varying mycorrhizal dependency for P uptake. Biology and Fertility of Soils. 2014;**50**:405-414. DOI: 10.1007/s00374-013-0847-x

[28] Osakabe Y, Osakabe K, Shinozaki K, Tran LS. Response of plants to water stress. Frontiers in Plant Science. 2014;**13**(5):86. DOI: 10.3389/fpls.2014.00086

[29] Seleiman MF, Al-Suhaibani N, Ali N, Akmal M, Alotaibi M, Refay Y, et al. Drought stress impacts on plants and different approaches to alleviate its adverse effects. Plants. 2021;**10**:259. DOI: 10.3390/plants10020259

[30] Begum N, Ahanger MA, Su Y, Lei Y, Mustafa NSA, Ahmad P, et al. Improved drought tolerance by AMF inoculation in maize (Zea mays) involves physiological and biochemical implications. Plants. 2019;**8**(12):579. DOI: 10.3390/plants8120579

[31] Igiehon ON, Babalola OO. Rhizobium and Mycorrhizal fungal species improved soybean yield under drought stress conditions. Current Microbiology. 2021;**78**:1615-1627. DOI: 10.1007/s00284-021-02432-w

[32] Sheteiwy MS, Ali DFI, Xiong YC, Brestic M, Skalicky M, Hamoud YA, et al. Physiological and biochemical responses of soybean plants inoculated with Arbuscular mycorrhizal fungi and Bradyrhizobium under drought stress. BMC Plant Biology. 2021;**21**:195. DOI: 10.1186/s12870-021-02949-z

[33] Vidal C, González F, Santander C, Pérez R, Gallardo V, Santos C, et al. Management of Rhizosphere Microbiota and Plant Production under drought stress: A comprehensive review. Plants. 2022;**11**(18):2437. DOI: 10.3390/plants11182437

[34] Arun KD, Sabarinathan KG, Gomathy M, Kannan R, Balachandar D. Mitigation of drought stress in rice crop with plant growth-promoting abiotic stress-tolerant rice phyllosphere bacteria. Journal of Basic Microbiology. 2020;**60**(9):768-786. DOI: 10.1002/jobm.202000011

[35] Win KT, Tanaka F, Okazaki K, Ohwaki Y. The ACC deaminase expressing endophyte Pseudomonas spp. enhances NaCl stress tolerance by reducing stress-related ethylene production, resulting in improved growth, photosynthetic performance, and ionic balance in tomato plants. Plant Physiology and Biochemistry. 2018;**127**:599-607. DOI: 10.1016/j.plaphy.2018.04.038

[36] Afridi MS, Amna S, Mahmood T, Salam A, Mukhtar T, Mehmood S, et al. Induction of tolerance to salinity in wheat genotypes by plant growth promoting endophytes: Involvement of ACC deaminase and antioxidant enzymes. Plant Physiology and

Biochemistry. 2019;**139**:569-577. DOI: 10.1016/j.plaphy.2019.03.041

[37] Lata R, Chowdhury S, Gond SK, White JF Jr. Induction of abiotic stress tolerance in plants by endophytic microbes. Letters in Applied Microbiology. 2018;**66**(4):268-276. DOI: 10.1111/lam.1285

[38] Jan R, Khan MA, Asaf S, Lubna LIJ, Kim KM. Metal resistant endophytic bacteria reduces cadmium, nickel toxicity, and enhances expression of metal stress related genes with improved growth of *Oryza sativa*, via regulating its antioxidant machinery and endogenous hormones. Plants. 2019;**8**(10):363. DOI: 10.3390/plants8100363

[39] Wang X, Mavrodi DV, Ke L, Mavrodi OV, Yang M, Thomashow LS, et al. Biocontrol and plant growth-promoting activity of rhizobacteria from Chinese fields with contaminated soils. Microbial Biotechnology. 2015;**8**(3):404-418. DOI: 10.1111/1751-7915.12158

[40] Goswami D, Thakker JN, Dhandhukia PC. Portraying mechanics of plant growth promoting rhizobacteria (PGPR): A review. Cogent Food & Agriculture. 2016;**2**:1. DOI: 10.1080/23311932.2015.1127500

[41] Marag PS, Suman A. Growth stage and tissue specific colonization of endophytic bacteria having plant growth promoting traits in hybrid and composite maize (*Zea mays* L.). Microbiological Research. 2018;**214**:101-113. DOI: 10.1016/j.micres.2018.05.016

[42] Kandel SL, Joubert PM, Doty SL. Bacterial endophyte colonization and distribution within plants. Microorganisms. 2017;**5**(4):77. DOI: 10.3390/microorganisms5040077

[43] Li X, Rui J, Mao Y, Yannarell A, Mackie R. Dynamics of the bacterial community structure in the rhizosphere of a maize cultivar. Soil Biology and Biochemistry. 2014;**68**:392-401. DOI: 10.1016/j.soilbio.2013.10.017

[44] Diédhiou I, Ramírez-Tobias HM, Fortanelli-Martinez J, Flores-Ramírez R. Maize intercropping in the traditional "Milpa" system. Physiological, morphological, and agronomical parameters under induced warming: Evidence of related effect of climate change in San Luis Potosí (Mexico). Life. 2022;**12**(10):1589. DOI: 10.3390/life12101589

[45] Ali S, Charles TC, Glick BR. Amelioration of high salinity stress damage by plant growth promoting bacterial endophytes that contain ACC deaminase. Plant Physiology and Biochemistry. 2014;**80**:160-167. DOI: 10.1016/j.plaphy.2014.04.003

[46] Siddikee MA, Chauhan PS, Anandham R, Han GH, Sa T. Isolation, characterization, and use for plant growth promotion under salt stress, of ACC deaminase-producing halotolerant bacteria derived from coastal soil. Journal of Microbiology and Biotechnology. 2017;**27**(9):1724. DOI: 10.4014/jmb.1007.07011

[47] Lephatsi M, Nephali L, Meyer V, Piater LA, Buthelezi N, Dubery IA, et al. Molecular mechanisms associated with microbial biostimulant-mediated growth enhancement, priming and drought stress tolerance in maize plants. Scientific Reports. 2022;**12**(1):10450. DOI: 10.1038/s41598-022-14570-7